MATHEMATICAL MODELS IN BIOLOGY
AN INTRODUCTION

To J., R., and K.,
may reality live up to the model

MATHEMATICAL MODELS IN BIOLOGY
AN INTRODUCTION

ELIZABETH S. ALLMAN

Department of Mathematics and Statistics,
University of Southern Maine

JOHN A. RHODES

Department of Mathematics,
Bates College

CAMBRIDGE
UNIVERSITY PRESS

PUBLISHED BY THE PRESS SYNDICATE OF THE UNIVERSITY OF CAMBRIDGE
The Pitt Building, Trumpington Street, Cambridge, United Kingdom

CAMBRIDGE UNIVERSITY PRESS
The Edinburgh Building, Cambridge CB2 2RU, UK
40 West 20th Street, New York, NY 10011-4211, USA
477 Williamstown Road, Port Melbourne, VIC 3207, Australia
Ruiz de Alarcón 13, 28014 Madrid, Spain
Dock House, The Waterfront, Cape Town 8001, South Africa

http://www.cambridge.org

First published 2004

Printed in the United States of America

Typeface Times 10.25/13 pt.   *System* LaTeX 2_ε [TB]

A catalog record for this book is available from the British Library.

Library of Congress Cataloging in Publication Data
Allman, Elizabeth Spencer, 1965–
Mathematical models in biology : an introduction / Elizabeth S. Allman, John A. Rhodes.
p.   cm.
Includes bibliographical references (p.   ).
ISBN 0-521-81980-6 (hb.) – ISBN 0-521-52586-1 (pbk.)
1. Biology – Mathematical models.   I. Rhodes, John A. (John Anthony), 1960–   II. Title.
QH323.5.A44   2003
570′.1′5118 – dc21   2003043929

ISBN 0 521 81980 6 hardback
ISBN 0 521 52586 1 paperback

Contents

Preface   *page* vii
Note on MATLAB   xi

1. Dynamic Modeling with Difference Equations   1
    1.1.   The Malthusian Model   2
    1.2.   Nonlinear Models   11
    1.3.   Analyzing Nonlinear Models   20
    1.4.   Variations on the Logistic Model   33
    1.5.   Comments on Discrete and Continuous Models   39
2. Linear Models of Structured Populations   41
    2.1.   Linear Models and Matrix Algebra   41
    2.2.   Projection Matrices for Structured Models   53
    2.3.   Eigenvectors and Eigenvalues   65
    2.4.   Computing Eigenvectors and Eigenvalues   78
3. Nonlinear Models of Interactions   85
    3.1.   A Simple Predator–Prey Model   86
    3.2.   Equilibria of Multipopulation Models   94
    3.3.   Linearization and Stability   99
    3.4.   Positive and Negative Interactions   105
4. Modeling Molecular Evolution   113
    4.1.   Background on DNA   114
    4.2.   An Introduction to Probability   116
    4.3.   Conditional Probabilities   130
    4.4.   Matrix Models of Base Substitution   138
    4.5.   Phylogenetic Distances   155
5. Constructing Phylogenetic Trees   171
    5.1.   Phylogenetic Trees   172
    5.2.   Tree Construction: Distance Methods – Basics   180
    5.3.   Tree Construction: Distance Methods – Neighbor
           Joining   191

5.4.	Tree Construction: Maximum Parsimony	198
5.5.	Other Methods	206
5.6.	Applications and Further Reading	208
6.	Genetics	215
6.1.	Mendelian Genetics	215
6.2.	Probability Distributions in Genetics	228
6.3.	Linkage	244
6.4.	Gene Frequency in Populations	261
7.	Infectious Disease Modeling	279
7.1.	Elementary Epidemic Models	280
7.2.	Threshold Values and Critical Parameters	286
7.3.	Variations on a Theme	296
7.4.	Multiple Populations and Differentiated Infectivity	307
8.	Curve Fitting and Biological Modeling	315
8.1.	Fitting Curves to Data	316
8.2.	The Method of Least Squares	325
8.3.	Polynomial Curve Fitting	335
A.	Basic Analysis of Numerical Data	345
A.1.	The Meaning of a Measurement	345
A.2.	Understanding Variable Data – Histograms and Distributions	348
A.3.	Mean, Median, and Mode	352
A.4.	The Spread of Data	355
A.5.	Populations and Samples	359
A.6.	Practice	360
B.	For Further Reading	362
References		365
Index		367

Preface

Interactions between the mathematical and biological sciences have been increasing rapidly in recent years. Both traditional topics, such as population and disease modeling, and new ones, such as those in genomics arising from the accumulation of DNA sequence data, have made biomathematics an exciting field. The best predictions of numerous individuals and committees have suggested that the area will continue to be one of great growth.

We believe these interactions should be felt at the undergraduate level. Mathematics students gain from seeing some of the interesting areas open to them, and biology students benefit from learning how mathematical tools might help them pursue their own interests. The image of biology as a non-mathematical science, which persists among many college students, does a great disservice to those who hold it. This text is an attempt to present some substantive topics in mathematical biology at the early undergraduate level. We hope it may motivate some to continue their mathematical studies beyond the level traditional for biology students.

The students we had in mind while writing it have a strong interest in biological science and a mathematical background sufficient to study calculus. We do not assume any training in calculus or beyond; our focus on modeling through difference equations enables us to keep prerequisites minimal. Mathematical topics ordinarily spread through a variety of mathematics courses are introduced as needed for modeling or the analysis of models.

Despite this organization, we are aware that many students will have had calculus and perhaps other mathematics courses. We therefore have not hesitated to include comments and problems (all clearly marked) that may benefit those with additional background. Our own classes using this text have included a number of students with extensive mathematical backgrounds, and they have found plenty to learn. Much of the material is also appealing to students in other disciplines who are simply curious. We believe the text can be used productively in many ways, for both classes and independent study, and at many levels.

Our writing style is intentionally informal. We have not tried to offer definitive coverage of any topic, but rather draw students into an interesting field. In particular, we often only introduce certain models and leave their analysis to exercises. Though this would be an inefficient way to give encyclopedic exposure to topics, we hope it leads to deeper understanding and questioning.

Because computer experimentation with models can be so informative, we have supplemented the text with a number of MATLAB programs. MATLAB's simple interface, its widespread availability in both professional and student versions, and its emphasis on numerical rather than symbolic computation have made it well-suited to our goals. We suggest appropriate MATLAB commands within problems, so that effort spent teaching its syntax should be minimal. Although the computer is a tool students should use, it is by no means a focus of the text.

In addition to many exercises, a variety of projects are included. These propose a topic of study and suggest ways to investigate it, but they are all at least partially open-ended. Not only does this allow students to work at different levels, it also is more true to the reality of mathematical and scientific work.

Throughout the text are questions marked with "▶." These are intended as gentle prods to prevent passive reading. Answers should be relatively clear after a little reflection, or the issue will be discussed in the text afterward. If you find such nagging annoying, please feel free to ignore them.

There is more material in the text than could be covered in a semester, offering instructors many options. The topics of Chapters 1, 2, 3, and 7 are perhaps the most standard for mathematical biology courses, covering population and disease models, both linear and nonlinear. Chapters 4 and 5 offer students an introduction to newer topics of molecular evolution and phylogenetic tree construction that are both appealing and useful. Chapter 6, on genetics, provides a glimpse of another area in which mathematics and biology have long been intertwined. Chapter 8 and the Appendix give a brief introduction to the basic tools of curve fitting and statistics.

In terms of logical development, mathematical topics are introduced as they are needed in addressing biological topics. Chapter 1 introduces the concepts of dynamic modeling through one-variable difference equations, including the key notions of equilibria, linearization, and stability. Chapter 2 motivates matrix algebra and eigenvector analysis through two-variable linear models. These chapters are a basis for all that follows.

An introduction to probability appears in two sections of Chapter 4, in order to model molecular evolution, and is then extended in Chapter 6 for genetics applications. Chapter 5, which has an algorithmic flavor different from the rest of the text, depends in part on the distance formulas derived in Chapter 4. Chapter 8's treatment of infectious disease models naturally depends on Chapter 3's introduction to models of interacting populations.

The development of this course began in 1994, with support from a Hughes Foundation Grant to Bates College. Within a few years, brief versions of a few chapters written by the second author had evolved. The first author supplemented these with additional chapters, with support provided by the American Association of University Women. After many additional joint revisions, the course notes reached a critical mass where publishing them for others to use was no longer frightening. A Phillips Grant from Bates and a professional leave from the University of Southern Maine aided the completion.

We thank our many colleagues, particularly those in the biological sciences, who aided us over the years. Seri Rudolph, Karen Rasmussen, and Melinda Harder all helped outline the initial course, and Karen provided additional consultations until the end. Many students helped, both as assistants and classroom guinea pigs, testing problems and text and asking many questions. A few who deserve special mention are Sarah Baxter, Michelle Bradford, Brad Cranston, Jamie McDowell, Christopher Hallward, and Troy Shurtleff. We also thank Cheryl McCormick for informal consultations.

Despite our best intentions, errors are sure to have slipped by us. Please let us know of any you find.

Elizabeth Allman
eallman@maine.edu
Portland, Maine

John Rhodes
jrhodes@bates.edu
Turner, Maine

Note on MATLAB

Many of the exercises and projects refer to the computer package MATLAB. Learning enough of the basic MATLAB commands to use it as a high-powered calculator is both simple and worthwhile. When the text requires more advanced commands for exercises, examples are generally given within the statements of the problems. In this way, facility with the software can be built gradually.

MATLAB is in fact a complete programming language with excellent graphical capabilities. We have taken advantage of these features to provide a few programs, making investigating the models in this text easier for the MATLAB beginner. Both exercises and projects refer to some of the programs (called m-files) or data files (called `mat`-files) below.

The m-files have been written to minimize necessary background knowledge of MATLAB syntax. To run most of the m-files below, say `onepop.m`, be sure it is in your current MATLAB directory or path and type `onepop`. You will then be asked a series of questions about models and parameters. The command `help onepop` also provides a brief description of the program's function. Since m-files are text files, they can be read and modified by anyone interested.

Some of the m-files define functions, which take arguments. For instance, a command like `compseq(seq1,seq2)` runs the program `compseq.m` to compare the two DNA sequences `seq1` and `seq2`. Typing `help compseq` prints an explanation of the syntax of such a function.

A `mat`-file contains data that may only be accessed from within MATLAB. To load such a file, say `seqdata.mat`, type `load seqdata`. The names of any new variables this creates can be seen by then typing `who`, while values stored in those variables can be seen by typing the variable name.

Some data files have been given in the form of m-files, so that supporting comments and explanations could be saved with the data. For these, running the m-file creates variables, just as loading a `mat`-file would. The comments can be read with any editor.

The MATLAB files made available with the text are:

- `aidsdata.m` – contains data from the Centers for Disease Control and Prevention on acquired immune deficiency syndrome (AIDS) cases in the United States
- `cobweb.m`, `cobweb2.m` – produce cobweb diagram movies for iterations of a one-population model; the first program leaves all web lines that are drawn, and the second program gradually erases them
- `compseq.m` – compares two DNA sequences, producing a frequency table of the number of sites with each of the possible base combinations
- `distances.m` – computes Jukes-Cantor, Kimura 2-parameter, and log-det (paralinear) distances between all pairs in a collection of DNA sequences
- `distJC.m`, `distK2.m`, `distLD.m` – compute Jukes-Cantor, Kimura 2-parameter, or log-det (paralinear) distance for one pair of sequences described by a frequency array of sites with each base combination
- `flhivdata.m` – contains DNA sequences of the envelope gene for human immunodeficiency virus (HIV) from the "Florida dentist case"
- `genemap.m` – simulates testcross data for a genetic mapping project, using either fly or mouse genes
- `genesim.m` – produces a time plot of allele frequency of a gene in a population of fixed size; relative fitness values for genotypes can be set to model natural selection
- `informative.m` – locates sites in aligned DNA sequences that are informative for the method of maximum parsimony
- `longterm.m` – draws a bifurcation diagram for a one-population model, showing long-term behavior as one parameter value varies
- `markovJC.m`, `markovK2.m` – produce a Markov matrix of Jukes-Cantor or Kimura 2-parameter form with specified parameter values
- `mutate.m`, `mutatef.m` – simulate DNA sequence mutation according to a Markov model of base substitution; the second program is a function version of the first
- `nj.m` – performs the Neighbor Joining algorithm to construct a tree from a distance array
- `onepop.m` – displays time plots of iterations of a one-population model
- `primatedata.m` – contains mitochondrial DNA sequences from 12 primates, as well as computed distances between them
- `seqdata.mat` – contains simulated DNA sequence data

- `seqgen.m` – generates DNA sequences with specified length and base distribution
- `sir.m` – displays iterations of an *SIR* epidemic model, including time and phase plane plots
- `twopop.m` – displays iterations of a two-population model, including time and phase plane plots

Of the above programs, `compseq`, `distances`, `distJC`, `distK2`, `distLD`, `informative`, `markovJC`, `markovK2`, `mutatef`, `nj`, and `seqgen` are functions requiring arguments.

All these files can be found on the web site www.cup.org/titles/0521525861

1

Dynamic Modeling with Difference Equations

Whether we investigate the growth and interactions of an entire population, the evolution of DNA sequences, the inheritance of traits, or the spread of disease, biological systems are marked by change and adaptation. Even when they appear to be constant and stable, it is often the result of a balance of tendencies pushing the systems in different directions. A large number of interactions and competing tendencies can make it difficult to see the full picture at once.

How can we understand systems as complicated as those arising in the biological sciences? How can we test whether our supposed understanding of the key processes is sufficient to describe how a system behaves? Mathematical language is designed for precise description, and so describing complicated systems often requires a *mathematical model*.

In this text, we look at some ways mathematics is used to model dynamic processes in biology. Simple formulas relate, for instance, the population of a species in a certain year to that of the following year. We learn to understand the consequences an equation might have through mathematical analysis, so that our formulation can be checked against biological observation. Although many of the models we examine may at first seem to be gross simplifications, their very simplicity is a strength. Simple models show clearly the implications of our most basic assumptions.

We begin by focusing on modeling the way populations grow or decline over time. Since mathematical models should be driven by questions, here are a few to consider: Why do populations sometimes grow and sometimes decline? Must populations grow to such a point that they are unsustainably large and then die out? If not, must a population reach some equilibrium? If an equilibrium exists, what factors are responsible for it? Is such an equilibrium so delicate that any disruption might end it? What determines whether a given population follows one of these courses or another?

To begin to address these questions, we start with the simplest mathematical model of a changing population.

1.1. The Malthusian Model

Suppose we grow a population of some organism, say flies, in the laboratory. It seems reasonable that, on any given day, the population will change due to new births, so that it increases by the addition of a certain multiple f of the population. At the same time, a fraction d of the population will die.

Even for a human population, this model might apply. If we assume humans live for 70 years, then we would expect that from a large population roughly $1/70$ of the population will die each year; so, $d = 1/70$. If, on the other hand, we assume there are about four births in a year for every hundred people, we have $f = 4/100$. Note that we have chosen *years* as units of time in this case.

▶ Explain why, for any population, d must be between 0 and 1. What would $d < 0$ mean? What would $d > 1$ mean?

▶ Explain why f must be at least 0, but could be bigger than 1. Can you name a real organism (and your choice of units for time) for which f would be bigger than 1?

▶ Using *days* as your unit of time, what values of f and d would be in the right ballpark for elephants? Fish? Insects? Bacteria?

To track the population P of our laboratory organism, we focus on ΔP, the *change* in population over a single day. So, in our simple conception of things,

$$\Delta P = fP - dP = (f - d)P.$$

What this means is simply that given a current population P, say $P = 500$, and the fecundity and death rates f and d, say $f = .1$ and $d = .03$, we can predict the change in the population $\Delta P = (.1 - .03)500 = 35$ over a day. Thus, the population at the beginning of the next day is $P + \Delta P = 500 + 35 = 535$.

Some more notation will make this simpler. Let

$$P_t = P(t) = \text{the size of the population measured on day } t,$$

so

$$\Delta P = P_{t+1} - P_t$$

Table 1.1. *Population Growth According to a Simple Model*

Day	Population
0	500
1	$(1.07)500 = 535$
2	$(1.07)^2 500 = 572.45$
3	$(1.07)^3 500 \approx 612.52$
4	$(1.07)^4 500 \approx 655.40$
$\vdots$	$\vdots$

is the *difference* or change in population between two consecutive days. (If you think there should be a subscript t on that ΔP, because ΔP might be different for different values of t, you are right. However, it's standard practice to leave it off.)

Now what we ultimately care about is understanding the population P_t, not just ΔP. But

$$P_{t+1} = P_t + \Delta P = P_t + (f - d)P_t = (1 + f - d)P_t.$$

Lumping some constants together by letting $\lambda = 1 + f - d$, our model of population growth has become simply

$$P_{t+1} = \lambda P_t.$$

Population ecologists often refer to the constant λ as the *finite growth rate* of the population. (The word "finite" is used to distinguish this number from any sort of instantaneous rate, which would involve a derivative, as you learn in calculus.)

For the values $f = .1$, $d = .03$, and $P_0 = 500$ used previously, our entire model is now

$$P_{t+1} = 1.07P_t, \quad P_0 = 500.$$

The first equation, relating P_{t+1} and P_t, is referred to as a *difference equation* and the second, giving P_0, is its *initial condition*. With the two, it is easy to make a table of values of the population over time, as in Table 1.1.

From Table 1.1, it's even easy to recognize an explicit formula for P_t,

$$P_t = 500(1.07)^t.$$

For this model, we can now easily predict populations at any future times.

It may seem odd to call $P_{t+1} = (1 + f - d)P_t$ a difference equation, when the difference ΔP does not appear. However, the equations

$$P_{t+1} = (1 + f - d)P_t$$

and

$$\Delta P = (f - d)P$$

are mathematically equivalent, so either one is legitimately referred to by the same phrase.

Example. Suppose that an organism has a very rigid life cycle (which might be realistic for an insect), in which each female lays 200 eggs, then all the adults die. After the eggs hatch, only 3% survive to become adult females, the rest being either dead or males. To write a difference equation for the females in this population, where we choose to measure t in generations, we just need to observe that the death rate is $d = 1$, while the effective fecundity is $f = .03(200) = 6$. Therefore,

$$P_{t+1} = (1 + 6 - 1)P_t = 6P_t.$$

▶   Will this population grow or decline?

▶   Suppose you don't know the effective fecundity, but do know that the population is stable (unchanging) over time. What must the effective fecundity be? (*Hint*: What is $1 + f - d$ if the population is stable?) If each female lays 200 eggs, what fraction of them must hatch and become females?

Notice that in this last model we ignored the males. This is actually a quite common approach to take and simplifies our model. It does mean we are making some assumptions, however. For this particular insect, the precise number of males may have little effect on how the population grows. It might be that males are always found in roughly equal numbers to females so that we know the total population is simply double the female one. Alternately, the size of the male population may behave differently from the female one, but whether there are few males or many, there are always enough that female reproduction occurs in the same way. Thus, the female population is the important one to track to understand the long-term growth or decline of the population.

▶   Can you imagine circumstances in which ignoring the males would be a bad idea?

What is a difference equation? Now that you have seen a difference equation, we can attempt a definition: a difference equation is a formula expressing values of some quantity Q in terms of previous values of Q. Thus, if $F(x)$ is any function, then

$$Q_{t+1} = F(Q_t)$$

is called a difference equation. In the previous example, $F(x) = \lambda x$, but often F will be more complicated.

In studying difference equations and their applications, we will address two main issues: 1) How do we find an appropriate difference equation to model a situation? 2) How do we understand the behavior of the difference equation model once we have found it?

Both of these things can be quite hard to do. You learn to model with difference equations by looking at ones other people have used and then trying to create some of your own. To be honest, though, this will not necessarily make facing a new situation easy. As for understanding the behavior a difference equation produces, usually we cannot hope to find an explicit formula like we did for P_t describing the insect population. Instead, we develop techniques for getting less precise qualitative information from the model.

The particular difference equation discussed in this section is sometimes called an *exponential* or *geometric model*, since the model results in exponential growth or decay. When applied to populations in particular, it is associated with the name of Thomas Malthus. Mathematicians, however, tend to focus on the form of the equation $P_{t+1} = \lambda P_t$ and say the model is *linear*. This terminology can be confusing at first, but it will be important; *a linear model produces exponential growth or decay.*

Problems

1.1.1. A population is originally 100 individuals, but because of the combined effects of births and deaths, it triples each hour.

    a. Make a table of population size for $t = 0$ to 5, where t is measured in hours.

    b. Give two equations modeling the population growth by first expressing P_{t+1} in terms of P_t and then expressing ΔP in terms of P_t.

    c. What, if anything, can you say about the birth and death rates for this population?

1.1.2. In the early stages of the development of a frog embryo, cell division occurs at a fairly regular rate. Suppose you observe that all cells divide, and hence the number of cells doubles, roughly every half-hour.

a. Write down an equation modeling this situation. You should specify how much real-world time is represented by an increment of 1 in t and what the initial number of cells is.

b. Produce a table and graph of the number of cells as a function of t.

c. Further observation shows that, after 10 hours, the embryo has around 30,000 cells. Is this roughly consistent with your model? What biological conclusions and/or questions does this raise?

1.1.3. Using a hand calculator, make a table of population values at times 0 through 6 for the following population models. Then graph the tabulated values.

a. $P_{t+1} = 1.3P_t$, $P_0 = 1$

b. $N_{t+1} = .8N_t$, $N_0 = 10$

c. $\Delta Z = .2Z$, $Z_0 = 10$

1.1.4. Redo Problem 1.1.3(a) using MATLAB by entering a command sequence like:

```
p=1
x=p
p=1.3*p
x=[x p]
p=1.3*p      (Because this repeats an earlier command, you can save
x=[x p]      some typing by hitting the "↑" key twice.)
⋮
```

Explain how this works.
Now redo the problem again by a command sequence like:

```
p=1
x=1
for i=1:10
    p=1.3*p      (The indentation is not necessary, but helps make
    x=[x p]      the for-end loop clearer to read.)
end
```

Explain how this works as well.

Graph your data with:

```
plot([0:10],x)
```

1.1.5. For the model in Problem 1.1.3(a), how much time must pass before the population exceeds 10, exceeds 100, and exceeds 1,000? (Use MATLAB to do this experimentally, and then redo it using logarithms and the fact that $P_t = 1.3^t$.) What do you notice about the difference between these times? Explain why this pattern holds.

1.1.6. If the data in Table 1.2 on population size were collected in a laboratory experiment using insects, would it be consistent with a geometric model? Would it be consistent with a geometric model for at least some range of times? Explain.

1.1.7. Complete the following:
   a. The models $P_t = k P_{t-1}$ and $\Delta P = rP$ represent *growing* populations when k is any number in the range ____ and when r is any number in the range ____.
   b. The models $P_t = k P_{t-1}$ and $\Delta P = rP$ represent *declining* populations when k is any number in the range ____ and when r is any number in the range ____.
   c. The models $P_t = k P_{t-1}$ and $\Delta P = rP$ represent *stable* populations when k is any number in the range ____ and when r is any number in the range ____.

1.1.8. Explain why the model $\Delta Q = rQ$ cannot be biologically meaningful for describing a population when $r < -1$.

1.1.9. Suppose a population is described by the model $N_{t+1} = 1.5 N_t$ and $N_5 = 7.3$. Find N_t for $t = 0, 1, 2, 3$, and 4.

1.1.10. A model is said to have a *steady state* or *equilibrium point* at P^* if whenever $P_t = P^*$, then $P_{t+1} = P^*$ as well.
   a. Rephrase this definition as: A model is said to have a *steady state* at P^* if whenever $P = P^*$, then $\Delta P = \ldots$.
   b. Rephrase this definition in more intuitive terms: A model is said to have a *steady state* at P^* if $\ldots$.
   c. Can a model described by $P_{t+1} = (1 + r)P_t$ have a steady state? Explain.

Table 1.2. *Insect Population Values*

t	0	1	2	3	4	5	6	7	8	9	10
P_t	.97	1.52	2.31	3.36	4.63	5.94	7.04	7.76	8.13	8.3	8.36

Table 1.3. *U.S. Population Estimates*

Year	Population (in 1,000s)
1920	106,630
1925	115,829
1930	122,988
1935	127,252
1940	131,684
1945	131,976
1950	151,345
1955	164,301
1960	179,990

1.1.11. Explain why the model $\Delta P = rP$ leads to the formula $P_t = (1 + r)^t P_0$.

1.1.12. Suppose the size of a certain population is affected only by birth, death, immigration, and emigration – each of which occurs in a yearly amount proportional to the size of a population. That is, if the population is P, within a time period of 1 year, the number of births is bP, the number of deaths is dP, the number of immigrants is iP, and the number of emigrants is eP, for some b, d, i, and e. Show the population can still be modeled by $\Delta P = rP$ and give a formula for r.

1.1.13. As limnologists and oceanographers are well aware, the amount of sunlight that penetrates to various depths of water can greatly affect the communities that live there. Assuming the water has uniform *turbidity*, the amount of light that penetrates through a 1-meter column of water is proportional to the amount entering the column.

   a. Explain why this leads to a model of the form $L_{d+1} = kL_d$, where L_d denotes the amount of light that has penetrated to a depth of d meters.
   b. In what range must k be for this model to be physically meaningful?
   c. For $k = .25$, $L_0 = 1$, plot L_d for $d = 0, 1, \ldots, 10$.
   d. Would a similar model apply to light filtering through the canopy of a forest? Is the "uniform turbidity" assumption likely to apply there?

1.1.14. The U.S. population data in Table 1.3 is from (Keyfitz and Flieger, 1968).

   a. Graph the data. Does this data seem to fit the geometric growth model? Explain why or why not using graphical and numerical evidence. Can you think of factors that might be responsible for any deviation from a geometric model?

b. Using the data only from years 1920 and 1925 to estimate a growth rate for a geometric model, see how well the model's results agree with the data from subsequent years.

c. Rather than just using 1920 and 1925 data to estimate a growth parameter for the U.S. population, find a way of using all the data to get what (presumably) should be a better geometric model. (Be creative. There are several reasonable approaches.) Does your new model fit the data better than the model from part (b)?

1.1.15. Suppose a population is modeled by the equation $N_{t+1} = 2N_t$, when N_t is measured in *individuals*. If we choose to measure the population in *thousands of individuals*, denoting this by P_t, then the equation modeling the population *might* change. Explain why the model is still just $P_{t+1} = 2P_t$. (*Hint*: Note that $N_t = 1000P_t$.)

1.1.16. In this problem, we investigate how a model must be changed if we change the amount of time represented by an increment of 1 in the time variable t. It is important to note that this is not always a biologically meaningful thing to do. For organisms like certain insects, generations do not overlap and reproduction times are regularly spaced, so using a time increment of less than the span between two consecutive birth times would be meaningless. However, for organisms like humans with overlapping generations and continual reproduction, there is no natural choice for the time increment. Thus, these populations are sometimes modeled with an "infinitely small" time increment (i.e., with differential equations rather than difference equations). This problem illustrates the connection between the two types of models.

A population is modeled by $N_{t+1} = 2N_t$, $N_0 = A$, where each increment of t by 1 represents a passage of 1 year.

a. Suppose we want to produce a new model for this population, where each time increment of t by 1 now represents 0.5 years, and the population size is now denoted P_t. We want our new model to produce the same populations as the first model at 1-year intervals (so $P_{2t} = N_t$). Thus, we have Table 1.4. Complete the table for P_t so that the growth is still geometric. Then give an equation of the model relating P_{t+1} to P_t.

b. Produce a new model that agrees with N_t at 1-year intervals, but denote the population size by Q_t, where each time increment of

Table 1.4. *Changing Time Steps in a Model*

t	0		1		2		3
N_t	A		2A		4A		8A
t	0	1	2	3	4	5	6
P_t	A		2A		4A		8A

t by 1 represents 0.1 years (so, $Q_{10t} = N_t$). You should begin by producing tables similar to those in part (a).

c. Produce a new model that agrees with N_t at 1-year intervals, but denote the population size by R_t, where each time increment of t by 1 represents h years (so $R_{\frac{1}{h}t} = N_t$). (h might be either bigger or smaller than 1; the same formula describes either situation.)

d. Generalize parts (a–c), writing a paragraph to explain why, if our original model uses a time increment of 1 year and is given by $N_{t+1} = kN_t$, then a model producing the same populations at 1-year intervals, but that uses a time increment of h years, is given by $P_{t+1} = k^h P_t$.

e. (Calculus) If we change the name of the time interval h to Δt, part (d) shows that

$$\frac{\Delta P}{\Delta t} = \frac{k^h - 1}{h} P.$$

If $\Delta t = h$ is allowed to become *infinitesimally small*, this means

$$\frac{dP}{dt} = \lim_{h \to 0} \frac{k^h - 1}{h} P.$$

Illustrate that

$$\lim_{h \to 0} \frac{k^h - 1}{h} = \ln k$$

by choosing a few values of k and a very small h and comparing the values of $\ln k$ and $\frac{k^h - 1}{h}$.

This result is formally proved by:

$$\lim_{h \to 0} \frac{k^h - 1}{h} = \lim_{h \to 0} \frac{k^{0+h} - k^0}{h} = \frac{d}{dx} k^x \Big|_{x=0} = \ln k \, k^x \Big|_{x=0} = \ln k.$$

f. (Calculus) Show the solution to $\frac{dP}{dt} = \ln k P$ with initial value $P(0) = P_0$ is

$$P(t) = P_0 e^{t \ln k} = P_0 k^t.$$

How does this compare to the formula for N_t, in terms of N_0 and k, for the difference equation model $N_{t+1} = kN_t$? Ecologists often refer to the k in either of these formulas as the *finite growth rate* of the population, while $\ln k$ is referred to as the *intrinsic growth rate*.

1.2. Nonlinear Models

The Malthusian model predicts that population growth will be exponential. However, such a prediction cannot really be accurate for very long. After all, exponential functions grow quickly and without bound; and, according to such a model, sooner or later there will be more organisms than the number of atoms in the universe. The model developed in the last section must be overlooking some important factor. To be more realistic in our modeling, we need to reexamine the assumptions that went into that model.

The main flaw is that we have assumed the fecundity and death rates for our population are the same regardless of the size of the population. In fact, when a population gets large, it might be more reasonable to expect a higher death rate and a lower fecundity. Combining these factors, we could say that, as the population size increases, the finite growth rate should decrease. We need to somehow modify our model so that the growth rate depends on the size of the population; that is, the growth rate should be *density dependent*.

▶   What biological factors might be the cause of the density dependence? Why might a large population have an increased death rate and/or decreased birth rate?

Creating a nonlinear model. To design a better model, it's easiest to focus on $\dfrac{\Delta P}{P}$, the change in population per individual, or the *per-capita growth rate* over a single time step. Once we have understood the per-capita growth rate and found a formula to describe it, we will be able to obtain a formula for ΔP from that.

For small values of P, the per-capita growth rate should be large, since we imagine a small population with lots of resources available in its environment to support further growth. For large values of P, however, per-capita growth should be much smaller, as individuals compete for both food and space. For even larger values of P, the per-capita growth rate should be negative, since that would mean the population will decline. It is reasonable then to assume $\Delta P / P$, as a function of P, has a graph something like that in Figure 1.1.

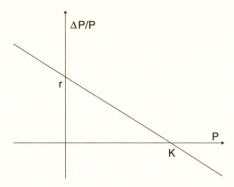

Figure 1.1. Per-capita growth rate as a function of population size.

Of course we cannot say exactly what a graph of $\Delta P/P$ should look like without collecting some data. Perhaps the graph should be concave for instance. However, this is a good first attempt at creating a better model.

▶   Graph the per-capita growth rate for the Malthusian model. How is your graph different from Figure 1.1?

For the Malthusian model $\Delta P/P = r$, so that the graph of the per-capita growth rate is a horizontal line – there is no decrease in $\Delta P/P$ as P increases.

In contrast, the sloping line of Figure 1.1 for an improved model leads to the formula $\Delta P/P = mP + b$, for some $m < 0$ and $b > 0$. It will ultimately be clearer to write this as

$$\frac{\Delta P}{P} = r\left(1 - \frac{P}{K}\right)$$

so that K is the horizontal intercept of the line, and r is the vertical intercept. Note that both K and r should be positive. With a little algebra, we get

$$P_{t+1} = P_t\left(1 + r\left(1 - \frac{P_t}{K}\right)\right)$$

as our difference equation. This model is generally referred to as the *discrete logistic model*, though, unfortunately, other models also go by that name as well.

The parameters K and r in our model have direct biological interpretations. First, if $P < K$, then $\Delta P/P > 0$. With a positive per-capita growth rate, the population will increase. On the other hand, if $P > K$, then $\Delta P/P < 0$. With a negative per-capita growth rate, the population will decrease. K is therefore called the *carrying capacity* of the environment, because it represents the maximum number of individuals that can be supported over a long period.

However, when the population is small (i.e., P is much smaller than K), the factor $(1 - P/K)$ in the per-capita growth rate should be close to 1. Therefore, for small values of P, our model is approximately

$$P_{t+1} \approx (1 + r)P_t.$$

In other words, r plays the role of $f - d$, the fecundity minus the death rate, in our earlier linear model. The parameter r simply reflects the way the population would grow or decline in the absence of density-dependent effects – when the population is far below the carrying capacity. The standard terminology for r is that it is the *finite intrinsic growth rate*. "Intrinsic" refers to the absence of density-dependent effects, whereas "finite" refers to the fact that we are using time steps of finite size, rather than the infinitesimal time steps of a differential equation.

▶   What are ballpark figures you might expect for r and K, assuming you want to model your favorite species of fish in a small lake using a time increment of 1 year?

As you will see in the problems, there are many ways different authors choose to write the logistic model, depending on whether they look at ΔP or P_{t+1} and whether they multiply out the different factors. A key point to help you recognize this model is that both ΔP and P_{t+1} are expressed as quadratic polynomials in terms of P_t. Furthermore, these polynomials have no constant term (i.e., no term of degree zero in P). Thus, the logistic model is about the simplest nonlinear model we could develop.

Iterating the model. As with the linear model, our first step in understanding this model is to choose some particular values for the parameters r and K, and for the initial population P_0, and compute future population values. For example, choosing K and r so that $P_{t+1} = P_t(1 + .7(1 - P_t/10))$ and $P_0 = 0.4346$, we get Table 1.5.

▶   How can it make sense to have populations that are not integers?

Table 1.5. *Population Values from a Nonlinear Model*

t	0	1	2	3	4	5	6
P_t	.4346	.7256	1.1967	1.9341	3.0262	4.5034	6.2362

t	7	8	9	10	11	12	...
P_t	7.8792	9.0489	9.6514	9.8869	9.9652	9.9895	...

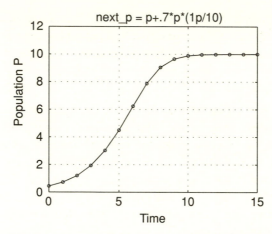

Figure 1.2. Population values from a nonlinear model.

If we measure population size in units such as thousands, or millions of individuals, then there is no reason for populations to be integers. For some species, such as commercially valuable fish, it might even be appropriate to use units of mass or weight, like tons.

Another reason that noninteger population values are not too worrisome, even if we use units of individuals, is that we are only attempting to approximately describe a population's size. We do not expect our model to give exact predictions. As long as the numbers are large, we can just ignore fractional parts without a significant loss.

In the table, we see the population increasing toward the carrying capacity of 10 as we might have expected. At first this increase seems slow, then it speeds up and then it slows again. Plotting the population values in Figure 1.2 shows the sigmoid-shaped pattern that often appears in data from carefully controlled laboratory experiments in which populations increase in a limited environment. (The plot shows the population values connected by line segments to make the pattern clearer, even though the discrete time steps of our model really give populations only at integer times.) Biologically, then, we have made some progress; we have a more realistic model to describe population growth.

Mathematically, things are not so nice, though. Unlike with the linear model, there is no obvious formula for P_t that emerges from our table. In fact, the only way to get the value of P_{100} seems to be to create a table with a hundred entries in it. We have lost the ease with which we could predict future populations.

This is something we simply have to learn to live with: Although nonlinear models are often more realistic models to use, we cannot generally get explicit formulas for solutions to nonlinear difference equations. Instead, we must rely more on graphical techniques and numerical experiments to give us insight into the models' behaviors.

Cobwebbing. Cobwebbing is the basic graphical technique for understanding a model such as the discrete logistic equation. It's best illustrated by an example. Consider again the model

$$P_0 = 2.3, \quad P_{t+1} = P_t \left(1 + .7 \left(1 - \frac{P_t}{10} \right) \right).$$

Begin by graphing the parabola defined by the equation giving P_{t+1} in terms of P_t, as well as the diagonal line $P_{t+1} = P_t$, as shown in Figure 1.3. Since the population begins at $P_0 = 2.3$, we mark that on the graph's horizontal axis. Now, to find P_1, we just move vertically upward to the graph of the parabola to find the point (P_0, P_1), as shown in the figure.

We would like to find P_2 next, but to do that we need to mark P_1 on the horizontal axis. The easiest way to do that is to move horizontally from the point (P_0, P_1) toward the diagonal line. When we hit the diagonal line, we will be at (P_1, P_1), since we've kept the same second coordinate, but changed the first coordinate. Now, to find P_2, we just move vertically back

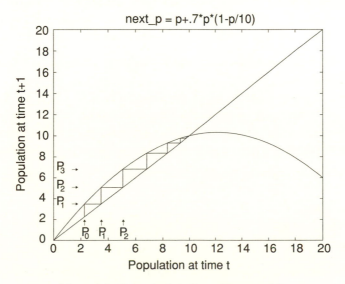

Figure 1.3. Cobweb plot of a nonlinear model.

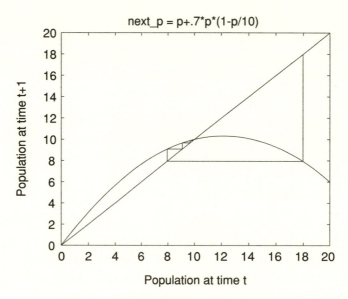

Figure 1.4. Cobweb plot of a nonlinear model.

to the parabola to find the point (P_1, P_2). Now it's just a matter of repeating these steps forever: Move vertically to the parabola, then horizontally to the diagonal line, then vertically to the parabola, then horizontally to the diagonal line, and so on.

It should be clear from this graph that if the initial population P_0 is anything between 0 and $K = 10$, then the model with $r = .7$ and $K = 10$ will result in an always increasing population that approaches the carrying capacity.

If we keep the same values of r and K, but let $P_0 = 18$, the cobweb looks like that in Figure 1.4.

Indeed, it becomes clear that if P_0 is any value above $K = 10$, then we see an immediate drop in the population. If this drop is to a value below the carrying capacity, there will then be a gradual increase back toward the carrying capacity.

▶   Find the positive population size that corresponds to where the parabola crosses the horizontal axis for the model $P_{t+1} = P_t(1 + .7(1 - P_t/10))$ by setting $P_{t+1} = 0$.

▶   What happens if P_0 is higher than the value you found in the last question?

If the population becomes negative, then we should interpret that as extinction.

At this point, you can learn a lot more from exploring the logistic model with a calculator or computer than you can by reading this text. The exercises will guide you in this. In fact, you will find that the logistic model has some surprises in store that you might not expect.

Problems

1.2.1. With a hand calculator, make a table of population values for $t = 0, 1, 2, \ldots, 10$ with $P_0 = 1$ and $\Delta P = 1.3P(1 - P/10)$. Graph your results.

1.2.2. In the model $\Delta P = 1.3P(1 - P/10)$, what values of P will cause ΔP to be positive? Negative? Why does this matter biologically?

1.2.3. Repeat problem 1 using MATLAB commands like:

```
p=1; x=p
for i=1:10; p=p+1.3*p*(1-p/10); x=[x p]; end
plot([0:10], x)
```

Explain why this works.

1.2.4. Using the MATLAB program onepop and many different values for P_0, investigate the long-term behavior of the model $\Delta P = rP(1 - P/10)$ for $r = .2, .8, 1.3, 2.2, 2.5, 2.9,$ and 3.1. (You may have to vary the number of time steps that you run the model to study some of these.)

1.2.5. Four of the many common ways of writing the discrete logistic growth equation are:

$$\Delta P = rP(1 - P/K), \qquad \Delta P = sP(K - P),$$
$$\Delta P = tP - uP^2, \qquad P_{t+1} = vP_t - wP_t^2.$$

Write each of the following in all four of these forms.
a. $P_{t+1} = P_t + .2P_t(10 - P_t)$
b. $P_{t+1} = 2.5P_t - .2P_t^2$

1.2.6. For the model $\Delta P = .8P(1 - P/10)$
a. Graph ΔP as a function of P using MATLAB by entering:

```
x=[0:.1:12]
y=.8*x.*(1-x/10)
plot(x,y)
```

b. Graph P_{t+1} as a function of P_t by modifying the MATLAB commands in part (a).

Table 1.6. *Insect Population Values*

t	0	1	2	3	4	5	6	7	8	9	10
P_t	.97	1.52	2.31	3.36	4.63	5.94	7.04	7.76	8.13	8.3	8.36

    c. Construct a table of values of P_t for $t = 0, 1, 2, 3, 4, 5$ starting with $P_0 = 1$. Then, on your graph from part (b), construct a cobweb beginning at $P_0 = 1$. (You can add the $y = x$ line to your graph by entering the commands `hold on, plot(x,y,x,x)`.) Does your cobweb match the table of values very accurately?

1.2.7. If the data in Table 1.6 on population size were collected in a laboratory experiment using insects, would it be at least roughly consistent with a logistic model? Explain. If it is consistent with a logistic model, can you estimate r and K in $\Delta N = rN(1 - N/K)$?

1.2.8. Suppose a population is modeled by the equation

$$N_{t+1} = N_t + .2N_t(1 - N_t/200000)$$

when N_t is measured in *individuals*.

    a. Find an equation of the same form, describing the same model, but with the population measured in *thousands of individuals*. (*Hint*: Let $N_t = 1000M_t$, $N_{t+1} = 1000M_{t+1}$, and find a formula for M_{t+1} in terms of M_t.)

    b. Find an equation of the same form, describing the same model, but with the population measured in units chosen so that the carrying capacity is 1 in those units. (To get started, determine the carrying capacity in the original form of the model.)

1.2.9. The technique of cobwebbing to study iterated models is not limited to just logistic growth. Graphically determine the populations for the next six time increments in each of the models of Figure 1.5 using the initial populations shown.

1.2.10. Give a formula for the graph appearing in part (a) of Figure 1.5. What is the name of this population model?

1.2.11. Some of the same modeling ideas and models used in population studies appear in very different scientific settings.

    a. Often, chemical reactions occur at rates proportional to the amount of raw materials present. Suppose we use a very small time interval to model such a reaction with a difference equation. Assume a fixed total amount of chemicals K, and that chemical 1, which initially

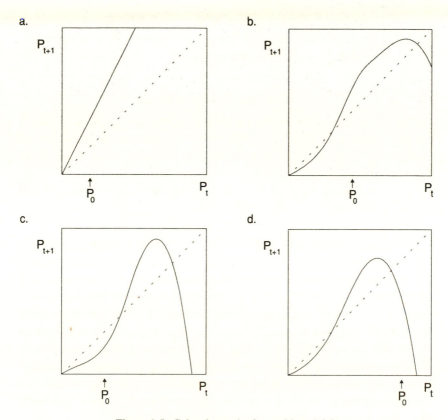

Figure 1.5. Cobweb graphs for problem 1.2.9.

occurs in amount K, is converted to chemical 2, which occurs in amount N_t at time t. Explain why $\Delta N = r(K - N)$. What values of r are reasonable? What is N_0? What does a graph of N_t as a function of t look like?

b. Chemical reactions are said to be *autocatalytic* if the rate at which they occur is proportional to both the amount of raw materials and to the amount of the product (i.e., the product of the reaction is a catalyst to the reaction). We can again use a very small time interval to model such a reaction with a difference equation. Assume a fixed total amount of chemicals K and that chemical 1 is converted to chemical 2, which occurs in amount N_t. Explain why $\Delta N = rN(K - N)$. If N_0 is small (but not 0), what will the graph of N_t as a function of t look like? If $N_0 = 0$, what will the graph of N_t as a function of t look like? Can you explain the shape of the graph intuitively? (Note that r will be very small, because we are using a small time interval.) The logistic growth model is sometimes also referred to as the *autocatalytic model*.

1.3. Analyzing Nonlinear Models

Unlike the simple linear model producing exponential growth, nonlinear models – such as the discrete logistic one – can produce an assortment of complicated behaviors. No doubt you noticed this while doing some of the exercises in the last section.

In this section, we will look at some of the different types of behavior and develop some simple tools for studying them.

Transients, equilibrium, and stability. It is helpful to distinguish several aspects of the behavior of a dynamic model. We sometimes find that regardless of our initial value, after many time steps have passed, the model seems to settle down into a pattern. The first few steps of the iteration, though, may not really be indicative of what happens over the long term. For example, with the discrete logistic model $P_{t+1} = P_t(1 + .7(1 - P_t/10))$ and most initial values P_0, the first few iterations of the model produce relatively large changes in P_t as it moves toward 10. This early behavior is thus called *transient*, because it is ultimately replaced with a different sort of behavior. However, that does not mean it is unimportant, since a real-world population may well experience disruptions that keep sending it back into transient behavior.

Usually, though, what we care about more is the long-term behavior of the model. The reason for this is we often expect the system we are studying to have been undisturbed long enough for transients to have died out. Often (but not always) the long-term behavior is independent of the exact initial population. In the model $P_{t+1} = P_t(1 + .7(1 - P_t/10))$, the long-term behavior for most initial values was for the population to stay very close to $K = 10$. Note that if $P_t = 10$ exactly, then $P_{t+1} = 10$ as well and the population never changes. Thus, $P_t = 10$ is an *equilibrium* (or *steady-state* or *fixed point*) of the model.

Definition. An *equilibrium* value for a model $P_{t+1} = F(P_t)$ is a value P^* such that $P^* = F(P^*)$. Equivalently, for a model $\Delta P = G(P_t)$, it is a value P^* such that $G(P^*) = 0$.

Finding equilibrium values is simply a matter of solving the equilibrium equation. For the model $P_{t+1} = P_t(1 + .7(1 - P_t/10))$, we solve $P^* =$

$P^*(1 + .7(1 - P^*/10))$ to see that there are precisely two equilibrium values, $P^* = 0$ or 10.

▶   Graphically, we can locate equilibria by looking for the intersection of the $P_{t+1} = F(P_t)$ curve with the diagonal line. Why does this work?

Equilibria can still have different qualitative features, though. In our example, $P^* = 0$ and 10 are both equilibria, but a population near 0 tends to move away from 0, whereas one near 10 tends to move toward 10. Thus, 0 is a *unstable* or repelling equilibrium, and 10 is a *stable* or attracting equilibrium.

Assuming our model comes close to describing a real population, stable equilibria are the ones that we would tend to observe in nature. Since any biological system is likely to experience small perturbations from our idealized model, even if a population was exactly at an equilibrium, we would expect it to be bounced at least a little away from it by factors left out of our model. If it is bounced a small distance from a stable equilibrium, though, it will move back toward it. On the other hand, if it is bounced away from an unstable equilibrium, it stays away. Although unstable equilibria are important for understanding the model as a whole, they are not population values we should ever really expect to observe for long in the real world.

Linearization. Our next goal is to determine what causes some equilibria to be stable and others to be unstable.

Stability depends on what happens close to an equilibrium; so, to focus attention near P^*, we consider a population $P_t = P^* + p_t$, where p_t is a very small number that tells us how far the population is from equilibrium. We call p_t the *perturbation* from equilibrium and are interested in how it changes. Therefore, we compute $P_{t+1} = P^* + p_{t+1}$ and use it to find p_{t+1}. If p_{t+1} is bigger than p_t in absolute value, then we know that P_{t+1} has moved away from P^*. If p_{t+1} is smaller than p_t in absolute value, then we know that P_{t+1} has moved toward P^*. Provided we can analyze how p_t changes for *all* small values of p_t, we'll be able to decide if the equilibrium is stable or unstable. A growing perturbation means instability, while a shrinking one means stability. (We are ignoring the sign of the perturbation here by considering its absolute value. Although the sign is worth understanding eventually, it is irrelevant to the question of stability.)

Example. Consider again the model $P_{t+1} = P_t(1 + .7(1 - P_t/10))$, which we know has equilibria $P^* = 0$ and 10. First, we'll investigate $P^* = 10$, which we believe is stable from numerical experiments. Substituting $P_t = 10 + p_t$

and $P_{t+1} = 10 + p_{t+1}$ into the equation for the model yields:

$$10 + p_{t+1} = (10 + p_t)(1 + .7(1 - (10 + p_t)/10))$$
$$10 + p_{t+1} = (10 + p_t)(1 + .7(-p_t/10))$$
$$10 + p_{t+1} = (10 + p_t)(1 - .07p_t)$$
$$10 + p_{t+1} = 10 + 0.3p_t - .07p_t^2$$
$$p_{t+1} = 0.3p_t - .07p_t^2.$$

But we are only interested in p_t being a very small number; so, p_t^2 is much smaller and negligible in comparison with p_t. Thus,

$$p_{t+1} \approx 0.3p_t.$$

This means that values of P_t close to the equilibrium will have their offset from the equilibrium compressed by a factor of about 0.3 with each time step. Small perturbations from the equilibrium therefore shrink, and $P^* = 10$ is indeed stable.

You should think of the number 0.3 as a "stretching factor" that tells how much perturbations from the equilibrium are increased. Here, because we stretch by a factor less than 1, we are really compressing.

The process performed in this example is called *linearization* of the model at the equilibrium, because we first focus attention near the equilibrium by our substitution $P_t = P^* + p_t$, and then ignore the terms of degree greater than 1 in p_t. What remains is just a linear model approximating the original model. Linear models, as we have seen, are easy to understand, because they produce either exponential growth or decay.

▶   Do a similar analysis for this model's other equilibrium to show it is unstable. What is the stretching factor by which distances from the equilibrium grow with each time step?

You should have found that linearization at $P^* = 0$ yields $p_{t+1} = 1.7p_t$. Therefore, perturbations from this equilibrium grow over time, so $P^* = 0$ is unstable. In general, when the stretching factor is greater than 1 in absolute value, the equilibrium is unstable. When it's less than 1 in absolute value, the equilibrium is stable.

A remark on calculus: If you know calculus, the linearization process might remind you of approximating the graph of a function by its tangent line. To develop this idea further, the stretching factor in the previous discussion could be expressed as the ratio $\dfrac{p_{t+1}}{p_t}$ for small values of p_t. But

$$\frac{p_{t+1}}{p_t} = \frac{P_{t+1} - P^*}{P_t - P^*} = \frac{F(P_t) - P^*}{P_t - P^*} = \frac{F(P_t) - F(P^*)}{P_t - P^*},$$

where $P_{t+1} = F(P_t)$ is the equation defining the model. (Note that we used $P^* = F(P^*)$ for the last equality.) Because we are interested only in values of P_t very close to P^*, this last expression is very close to

$$\lim_{P_t \to P^*} \frac{F(P_t) - F(P^*)}{P_t - P^*}.$$

But this limit is, by definition, nothing more than $F'(P^*)$, the derivative of the function defining our model. So, we have shown

Theorem. *If a model $P_{t+1} = F(P_t)$ has equilibrium P^*, then $|F'(P^*)| > 1$ implies P^* is unstable, while $|F'(P^*)| < 1$ implies P^* is stable. If $|F'(P^*)| = 1$, then this information is not enough to determine stability.*

Example. Using $P_{t+1} = P_t(1 + .7(1 - P_t/10))$ so $F(P) = P(1 + .7(1 - P/10))$, we compute $F'(P) = (1 + .7(1 - P/10)) + P(.7)(-1/10)$. Therefore, $F'(10) = 1 - .7 = 0.3$, and $P^* = 10$ is stable.

Note that, in this example, the value we found for $F'(10)$ was exactly the same as the value we found for the "stretching factor" in our earlier noncalculus approach. This had to happen, of course, because what lead us to the derivative initially was investigating this factor more thoroughly. The derivative can be interpreted, then, as a measure of how much a function "stretches out" values plugged into it.

Because we have taken a symbolic approach (i.e., writing down formulas and equations) in showing the connection between derivatives and stability, you should be sure to do problems 1.3.1 to 1.3.3 at the end of this section to see the connection graphically.

Why are both noncalculus and calculus approaches to stability presented here? The noncalculus one is the most intuitive and makes the essential ideas clearest, we think. It was even easy to do in the example. The weakness of it is that it only works for models involving simple algebraic formulas. If the model equation had exponentials or other complicated functions in it, the algebra simply would not have worked out. When things get complicated, calculus is a more powerful tool for analysis.

When linearizing to determine stability, it is vital that you are focusing on an equilibrium. Do not attempt to decide if a point is a stable or unstable equilibrium until *after* you have made sure it is an equilibrium; the analysis assumes that the point P^* satisfies $F(P^*) = P^*$. For example, if we tried to linearize F at 11 in the previous example, we could not conclude anything from the work, because 11 is not an equilibrium.

Finally, it is also important to realize that our analysis of stable and unstable equilibria has been a *local* one rather than a *global* one. What this terminology means is that we have considered what happens only in very small regions around an equilibrium. Although a stable equilibrium will attract values close to it, this does not mean that values far away must move toward it. Likewise, even though an equilibrium is unstable, we cannot say that values far away will not move toward (or even exactly to) it.

Oscillations, bifurcations, and chaos. In Problem 1.2.4 of the last section, you investigated the behavior of the logistic model $\Delta P = rP(1 - P/K)$ for $K = 10$ and a variety of values of r. In fact, the parameter K in the model is not really important; we can choose the units in which we measure the population so that the carrying capacity becomes 1. For example, if the carrying capacity is 10,000 organisms, we could choose to use units of 10,000 organisms, and then $K = 1$. This observation lets us focus more closely on how the parameter r affects the behavior of the model.

Setting $K = 1$, for any value of r the logistic model has two equilibria, 0 and 1, since those are the only values of P that make $\Delta P = 0$. As you will see in the problems section later, the "stretching factor" at $P^* = 0$ is $1 + r$, and at $P^* = 1$ is $1 - r$. $P^* = 0$, then is always an unstable equilibrium for $r > 0$.

$P^* = 1$ is much more interesting. First, when $0 < r \leq 1$, then $0 \leq 1 - r < 1$, so the equilibrium is stable. The formula $p_{t+1} \approx (1 - r)p_t$ shows that the sign of p_t will never change; although the perturbation shrinks, an initially positive perturbation remains positive and an initially negative one remains negative. The population simply moves toward equilibrium without ever overshooting it.

When r is increased so that $1 < r < 2$, then $-1 < 1 - r < 0$ and the equilibrium is still stable. Now, however, we see that because $p_{t+1} \approx (1 - r)p_t$, the sign of p_t will alternate between positive and negative as t increases. Thus, we should see oscillatory behavior above and below the equilibrium as our perturbation from equilibrium alternates in sign. The population therefore approaches the equilibrium as a *damped oscillation*.

Think about why this oscillation might happen in terms of a population being modeled. If r, a measure of the reproduction rate, is sufficiently large, a population below the carrying capacity of the environment may in a single time step grow so much that it exceeds the carrying capacity. Once it exceeds the carrying capacity, the population dies off rapidly enough that by the next time step it is again below the carrying capacity. But then it will once again grow enough to overshoot the carrying capacity. It's as if the population overcompensates at each time step.

If the parameter r of the logistic model is even larger than the values just considered, the population no longer approaches an equilibrium. When r is increased beyond 2, we find $|1 - r|$ exceeds 1 and therefore the stable equilibrium at $P^* = 1$ becomes unstable. Thus, a dramatic qualitative behavior change occurs as the parameter is increased across the value 2. An interesting question arises as to what the possibilities are for a model that has two unstable equilibria and no stable ones. What long-term behavior can we expect?

A computer experiment shows that for values of r slightly larger than 2, the population falls into a 2-cycle, endlessly bouncing back and forth between a value above 1 and a value below 1. As r is increased further, the values in the 2-cycle change, but the presence of the 2-cycle persists until we hit another value of r, at which another sudden qualitative change occurs. This time we see the 2-cycle becoming a 4-cycle. Further increases in r produce an 8-cycle, then a 16-cycle, and so on.

Already, this model has lead to an interesting biological conclusion: It is possible for a population to exhibit cycles even though the environment is completely unchanging. Assuming our modeling assumptions are correct and a population has a sufficiently high value of r, it may never reach a single equilibrium value.

A good way of understanding the effect of changing r on this model is through the *bifurcation diagram* of Figure 1.6.

Figure 1.6 is produced as follows. For each value of r on the horizontal axis, choose some value of P_0, and iterate the model for many time steps, so that transient behavior is past. (In practice, this means iterate for as many times as you can stand and you think might be necessary.) Then continue iterating for lots of additional steps, but now plot all these values of P_t on the vertical axis above the particular r used.

To illustrate the process for the discrete logistic model, suppose $r = 1.5$. Then, regardless of P_0, after the first set of many iterations, P_t will be very close to the stable equilibrium 1. Thus, when we plot the next set of many iterations, we just repeatedly plot points that will look like they are at $P = 1$.

If we then think of this process for an r slightly bigger than 2, the first set of iterations sends the population into a 2-cycle, and then when we plot the

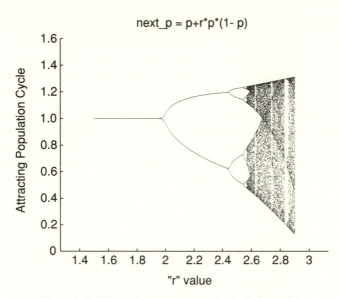

Figure 1.6. Bifurcation diagram for the logistic model.

next set of iterations, we plot points that bounce back and forth between the two values in the cycle, so that it appears that we have just plotted two points.

From this diagram, we notice several things. First the interval of r values over which we get a 2^{n+1}-cycle is shorter than that for a 2^n-cycle. Thus, once r is large, small additional increases can have more drastic effects.

Second, if r is increased past a certain point ($\approx 2.692\ldots$), all the bifurcations into 2^n-cycles have already taken place and a new type of behavior emerges. It appears as if the model values are changing more or less at random. However, the behavior is certainly *not* random – there is a completely deterministic formula producing it. The technical terminology for what has happened is that the model's behavior has become *chaotic*. The choice of the word "chaos" to describe this is perhaps unfortunate, since it calls up images of randomness and primordial confusion that are really irrelevant.

Chaos actually has a rather precise technical definition that we will not give. Instead, we merely informally point out two of the requirements mathematicians impose on the use of the word: 1) the model must be deterministic – that is, there can be no randomness in it; and 2) the predictions of the model are extremely sensitive to initial conditions.

To see that the discrete logistic model is in fact chaotic for $r = 2.75$, for example, we need to look into condition (2) a bit more. The plot in Figure 1.7 shows population values arising from two different, but very close, values of P_0.

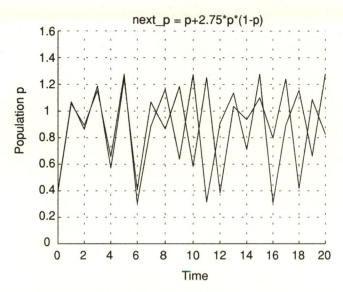

Figure 1.7 Populations resulting from two nearby initial values; logistic model $r = 2.75$.

Note that although the populations change similarly for the first few time steps, after a while they seem to be changing in completely different ways. For these initial values, then, we seem to have observed an extreme sensitivity of the model to the initial conditions. Of course, this is no proof of anything and it's conceivable that this behavior was just an artifact of computer round-off errors. It has been proven, however, that this is genuine chaos.

The possibility of chaotic behavior in as simple a population model as the discrete logistic one created quite a stir in the 1970s when it was first publicized by May (May, 1978). If such a simple model could produce such complicated behavior, then the natural view that complicated population dynamics can arise only from complicated interactions and environmental fluctuations would have to be abandoned.

Further work by May and others on determining appropriate values for parameters such as r in models of both laboratory and real-world insect populations led them to doubt that chaotic behavior was actually seen in real population dynamics. However, one examination of measles epidemics in New York City did suggest the possibility of chaos. Mumps and chickenpox, however, did not seem to behave chaotically. Although work is still being done, there is little data of high enough quality and long enough duration to really test the idea. More recent focus has been on population models more complex than the logistic one. In fact, in 1996, Cushing et al. announced the first unequivocal discovery of a real population, a laboratory population of the flour beetle *tribolium*, that exhibits chaotic dynamics (see (Cushing *et al.*, 2001)).

Problems

1.3.1. The equilibrium points of a model are located where the graph of P_{t+1} as a function of P_t crosses the line $P_{t+1} = P_t$. Suppose we focus on a section of the graph around an equilibrium point and zoom in so that the graph of P_{t+1} as a function of P_t appears to be a straight line. In each of the models of Figure 1.8, decide whether the equilibrium is stable or unstable by choosing a P_0 close to the steady state and then cobwebbing.

1.3.2. Reasoning from the preceding problem, in what range must the slope of the graph of P_{t+1} vs. P_t be at an equilibrium point to produce stability? Instability? (*Hint*: You might want to think about the special cases where the slope is first -1 and then 1.)

1.3.3. (Calculus) Phrase your answer to the preceding problem by using derivatives: If P^* is an equilibrium point of the model $P_{t+1} = f(P_t)$, then it is stable if . . . .

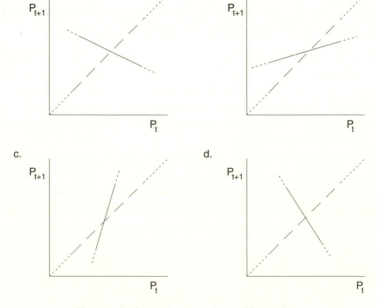

Figure 1.8. Cobweb graphs for problem 1.3.1.

1.3.4. Mathematically, when dealing with the logistic growth model $\Delta N = rN(1 - N/K)$, we can always choose the units in which N is measured so that $K = 1$. Thus, we need only consider $\Delta N = rN(1 - N)$, which has only one parameter, r, rather than two. Carefully investigate the long-term behavior of this model for various values of r starting at .5 and gradually increasing, by using the MATLAB program `onepop`. For what values of r do you see a simple approach to equilibrium without oscillations? An approach to equilibrium with oscillations? 2-cycle behavior? 4-cycle behavior?

1.3.5. In the preceding exercise, you discovered that as r is increased past 2, the population will stop tending to $K = 1$ and instead fall into a 2-cycle.

   a. Show that, regardless of the fact that the model falls into a 2-cycle, the only equilibrium points are still $N^* = 0$ and 1.

   b. If N_t falls into a 2-cycle, then $N_{t+2} \approx N_t$. Therefore, it may be worthwhile to find a formula for N_{t+2} in terms of N_t. Do it for $K = 1, r = 2.2$. Your answer should be a fourth-degree polynomial.

   c. Can you use your work in part (b) to find formulas for the points in the 2-cycle by setting $N_{t+2} = N_t$? Try it. Things may not work out nicely, but at least explain the difficulty.

1.3.6. For each of the following, determine the equilibrium points.

   a. $P_{t+1} = 1.3P_t - .02P_t^2$
   b. $P_{t+1} = 3.2P_t - .05P_t^2$
   c. $\Delta P = .2P(1 - P/20)$
   d. $\Delta P = aP - bP^2$
   e. $P_{t+1} = cP_t - dP_t^2$.

1.3.7. For (a–e) of the preceding problem, algebraically linearize the model first about the steady state 0 and then about the other steady state to determine their stability.

1.3.8. Compute the equilibrium points of the model $P_{t+1} = P_t + rP_t(1 - P_t)$. Then use only algebra to linearize at each of these points to determine when they are stable or unstable.

1.3.9. (Calculus) Redo the preceding problem, but use derivatives to determine the stability of the equilibria of $P_{t+1} = P_t + rP_t(1 - P_t)$. You should, of course, get the same answers.

1.3.10. (Calculus) Here is a slightly different approach to the relationship between derivatives and stability: Find the tangent line approximations to $f(P) = P + rP(1 - P)$ at the equilibria $P^* = 0$ and 1. Then replace $P + rP(1 - P)$ by these approximations in $P_{t+1} = P_t + rP_t(1 - P_t)$. Use this to determine the stability of the equilibria. Your answer should agree with the preceding two problems.

1.3.11. Many biological processes involve *diffusion*. A simple example is the passage of oxygen from the lung into the bloodstream (and the passage of carbon dioxide in the opposite direction). A simple model views the lung as a single compartment with oxygen concentration L and the bloodstream as an adjoining compartment with oxygen concentration B. If, for simplicity, we assume the compartments both have volume 1, then in the time span of a single breath the total oxygen $K = L + B$ is constant. If we think of a *very small* fixed-time interval, then the increase in B over this time interval will be proportional to the difference between L and B. That is,

$$\Delta B = r(L - B).$$

(This experimental fact is sometimes called *Fick's law*.)
   a. In what range must the parameter r be for this model to be meaningful?
   b. Use the fact that $L + B = K$ to write the model using only two parameters, r and K, to describe ΔB in terms of B.
   c. For $r = .1$, $K = 1$, and a variety of choices of B_0, investigate the model using the MATLAB program onepop. How do things change if a different value of r is used?
   d. Algebraically, find the equilibrium point B^* (in terms of r and K) for this model. Does this fit with what you saw in part (c)? Can you explain the result intuitively?
   e. Let $b = B - B^*$, and rewrite the model in terms of b, the offset from equilibrium, by substituting in $B = B^* + b$ and simplifying.
   f. Use part (e) to find a formula for b_t and then one for B_t. Make sure your formula gives the same results as onepop.
   g. Can you modify the model to deal with two compartments of unequal size?

Projects

1.   Suppose we know that, when undisturbed by humans, a commercially valuable population (e.g., a particular species of fish) has dynamics

modeled well by the discrete logistic difference equation

$$\Delta P = r P (1 - P/K).$$

Of course, the dynamics of the population will depend on the value of r, but by choosing appropriate units, we may assume $K = 1$.

Investigate the effect of regular harvesting of the population under two different types of assumptions.

a. $\Delta P = r P (1 - P/K) - H$, where H is some fixed number of individuals harvested at each time step.

b. $\Delta P = r P (1 - P/K) - hP$, where h is some fixed percentage of the population harvested at each time step (so, $0 \le h \le 1$).

Suggestions
- To get a feel for the models, investigate them numerically with onepop for lots of reasonable choices of the parameters. Make a note of any unusual behavior and try to explain it.
- Calculate analytically the equilibria (which may be in terms of r and H or h) and the stability of these equilibria (which may also depend on r and H or h).
- Explain the equilibria and stability in terms of cobweb diagrams. What effect does subtracting the harvesting terms H and hP have on the cobweb diagram of the logistic model?
- Try to find the largest H or h can be so that there is still a stable equilibrium. If h or H is chosen to be as large as possible so that there is still a stable equilibrium (which might be economically desirable), what happens to the unstable equilibrium?
- If you were responsible for managing the population, would you be comfortable if the stable equilibrium was close to the unstable one?
- Are there values of r for which H can be larger than K? Does this make sense biologically?
- If, in the absence of harvesting, a population has no stable equilibrium, can imposing harvesting lead to stability? Does this make sense biologically?
- Use the program longterm to create diagrams showing changes in long-term behavior as the parameters of the model are varied.

2. For an insect with a generation time significantly shorter than 1 year, it may be inappropriate to think of the carrying capacity as a constant. Investigate what happens if the carrying capacity varies sinusoidally. To

get started, try the MATLAB commands:

```
t=[0:50]
K=5+sin((2*pi/12)*t)
p=.1; pops=p
for i=1:50 p=p+.2*p*(1-p/K(i)); pops=[pops p]; end
plot(t,K,t,pops)
```

Suggestions
- Explain why a sinusoidally varying carrying capacity might be biologically reasonable under some circumstances.
- Investigate this situation for a variety of choices of r and P_0. Does P_t oscillate with K? Pay particular attention to when the population peaks and what the average population is in the long run. Do the results fit your biological intuition?
- What happens if the frequency of oscillation of the carrying capacity is changed? (Try replacing the "2*pi/12" in the previous command with "2*pi/N" for different N.)
- As r increases, does this model exhibit bifurcations? Chaos?

3. Investigate what happens if the carrying capacity varies randomly in a logistic model, and, in particular, the effect of such a carrying capacity on small populations. You will need to know that the MATLAB command rand(1) produces a random number between 0 and 1 with a uniform distribution, and that randn(1) produces a random number from a normal distribution with mean 0 and standard deviation 1. You might begin with using onepop with an expression like 10+rand(1) as the carrying capacity in the logistic model.

Suggestions
- Perhaps 10*rand(1) or 10+2*randn(1) would be a better form for the carrying capacity. Describe the qualitative differences between the biological situations these different expressions might describe.
- For your chosen carrying capacity expression, investigate the behavior of the model for a variety of choices of r and P_0. How does P_t behave? What is the average population in the long run? Do the results fit your biological intuition?
- As r increases, does this model exhibit bifurcations? Chaos?
- Investigate what happens if we have a small population, that must, of course, be integer valued. You will need to know that the MATLAB command floor(p) returns the largest integer less than or equal to

p. Your model should be something like

$$P_{t+1} = \text{floor}(P_t + r\,P_t(1 - P_t/K)),$$

where K is first a constant and then is made to vary randomly.

1.4. Variations on the Logistic Model

In presenting the discrete logistic model in previous sections, we have tried to keep the model as simple as possible to focus on developing the main ideas. Now that the concepts of equilibria and stability and the technique of cobwebbing have been developed, we can pay more attention to producing a more realistic model.

In looking at the graph of P_{t+1} as a function of P_t in Figure 1.9, for the model $P_{t+1} = P_t(1 + r(1 - P_t/K))$, one immediately obvious feature that is unrealistic is the fact that the parabola drops below the horizontal axis as we move off to the right. This means that large populations P_t become negative at the next time step. Although we can interpret a negative population as extinct, this may not be the behavior that would actually happen and that we would like our model to describe.

Perhaps a more reasonable model would have large values of P_t produce very small (but still positive) values of P_{t+1}. Thus, a population well over the carrying capacity might immediately crash to very low levels, but at least some of the population would survive. Graphically, P_{t+1} should depend on P_t as shown in Figure 1.10.

A function producing such a graph is $F(P) = Pe^{r(1-P/K)}$. The exponential in this formula produces the exponential-like decay as we move horizontally out on the graph, while the factor of P causes the initial rise in the graph near the origin.

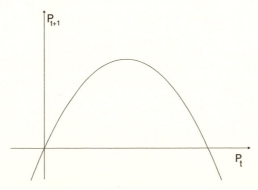

Figure 1.9. Model with unrealistic $P_{t+1} < 0$ for large P_t.

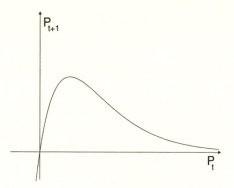

Figure 1.10. New model with $P_{t+1} > 0$.

The model $P_{t+1} = Pe^{r(1-P_t/K)}$ is sometimes called the discrete logistic model and is sometimes referred to as the Ricker model, after its first user (Ricker, 1954). As you can easily compute, the equilibria for this model are $P^* = 0$ and $P^* = K$. You can analyze this model by drawing cobweb diagrams and computing the stability of the equilibria, just as in the last sections.

You might object to this rabbit-out-of-the-hat approach to modeling; we have not quite explained where the equation for the Ricker model came from. Although we will give one explanation shortly, it is important to realize that what really matters about the formula is that it produces the qualitative graphical features we think are realistic. If a strange formula gives us the kind of graph we think we need, that is enough justification for using it.

To motivate more fully the Ricker model, let's return to the graph of the per-capita population change $\Delta P/P$ as a function of P that first motivated our development of the logistic model. Our sole reason for choosing the formula $\Delta P/P = r(1 - P/K)$ was to produce the downward trend shown in Figure 1.11.

How can we improve this? First, it is impossible for the per-capita population change to be less than -1, because that would mean more than one death per capita. That means our graph should really look more like Figure 1.12.

Since this looks like an exponential decay curve moved down 1 unit, that leads us to a formula such as:

$$\frac{\Delta P}{P} = ae^{-bP} - 1$$

for some positive values of a and b. To get the traditional form of the Ricker

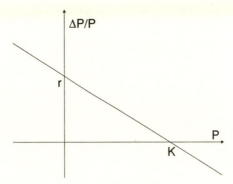

Figure 1.11. Per-capita growth rate for the logistic model.

model, we make some variable substitutions. Letting $b = \frac{r}{K}$ and $a = e^r$, in terms of the new parameters r and K, the model becomes

$$\frac{\Delta P}{P} = e^r e^{-rP/K} - 1 = e^{r(1-P/K)} - 1.$$

Now straightforward algebra leads to the Ricker formula

$$P_{t+1} = P_t e^{r(1-P_t/K)}.$$

In this formula, K should still be interpreted as the carrying capacity, because if $P > K$, then $\Delta P < 0$; and if $P < K$, then $\Delta P > 0$. The finite intrinsic rate of growth, however, is $e^r - 1$ rather than just r, although for small r these quantities are approximately the same.

Of course, the curve for $\Delta P/P$ does not have to be an exponential decay curve exactly. To model a population accurately, we would need to collect

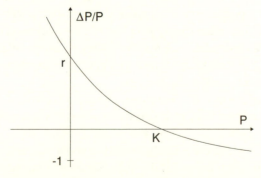

Figure 1.12. Per-capita growth rate for a new model.

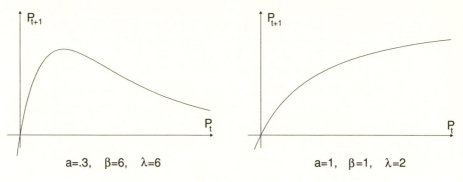

$$a=.3, \quad \beta=6, \quad \lambda=6 \qquad\qquad a=1, \quad \beta=1, \quad \lambda=2$$

Figure 1.13. Two models of the form $P_{t+1} = \frac{\lambda P_t}{(1+aP_t)^\beta}$.

data on how the population at time $t + 1$ depends on the size of the population at time t. We could then plot these points (P_t, P_{t+1}) and look for a formula that fits them well. Because the Ricker model has two parameters, r and K, by varying them, we may be able to make the curve fit the data reasonably well.

Another model often used is

$$P_{t+1} = \frac{\lambda P_t}{(1 + aP_t)^\beta}.$$

Although the meaning of the numbers λ, a, and β in this model may not be clear in biological terms, having three parameters simply allows more freedom in the shape of the curve to fit the data.

The graphs in Figure 1.13 show $P_{t+1} = \frac{\lambda P_t}{(1+aP_t)^\beta}$ for two different choices of values of the parameters. These two graphs describe drastically different population dynamics. The graph on the left that decays toward the horizontal axis represents a pure *scramble* competition for resources between individuals, where each individual simply gets less of the resources if the population is large. Thus, all individuals are hurt by having a large population around. A large value of P_t is thus likely to lead to a much smaller value for P_{t+1}; and, the larger P_t is, the smaller P_{t+1} will be.

The graph on the right that levels out above the horizontal axis represents a pure *contest* competition, in which if the population exceeds the carrying capacity, some individuals get all the resources, and others get none. Any large value of P_t is therefore likely to lead to about the same value of P_{t+1}. Of course, many populations exhibit behavior combining aspects of both of these competition types and so might be described by graphs somewhere in between.

Problems

1.4.1. For a discrete population model, the *relative growth rate* is defined as $\frac{P_{t+1}}{P_t}$.

    a. Complete: For a particular value of P_t, if the relative growth rate is larger than 1, then the population will _____ over the next time interval, whereas if it is smaller than 1, the population will _____.

    b. Does it make sense for the relative growth rate to be zero? Negative?

    c. Give expressions for the relative growth rates for the geometric and logistic population models, as well as the models of this section.

    d. Graph each of the relative growth rates you found in part (c) as functions of P_t. You may have to pick a few values of the parameters to draw the graphs.

1.4.2. Graphs (b), (c), and (d) of Problem 1.2.9 of Section 1.2 show $P_{t+1} = F(P_t) < P_t$ when P_t is small. Explain the affect of this feature on population dynamics. Why might this be a biologically important feature? (The resulting behavior is sometimes known as an *Allee effect*.)

1.4.3. Construct a simple model showing an Allee effect as follows.

    a. Explain why for some $0 < L < K$, the per-capita growth rate should be

$$\frac{\Delta P}{P} < 0, \quad \text{when} \quad 0 < P < L \text{ or } P > K,$$

$$\frac{\Delta P}{P} > 0, \quad \text{when} \quad L < P < K.$$

    Sketch a possible graph of $\Delta P/P$ vs. P.

    b. Explain why $\Delta P/P = P(K - P)(P - L)$ has the qualitative features desired.

    c. Investigate the resulting model using onepop and cobweb for some choices of K and L. Is the behavior as expected?

    d. What features of this modeling equation are unrealistic? How might the model be improved?

Projects

1. Investigate the Ricker model (Ricker, 1954)

$$N_{t+1} = N_t e^{r(1-N_t/K)}$$

more completely.

Suggestions
- Use a calculator or computer to graph N_{t+1} as a function of N_t (for several choices of r and K) and compare it with the corresponding graph for the logistic model. What are the qualitative similarities and differences between the graphs?
- Find all equilibria of the model.
- Use the MATLAB program `onepop` to investigate this model's dynamic behavior for $K = 1$ and a number of different r. Do you find stable equilibria? 2-cycles? 4-cycles? Chaotic behavior?
- Use the MATLAB program `longterm` to produce a bifurcation diagram for this model as r varies.

2. Repeat the last project for the model

$$X_{t+1} = \lambda X_t (1 + aX_t)^{-\beta},$$

which is frequently used for modeling insect populations. (For varying parameters, you might first let $\lambda = 6$ and $a = .3$ and vary β through positive values. Then hold λ and β fixed and vary a, etc.)

3. A famous model of the spruce budworm population proposed in (Ludwig *et al.*, 1978) (which used a differential equation) involved assuming logistic growth for the budworm, but introducing another term for predation due to birds. The predation term used was $g(N) = \frac{\beta^2 N^2}{\alpha^2 + N^2}$, where N denotes the number of budworms and α and β are two parameters than can be chosen to vary the graph to fit experimental data.

Suggestions
- Graph g and explain why its shape is reasonable to describe the number of budworms consumed by birds for various sizes of the budworm population. In particular, does it rise and level off as you think it should? How do the values of α and β affect the shape of the graph?
- Investigate the full model

$$N_{t+1} = N_t + rN_t(1 - N_t/K) - \frac{\beta^2 N_t^2}{\alpha^2 + N_t^2}$$

using MATLAB for a variety of parameter choices (but keep r small to avoid cycles or chaos in the logistic part of the model). Find parameter values that seem to give realistic behavior.
- What can you say about steady states and their stability?

1.5. Comments on Discrete and Continuous Models

In this chapter, we have discussed models using difference equations, which are built on discrete, finite (as opposed to infinitesimal) time steps. An alternative is to use differential equations, which assume things change continuously. Both difference and differential equations are used extensively for modeling throughout the sciences, and in many ways they have a parallel theory.

Differential equations are sometimes more amenable to analytic solution than difference equations. For example, the logistic differential equation does in fact have an explicit solution (i.e., a formula giving the value of the population at all times). In the precomputer era, differential equations were the primary choice of modelers, because more progress could be made in understanding such models. For certain fields, such as physiology (modeling such things as blood flow through the heart) and most of physics, where things really do seem to change continuously, they are still the natural choice.

Difference equations are more appropriate in situations in which there are natural discrete time steps. An example would be in modeling insect populations, which tend to have rather rigid life histories, with well-defined development stages and life spans. Now that computers are readily available, difference equations can be studied through numerical experiments.

In fact, because most complicated differential equation models are not explicitly solvable, those who use them often resort to using computers to perform simulations as well. Since computers work discretely, the models must first be translated into a discrete form. This may mean using an approach like Euler's method to approximate the differential equations – thus essentially pretending the differential equation is a difference equation. In the end, both difference and differential equations are valuable tools for investigating biological systems. Courses in calculus and differential equations are necessary for future biological modelers.

Though conceptually simpler than differential equations, difference equations often exhibit more complicated behavior. For instance, the discrete logistic model can exhibit cyclic or chaotic behavior, but the continuous logistic model never does. One explanation of this is that the time lags inherent in a discrete time step often mean the quantity being modeled cannot "figure out" by how much it should change quickly enough, so that it overshoots its "goal." However, sufficiently complicated differential equation models can also produce cycles and chaotic behavior.

Problems

1.5.1. (Calculus) The logistic differential equation is

$$\frac{dN}{dt} = rN(1 - N/K).$$

a. Show that

$$N(t) = \frac{K}{1 + Ce^{-rt}} \quad \text{where} \quad C = \frac{K - N_0}{N_0}$$

is a solution with initial condition $N(0) = N_0$.

b. Graph $N(t)$ for $K = 1$ and a few choices of r and N_0.

c. How does increasing r affect the solution? Explain how this compares with the behavior of the logistic difference equation.

2

Linear Models of Structured Populations

In the previous chapter, we first discussed the linear difference equation model $P_{t+1} = \lambda P_t$, which results in exponential growth or decay. After criticizing this model for not being realistic enough, we looked at nonlinear models that could result in quite complicated dynamics.

However, there is another way our models in the last chapter were simplistic – they treated all individuals in a population identically. In most populations, there are actually many subgroups whose vital behavior can be quite different. For instance, in humans, the death rate for infants is often higher than for older children. Also, children before the age of puberty contribute nothing to the birth rate. Even among adults, death rates are not constant, but tend to rise with advancing age.

In nonhuman populations, the differences can be more extreme. Insects go through a number of distinct life stages, such as egg, larva, pupa, and adult. Death rates may vary greatly across these different stages, and only adults are capable of reproducing. Plants also may have various stages they pass through, such as dormant seed, seedling, nonflowering, and flowering. How can a mathematical model take into account the subgroup structure that we would expect to play a large role in determining the overall growth or decline of such populations?

To create such *structured models*, we will focus on linear models. Even without resorting to nonlinear formulas, we can gain insight into how populations with distinct age groups, or developmental stages, can behave. Ultimately, we see that the behavior of these new linear models is quite similar to the exponential growth and decay of the linear model in the last chapter, with some important and interesting twists.

2.1. Linear Models and Matrix Algebra

The main modeling idea we use is simple. Rather than lumping the size of the entire population we are tracking into one quantity, with no regard for age or developmental stage, we consider several different quantities, such as the number of adults and the number of young. However, we limit ourselves to using very simple equations.

Example. Suppose we consider a hypothetical insect with three life stages: egg, larva, and adult. Our insect is such that individuals progress from egg to larva over one time step, and from larva to adult over another. Finally, adults lay eggs and die in one more time step. To formalize this, let

$$E_t = \text{the number of eggs at time } t,$$
$$L_t = \text{the number of larvae at time } t,$$
$$A_t = \text{the number of adults at time } t.$$

Suppose we collect data and find that only 4% of the eggs survive to become larvae, only 39% of the larvae make it to adulthood, and adults on average produce 73 eggs each. This can be expressed by the three equations

$$E_{t+1} = 73A_t,$$
$$L_{t+1} = .04E_t,$$
$$A_{t+1} = .39L_t.$$

This system of three difference equations is a model of the insect population. Note that because the equations involve no terms more complicated than those that appear in the equation of a line, it is justifiable to refer to this as a linear model. Also note that, if we wish to use this model to predict future populations, we need three initial values, E_0, L_0, and A_0, one for each stage class. Because the three equations are *coupled* (the population of one developmental stage appears in the formula giving the future population of a different stage), this system of difference equations is slightly more complicated than the linear models in the last chapter.

▶ The above example could actually be studied by the model

$$A_{t+3} = (.39)(.04)(73)A_t = 1.1388A_t,$$

where A_t is the number of adults. Explain why.

Of course, if we realize that $A_{t+3} = 1.1388A_t$ describes our population, then we immediately know that the population will grow exponentially, by a factor of 1.1388 for each three time steps.

Example. Consider the example above, but suppose that rather than dying, 65% of the adults alive at any time survive for an additional time step. Then the model becomes

$$E_{t+1} = 73A_t,$$
$$L_{t+1} = .04E_t,$$
$$A_{t+1} = .39L_t + .65A_t.$$

Again, we call this a linear model since all terms are of degree one. Because of our modification, however, it is no longer clear how to express the population's growth in terms of a single equation. It should be intuitively clear that the change in our model should result in an even more rapidly growing population than before. The adults who survive longer can produce more eggs, producing even more adults that survive longer, and so on. However, the new growth rate is by no means obvious.

Example. Suppose we are interested in a forest that is composed of two species of trees, with A_t and B_t denoting the number of each species in the forest in year t. When a tree dies, a new tree grows in its place, but the new tree might be of either species. To be concrete, suppose the species A trees are relatively long lived, with only 1% dying in any given year. On the other hand, 5% of the species B trees die. Because they are rapid growers, the B trees, however, are more likely to succeed in winning a vacant spot left by a dead tree; 75% of all vacant spots go to species B trees, and only 25% go to species A trees. All this can be expressed by

$$A_{t+1} = (.99 + (.25)(.01))A_t + (.25)(.05)B_t,$$
$$B_{t+1} = (.75)(.01)A_t + (.95 + (.75)(.05))B_t.$$

(2.1)

▶ Explain the source of each of the terms in these equations.

After simplifying, the model is a system of two linear difference equations

$$A_{t+1} = .9925A_t + .0125B_t,$$
$$B_{t+1} = .0075A_t + .9875B_t.$$

Unlike in the previous two examples, there is no obvious guess as to how populations modeled by these equations will behave.

In order to try to get numerical insight, suppose that we begin with a populations of $A_0 = 10$ and $B_0 = 990$. These initial population values might describe the forest if most of the A trees were selectively logged in the past.

Table 2.1. *Forest Model Simulation*

Year	A_t	B_t
0	10	990
1	22.30	977.70
2	34.35	965.65
3	46.17	953.83
4	57.74	942.26
5	69.09	930.91
$\vdots$	$\vdots$	$\vdots$
10	122.50	877.50
$\vdots$	$\vdots$	$\vdots$
50	401.04	598.96
$\vdots$	$\vdots$	$\vdots$
100	543.44	456.56
$\vdots$	$\vdots$	$\vdots$
500	624.97	375.03
$\vdots$	$\vdots$	$\vdots$
1000	625	375
$\vdots$	$\vdots$	$\vdots$

What will happen to the populations over time? A computer calculation shows the results in Table 2.1.

This table shows rather interesting behavior; it appears that the forest approaches an equilibrium, with 625 trees of species A and 375 of species B. In fact, as you can see in Figure 2.1, if we had started with any other nonnegative choices of A_0 and B_0, numerical calculations would have shown a similar movement toward exactly the same ratio $\frac{625}{375} = \frac{5}{3}$ of A trees to B trees. That the forest would even approach a stable distribution of the two species of trees is not obvious from our equations. It is even less clear why the stable distribution is in this $\frac{5}{3}$ ratio. To begin to understand the behavior of models such as the one above, we need to develop some more mathematical tools.

Vectors and matrices. The most convenient mathematical language to express models of the type above is that of linear algebra. It involves several types of mathematical objects that may be new to you.

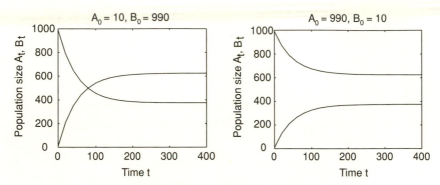

Figure 2.1. Two forest model simulations.

Definition. A *vector in* $\mathbb{R}^n$ is a list of n real numbers, usually written as a column.

Example. $\begin{pmatrix} 10 \\ 990 \end{pmatrix}$ and $\begin{pmatrix} 625 \\ 375 \end{pmatrix}$ are both vectors in $\mathbb{R}^2$; $\begin{pmatrix} 1 \\ -2 \\ 3 \end{pmatrix}$ is a vector in $\mathbb{R}^3$.

Vectors are usually denoted by small boldface letters; so, for instance, we might use $\mathbf{x}_t = \begin{pmatrix} A_t \\ B_t \end{pmatrix}$ to denote the tree distribution in year t in our last example, so that $\mathbf{x}_3 = \begin{pmatrix} 46.17 \\ 953.83 \end{pmatrix}$. As you can see, much space is being wasted on this page by insisting that vectors be written in columns. To remedy this, we will write things like $\mathbf{x}_3 = (46.17, 953.83)$ from now on, but we will always expect you to act as if we had written the numbers in a column.

Definition. An $m \times n$ *matrix* is a two-dimensional rectangular array of real numbers, with m rows and n columns.

Example. $\begin{pmatrix} .9925 & .0125 \\ .0075 & .9875 \end{pmatrix}$ is a 2×2 matrix and $\begin{pmatrix} 1 & -2 & 3 & -4 \\ 5 & -6 & -7 & 8 \\ -9 & 10 & -11 & 12 \end{pmatrix}$ is a 3×4 matrix.

If a matrix has the same number of rows as columns, it is said to be *square*. Note that there is not really any important difference between a vector in $\mathbb{R}^n$ and a $n \times 1$ matrix; they are written in an identical manner.

Matrices (the plural of "matrix") are usually denoted by capital letters, such as A, M, or P. For instance, we might say

$$P = \begin{pmatrix} .9925 & .0125 \\ .0075 & .9875 \end{pmatrix}$$

is the *projection*, or *transition*, matrix for our forest model above, because the entries in it are the numbers used to project future tree populations.

We now rewrite the forest model

$$\begin{aligned} A_{t+1} &= .9925A_t + .0125B_t \\ B_{t+1} &= .0075A_t + .9875B_t \end{aligned} \qquad (2.2)$$

in matrix notation as

$$\begin{pmatrix} A_{t+1} \\ B_{t+1} \end{pmatrix} = \begin{pmatrix} .9925 & .0125 \\ .0075 & .9875 \end{pmatrix} \begin{pmatrix} A_t \\ B_t \end{pmatrix} \qquad (2.3)$$

or $\mathbf{x}_{t+1} = P\mathbf{x}_t$. We've really gotten a bit ahead of ourselves here in our zeal to express the model in the simple form $\mathbf{x}_{t+1} = P\mathbf{x}_t$, which looks so much like the linear models we considered in the last chapter. What we have neglected to do is to make sure we know what we mean by writing $P\mathbf{x}_t$, a matrix times a vector.

We will define $P\mathbf{x}_t$ to be whatever is necessary, so that Eq. (2.3) means the same thing as Eq. (2.2). In other words, we need

$$\begin{pmatrix} .9925 & .0125 \\ .0075 & .9875 \end{pmatrix} \begin{pmatrix} A_t \\ B_t \end{pmatrix} = \begin{pmatrix} .9925A_t + .0125B_t \\ .0075A_t + .9875B_t \end{pmatrix}.$$

This leads us to define multiplication by:

Definition. The product of a 2×2 matrix and a vector in $\mathbb{R}^2$ is defined by

$$\begin{pmatrix} a & b \\ c & d \end{pmatrix} \begin{pmatrix} x \\ y \end{pmatrix} = \begin{pmatrix} ax + by \\ cx + dy \end{pmatrix}.$$

Rather than try to remember this formula, it is better to remember the process by which we multiply: Entries in the first row of the matrix are multiplied by the corresponding entries in the column vector and then all these products are added. This gives us the top entry in the product. The bottom entry is obtained the same way but using the bottom row of the matrix.

If we have a larger matrix than a 2×2, we proceed analogously. Note that to do this, each row of the matrix must have as many entries as the column vector. That means that if we have a vector in $\mathbb{R}^n$ and we try to multiply it on the left by a matrix, the matrix must have n entries in each row, and hence have n columns. Since we will be dealing primarily with square matrices, we will generally use $n \times n$ matrices to multiple vectors in $\mathbb{R}^n$.

Example.

$$\begin{pmatrix} 1 & -2 & 3 \\ -4 & 5 & -6 \\ 7 & -8 & 9 \end{pmatrix} \begin{pmatrix} 1 \\ 0 \\ -1 \end{pmatrix} = \begin{pmatrix} 1 \cdot 1 - 2 \cdot 0 + 3 \cdot -1 \\ -4 \cdot 1 + 5 \cdot 0 - 6 \cdot -1 \\ 7 \cdot 1 - 8 \cdot 0 + 9 \cdot -1 \end{pmatrix} = \begin{pmatrix} -2 \\ 2 \\ -2 \end{pmatrix}.$$

Think again of our forest with the two species of trees. Suppose the description above of the way the forest composition changes is what happens only in a wet year, so we rename the projection matrix

$$W = \begin{pmatrix} .9925 & .0125 \\ .0075 & .9875 \end{pmatrix}.$$

If, in dry years, we suppose species B dies at a greater rate, then a projection matrix for those years might be

$$D = \begin{pmatrix} .9925 & .0975 \\ .0075 & .9025 \end{pmatrix}.$$

▶ What is it about this matrix that suggests B trees have a higher mortality in dry years than in wet years?

In fact, all we have changed here is that the likelihood of a B tree dying in a dry year is now .39. All the other parameters are just as they were in Eq. (2.1).

▶ Verify that if the probability of a B tree dying is changed to .39, then the matrix D above results.

Suppose our initial populations are given by $x_0 = (10, 990)$ as before. Then, if the first year is dry,

$$x_1 = D x_0 = \begin{pmatrix} .9925 & .0975 \\ .0075 & .9025 \end{pmatrix} \begin{pmatrix} 10 \\ 990 \end{pmatrix}$$

$$= \begin{pmatrix} .9925 \cdot 10 + .0975 \cdot 990 \\ .0075 \cdot 10 + .9025 \cdot 990 \end{pmatrix} = \begin{pmatrix} 106.45 \\ 893.55 \end{pmatrix}.$$

Now suppose we have a dry year followed by a wet year; how should the populations change? Because we know $x_1 = D x_0$ and $x_2 = W x_1$, we see $x_2 =$

$W(D\mathbf{x}_0)$, which we could compute relatively easily by matrix multiplication:

$$\mathbf{x}_2 = \begin{pmatrix} .9925 & .0125 \\ .0075 & .9875 \end{pmatrix} \begin{pmatrix} 106.45 \\ 893.55 \end{pmatrix} \approx \begin{pmatrix} 116.82 \\ 883.18 \end{pmatrix}.$$

A more interesting question is can we find a *single* matrix that will tell us the cumulative effect on populations of a dry year followed by a wet year? Although we know $\mathbf{x}_2 = W(D\mathbf{x}_0)$, is there a matrix B so that $\mathbf{x}_2 = B\mathbf{x}_0$?

What we would like to do is simply move some parentheses in the equation $\mathbf{x}_2 = W(D\mathbf{x}_0)$, writing it as $\mathbf{x}_2 = (WD)\mathbf{x}_0$, and say the matrix that does what we want is WD. The problem with this is that we do not yet know how we could multiply the two matrices W and D to get a new matrix WD.

What should this matrix WD look like? Rather than worry about the particular numbers involved in our concrete example, let

$$D = \begin{pmatrix} a & b \\ c & d \end{pmatrix}, \quad W = \begin{pmatrix} e & f \\ g & h \end{pmatrix}, \quad \mathbf{x}_t = \begin{pmatrix} x_t \\ y_t \end{pmatrix}.$$

So

$$x_1 = ax_0 + by_0, \qquad x_2 = ex_1 + fy_1$$
$$y_1 = cx_0 + dy_0, \qquad y_2 = gx_1 + hy_1.$$

By substituting the left two equations into the right ones, we get

$$x_2 = e(ax_0 + by_0) + f(cx_0 + dy_0)$$
$$y_2 = (ax_0 + by_0) + h(cx_0 + dy_0),$$

or after rearranging,

$$x_2 = (ea + fc)x_0 + (eb + fd)y_0$$
$$y_2 = (ga + hc)x_0 + (gb + hd)y_0.$$

In matrix form, this becomes

$$\mathbf{x}_2 = \begin{pmatrix} ea + fc & eb + fd \\ ga + hc & gb + hd \end{pmatrix} \mathbf{x}_0.$$

This indicates how we should define the product of two matrices; we want

$$WD = \begin{pmatrix} e & f \\ g & h \end{pmatrix} \begin{pmatrix} a & b \\ c & d \end{pmatrix} = \begin{pmatrix} ea + fc & eb + fd \\ ga + hc & gb + hd \end{pmatrix}.$$

Notice the first column of our product on the right comes from multiplying W times the first column of D (treated as a vector), and the second column of the product comes from multiplying W times the second column of D.

Definition. The product of two matrices is a new matrix, whose columns are found by multiplying the matrix on the left times each of the columns of the matrix on the right.

This means that, in order to multiply two matrices, if the one on the right has n entries in each column, the one on the left must have n entries in each row.

Example.

$$\begin{pmatrix} 1 & 3 \\ -1 & 2 \end{pmatrix} \begin{pmatrix} 2 & 1 \\ -2 & 1 \end{pmatrix} = \begin{pmatrix} 1 \cdot 2 + 3 \cdot -2 & 1 \cdot 1 + 3 \cdot 1 \\ -1 \cdot 2 + 2 \cdot -2 & -1 \cdot 1 + 2 \cdot 1 \end{pmatrix} = \begin{pmatrix} -4 & 4 \\ -6 & 1 \end{pmatrix}.$$

An interesting thing happens if we multiply the above two matrices again, but with them written in the opposite order – we get a different result.

Example.

$$\begin{pmatrix} 2 & 1 \\ -2 & 1 \end{pmatrix} \begin{pmatrix} 1 & 3 \\ -1 & 2 \end{pmatrix} = \begin{pmatrix} 2 \cdot 1 + 1 \cdot -1 & 2 \cdot 3 + 1 \cdot 2 \\ -2 \cdot 1 + 1 \cdot -1 & -2 \cdot 3 + 1 \cdot 2 \end{pmatrix} = \begin{pmatrix} 1 & 8 \\ -3 & -4 \end{pmatrix}.$$

Warning: For most matrices A and B, $AB \neq BA$. Matrix multiplication is not commutative. The order within a product matters.

▶ Biologically, would you expect the effect on a forest of a dry year followed by a wet year to be exactly the same as that of a wet year followed by a dry year? What does this have to do with the warning?

Example. Note that, although a product like $\begin{pmatrix} .2 & .7 \\ 0 & .4 \end{pmatrix} \begin{pmatrix} 3.2 \\ 1.1 \end{pmatrix}$ makes sense, if the vector is placed on the left as $\begin{pmatrix} 3.2 \\ 1.1 \end{pmatrix} \begin{pmatrix} .2 & .7 \\ 0 & .4 \end{pmatrix}$, then the product does not make sense anymore. Because there is only one entry in each row of $\begin{pmatrix} 3.2 \\ 1.1 \end{pmatrix}$, and $\begin{pmatrix} .2 & .7 \\ 0 & .4 \end{pmatrix}$ has two entries in each column, the definition of matrix multiplication cannot be used. Since we are writing our vectors as columns, this means we must always put matrices to the left of vectors in products.

The fact that for matrices multiplication is not commutative – that order matters in a product – is a significant difference from the algebra of ordinary numbers. It is very important to always be aware of this when using matrices.

Fortunately, although we will not carefully show it here, matrix multiplication is associative: when multiplying three matrices, it is always true that $(AB)C = A(BC)$. You can regroup products however you wish, as long as you do not change the order. [A hint at why this turns out to be true: we defined the product of two matrices so that $A(BC) = (AB)C$ would hold in the special case when C is a vector. It only takes a little more thought to see that the definition then forces the same equality to be true when C is any matrix.]

Of course, it takes some practice to get comfortable with the algebra of matrices, but that is what the exercises are for. Most people use computers for performing matrix calculations, especially when the sizes of the matrices are large. Once you understand how to perform the work, the whole process becomes very tedious to do by hand. Nonetheless, you have to be able to do simple hand calculations to develop the understanding to use a computer effectively.

There are a few other terms and rules that are used in manipulating vectors and matrices.

Because we have names (vectors and matrices) for arrays of numbers, it is convenient to have a name for single numbers as well.

Definition. A *scalar* is a single number.

Definition. To multiply a vector or a matrix by a scalar, multiply every entry by that scalar.

Example. $3 \begin{pmatrix} 1 \\ 2 \\ 3 \end{pmatrix} = \begin{pmatrix} 3 \\ 6 \\ 9 \end{pmatrix}$ and $-.2 \begin{pmatrix} 1 & -1 \\ 2 & 1 \end{pmatrix} = \begin{pmatrix} -.2 & .2 \\ -.4 & -.2 \end{pmatrix}.$

Definition. To add two vectors together or to add two matrices together, add corresponding entries. The things being added must be the same size.

Example. $\begin{pmatrix} 1 \\ 2 \end{pmatrix} + \begin{pmatrix} -1 \\ 4 \end{pmatrix} = \begin{pmatrix} 0 \\ 6 \end{pmatrix}$ and $\begin{pmatrix} 1 & -1 \\ 2 & 1 \end{pmatrix} + \begin{pmatrix} 0 & 1 \\ -1 & 2 \end{pmatrix} = \begin{pmatrix} 1 & 0 \\ 1 & 3 \end{pmatrix}.$

Definition. A vector whose entries are all zero is called a *zero vector* and is denoted by **0**.

Vectors and matrices also obey several distributive laws of multiplication over addition such as

$$A(B + C) = AB + AC, (B + C)A = BA + CA, \quad \text{and}$$
$$A(\mathbf{x} + \mathbf{y}) = A\mathbf{x} + A\mathbf{y}.$$

Finally, we note that although matrices in products do not usually commute, it is valid to interchange the order of a matrix and a scalar; for instance, $Ac\mathbf{x} = cA\mathbf{x}$.

Problems

2.1.1. Without a computer, find the products

a. $\begin{pmatrix} 2 & -3 \\ 1 & 7 \end{pmatrix} \begin{pmatrix} 3 \\ 2 \end{pmatrix}$

b. $\begin{pmatrix} 1 & 3 & -2 \\ 4 & -3 & 1 \\ 0 & 1 & -4 \end{pmatrix} \begin{pmatrix} 3 \\ 2 \\ 5 \end{pmatrix}$

c. $\begin{pmatrix} 2 & -3 \\ 1 & 7 \end{pmatrix} \begin{pmatrix} 3 & 2 \\ 2 & 4 \end{pmatrix}$

d. $\begin{pmatrix} 1 & 3 & -2 \\ 4 & -3 & 1 \\ 0 & 1 & -4 \end{pmatrix} \begin{pmatrix} 3 & 1 & 0 \\ 2 & -1 & 3 \\ 5 & 0 & 1 \end{pmatrix}$

2.1.2. Explain why the product $\begin{pmatrix} 3 \\ 2 \end{pmatrix} \begin{pmatrix} 2 & -3 \\ 1 & 7 \end{pmatrix}$ cannot be calculated.

2.1.3. For $A = \begin{pmatrix} 1 & 2 \\ -1 & 1 \end{pmatrix}$, $B = \begin{pmatrix} 3 & -1 \\ -2 & 2 \end{pmatrix}$, and $C = \begin{pmatrix} -1 & 1 \\ -3 & 4 \end{pmatrix}$, find the following without a computer. Then check your answers with MATLAB. Matrices are entered as A=[1,2;-1,1].

a. $A + B$

b. AB

c. BA

d. $A^2 = AA$

e. $2A$

f. Show $(A + B)C = AC + BC$.

2.1.4. For $A = \begin{pmatrix} 1 & 0 & -1 \\ 2 & 1 & 0 \\ -1 & 1 & -2 \end{pmatrix}$, $B = \begin{pmatrix} 3 & 2 & -1 \\ -2 & 0 & 2 \\ 0 & -1 & 1 \end{pmatrix}$, and $C = \begin{pmatrix} 1 & 0 & 2 \\ -2 & 1 & 1 \\ 3 & -1 & 1 \end{pmatrix}$, find the following without a computer. Then check your answers with MATLAB.

a. $A + B$

b. AB

c. BA

d. $A^2 = AA$

e. $2A$

f. Show $C(A + B) = CA + CB$.

2.1.5. For $A = \begin{pmatrix} r & s \\ t & u \end{pmatrix}$ and $\mathbf{x} = \begin{pmatrix} x \\ y \end{pmatrix}$ and c a scalar, show $A(c\mathbf{x}) = c(A\mathbf{x})$ by computing each side.

2.1.6. For the matrix P in the text that models a forest succession, compute P^2, P^3, P^{500}. What is the biological meaning of each of these matrices? What is significant about the entries you see in P^{500}? (Use MATLAB for calculations.)

2.1.7. For the matrix P in the text that models a forest succession, produce a plot of the number of trees of each type over many years assuming $\mathbf{x}_0 = (10, 990)$. Use the MATLAB commands

```
P=[.9925 .0125;  .0075 .9875]
x=[10;  990]
pops=[x]
x=P*x
pops=[pops x]
x=P*x
pops=[pops x]
x=P*x
pops=[pops x]
        ⋮
plot(pops')
```

Repeat this process several more times using different initial vectors with entries adding to 1,000. Do all initial vectors ultimately lead to the same forest composition?

2.1.8. The first example of this section describes an insect model given by

$$E_{t+1} = 73A_t,$$
$$L_{t+1} = .04E_t,$$
$$A_{t+1} = .39L_t.$$

a. Express this model as $\mathbf{x}_{t+1} = P\mathbf{x}_t$ using a 3×3 matrix P. What is $\mathbf{x}_t$?
b. Compute P^2 and P^3 without the aid of a computer. What is the biological meaning of these matrices?
c. Your computation of P^3 should remind you of the equation

$$A_{t+3} = (.39)(.04)(73)A_t = 1.1388A_t$$

in the text. Explain the connection.

2.1.9. The second example of this section describes an insect model given by

$$E_{t+1} = 73A_t,$$
$$L_{t+1} = .04E_t,$$
$$A_{t+1} = .39L_t + .65A_t.$$

a. Express this model using a 3×3 matrix P.
b. Compute P^2 and P^3 without the aid of a computer.
c. Beginning with initial populations of $(E_0, L_0, A_0) = (10, 10, 10)$, produce a plot of the population sizes over time using a computer. You can modify the commands in Problem 2.1.7 to do this with MATLAB.

2.2. Projection Matrices for Structured Models

Although linear models have many applications beyond understanding population growth, there are several common applications of them in modeling populations. In this setting, the projection matrices often have a rather distinct form, because there are natural ways of breaking the population into subgroups by age or developmental stage.

The Leslie model. In some species, the amount of reproduction varies greatly with the age of individuals. For instance, consider two different human populations that have the same total size. If one is comprised primarily of those over 50 in age, while the other has mostly individuals in their 20s, we would expect quite different population growth from them. Clearly, the age structure of the population matters.

Humans progress through a relatively long period before puberty when no reproduction occurs. After puberty, various social factors discourage or encourage childbearing at certain ages. Finally, menopause limits reproduction by older women.

To capture the effects on population growth, we might begin modeling a human population by creating five age classes with:

$$x_1(t) = \text{no. of individuals age 0 through 14 at time } t,$$

$$x_2(t) = \text{no. of individuals age 15 through 29 at time } t,$$

$$x_3(t) = \text{no. of individuals age 30 through 44 at time } t,$$

$$x_4(t) = \text{no. of individuals age 45 through 59 at time } t,$$

$$x_5(t) = \text{no. of individuals age 60 through 75 at time } t.$$

Although this formulation makes the unrealistic assumption that no one survives past age 75, that shortcoming could of course be remedied by creating additional age classes. Using a time step of 15 years, we can describe the population through equations like:

$$
\begin{aligned}
x_1(t+1) &= f_1 x_1(t) &+ f_2 x_2(t) &+ f_3 x_3(t) &+ f_4 x_4(t) &+ f_5 x_5(t) \\
x_2(t+1) &= \tau_{1,2} x_1(t) \\
x_3(t+1) &= &\tau_{2,3} x_2(t) \\
x_4(t+1) &= &&\tau_{3,4} x_3(t) \\
x_5(t+1) &= &&&\tau_{4,5} x_4(t).
\end{aligned}
$$

Here, f_i denotes a birth rate (over a 15-year period) for parents in the ith age class, and $\tau_{i,i+1}$ denotes a survival rate for those in the ith age class passing into the $(i+1)$th. Because a single set of parents may be in different age groups, we should attribute half of their offspring to each in choosing values for f_i.

In matrix notation, the model is simply $\mathbf{x}_{t+1} = P\mathbf{x}_t$, where

$$\mathbf{x}_t = (x_1(t), x_2(t), x_3(t), x_4(t), x_5(t))$$

is the column vector of subpopulation sizes at time t and

$$
P = \begin{pmatrix}
f_1 & f_2 & f_3 & f_4 & f_5 \\
\tau_{1,2} & 0 & 0 & 0 & 0 \\
0 & \tau_{2,3} & 0 & 0 & 0 \\
0 & 0 & \tau_{3,4} & 0 & 0 \\
0 & 0 & 0 & \tau_{4,5} & 0
\end{pmatrix}
$$

is the projection matrix.

We might expect f_1 to be smaller than f_2, because fewer 0- to 15-year-olds are likely to give birth than 15- to 30-year-olds. However, remember that, in the course of a time step, the 0- to 15-year-olds age by 15 years; therefore, the birth rate to such parents is probably not as small as you might have thought.

It is also possible that some of the f_i are zero; for instance, the very old may not reproduce.

▶ If data were collected, which of the numbers f_i do you think would be largest? Which would be smallest? How might this vary depending on which particular human population was being modeled?

▶ What might be reasonable values for the $\tau_{i,i+1}$? Which are likely to be largest? Smallest?

Of course we might improve our model by using more age classes of smaller duration, say 5 years or even 1 year, and adding additional age classes for those over 75. For humans, age classes of 15 years are too long for much accuracy. Demographers often use 5-year classes and track individuals to age 85, which results in a 17×17 matrix.

With an improved model, our matrix would be larger, but it would still have the same form: The top row would have fecundity information, the *subdiagonal* would have survival information, and the rest of the matrix would have entries of 0. A model whose projection matrix has this form is called a *Leslie* model.

Example. A Leslie model describing the U.S. population in 1964 was formulated in (Keyfitz and Murphy, 1967). Tracking only females, and hence ignoring the births of any males in the computation of birth rates, it used 10 age groups of 5-year durations and a time step of 5 years. The top row of the matrix was

(.0000, .0010, .0878, .3487, .4761, .3377, .1833, .0761, .0174, .0010),

while the subdiagonal was

(.9966, .9983, .9979, .9968, .9961, .9947, .9923, .9987, .9831).

▶ What is the meaning of the fact that the first subdiagonal entry is smaller than the second? What are possible explanations for this?

▶ Why might the seventh subdiagonal entry be smaller than those to either side of it? What age group of females is this number describing?

▶ Why might it be reasonable to only include females up to age 50 in this model?

The Usher model. An Usher model is a slight variation on a Leslie model, in which there may be nonzero entries on the diagonal. For example, return to the 5×5 matrix model of humans above, and continue to use 15-year-long age classes, but make the time step only 5 years in duration. Then, while some of the individuals in a class will move up to the next class after a time step, many will stay where they are. This results in a matrix of the form

$$\begin{pmatrix} f_1 + \tau_{1,1} & f_2 & f_3 & f_4 & f_5 \\ \tau_{1,2} & \tau_{2,2} & 0 & 0 & 0 \\ 0 & \tau_{2,3} & \tau_{3,3} & 0 & 0 \\ 0 & 0 & \tau_{3,4} & \tau_{4,4} & 0 \\ 0 & 0 & 0 & \tau_{4,5} & \tau_{5,5} \end{pmatrix},$$

with the parameters $\tau_{i,i}$ describing the fraction of the ith age class that remains in that class in passage to the next time step. Note that the values of the entries $\tau_{i,i+1}$ and f_i will be different from what they were in the Leslie version above, because the time step size has been changed.

Perhaps a more natural example of an Usher model is one based on the developmental stages an organism passes through in its lifetime. For instance for a mammal such as a whale that takes several years to reach sexual maturity, and may also live past an age where it breeds, a three-stage model might be used, with immature, breeding, and postbreeding classes. The Usher matrix

$$\begin{pmatrix} \tau_{1,1} & f_2 & 0 \\ \tau_{1,2} & \tau_{2,2} & 0 \\ 0 & \tau_{2,3} & \tau_{3,3} \end{pmatrix}$$

could describe such a population.

▶ Why is there only one nonzero f_i in this matrix?

Other structured population models. Although Leslie and Usher models are natural and common ones for describing populations, mathematically there is little special about the particular matrix forms they use. If a species can be better modeled by a different matrix model, then there is no reason not to.

As an example, consider a plant that takes several years to mature to a flowering stage and that does not flower every year after reaching maturity. In addition, seeds may lie dormant for several years before germinating.

The life cycle of this plant could be modeled using time steps of a year and the classes

$$x_1(t) = \text{no. of ungerminated seeds at time } t,$$
$$x_2(t) = \text{no. of sexually immature plants at time } t,$$
$$x_3(t) = \text{no. of mature plants flowering at time } t,$$
$$x_4(t) = \text{no. of mature plants not flowering at time } t.$$

With $\mathbf{x}(t) = (x_1(t), x_2(t), x_3(t), x_4(t))$, the projection matrix for the model might have the form

$$\begin{pmatrix} \tau_{1,1} & 0 & f_{3,1} & 0 \\ \tau_{1,2} & \tau_{2,2} & f_{3,2} & 0 \\ 0 & \tau_{2,3} & \tau_{3,3} & \tau_{4,3} \\ 0 & 0 & \tau_{3,4} & \tau_{4,4} \end{pmatrix}.$$

Here, the parameter $\tau_{4,3}$ describes mature plants that did not flower in one season passing into the flowering class for the next. In addition, there are two parameters describing fertility – $f_{3,1}$ describes the production of seeds that do not germinate immediately, whereas $f_{3,2}$ describes the production of seedlings through new seeds that germinate by the next time step.

▶ Which parameter describes the seeds produced in previous years that again do not germinate, but may germinate in the future?

Example. For this plant model with the particular parameter choices given by

$$\begin{pmatrix} .02 & 0 & 11.9 & 0 \\ .05 & .12 & 5.7 & 0 \\ 0 & .14 & .21 & .32 \\ 0 & 0 & .43 & .11 \end{pmatrix}, \tag{2.4}$$

and an initial population vector of $\mathbf{x}_0 = (0, 50, 50, 0)$, the populations over the next 12 time steps are shown in Figure 2.2.. We see a clear growth trend in the sizes of all the classes, with some overlying oscillations for at least the first few time steps. Moreover, there is a roughly constant ratio between the sizes of the classes after a few steps.

The behavior exhibited in Figure 2.2. is typical of Leslie and Usher models as well, regardless of the number of classes involved. Generally, there is a dominant trend of growth or decay, although smaller scale fluctuations are often present also. The dominant trend appears similar to the exponential growth or decay of the Malthusian model. However, the class structure of the model produces more intricate behavior as well.

The forest model of Section 2.1 is another example of a linear model that is neither Leslie nor Usher. Because it tracks two types of trees, rather than an organism going through its life cycle, the projection matrix has a rather different form. It is an example of a *Markov* model, which we will develop further in Chapter 4. We saw, however, in Figure 2.1 that this model also

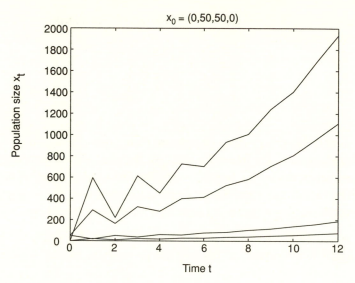

Figure 2.2. Simulation of plant model; on the right side of graph, classes are in order 1, 2, 3, and 4 from top.

showed a long-term trend, toward an equilibrium. We will develop a means of extracting information on the main trends produced by any linear model in the next section.

In modeling real populations' life stages, the decision to use a Leslie model, an Usher model, or a unnamed variant must take into account a number of factors. An understanding of the life cycle of the organism may make a natural choice of classes clear. However, the difficulty of finding good estimates of the parameters could also dictate choices, since if more classes are used, then more parameters appear in the model. Using very small age groups or many different stages should, in theory, produce a more accurate model. However, it also requires more detailed surveying to obtain reasonably accurate parameter values.

The identity matrix and matrix inverses. Having looked in more detail at the types of matrices used in linear population models, let's return to developing some mathematical tools for understanding them.

Suppose a linear population model uses only two classes, and hence has a 2×2 projection matrix P. If the population at time 1 is given by the vector $\mathbf{x}_1$, then computing the populations at the next time step just requires a multiplication

$$\mathbf{x}_2 = P\mathbf{x}_1.$$

But imagine that we are interested in deducing the populations at the previous time step. If we know $\mathbf{x}_1$ and P, how can we find $\mathbf{x}_0$? In other words, can we project populations *backward* in time if we only have a matrix P describing how they change *forward* in time?

If P were a scalar instead of a matrix, we would know how to do this. We would simply "divide" each side of the equation $\mathbf{x}_1 = P\mathbf{x}_0$ by P to solve for $\mathbf{x}_0$. Unfortunately, it is not clear what "dividing by a matrix" means.

A slightly better way to think of it is as follows: Can the equation $\mathbf{x}_1 = P\mathbf{x}_0$ be multiplied on each side by some matrix to remove the P from the right-hand side? Suppose we try this and pick some 2×2 matrix Q so that $\mathbf{x}_1 = P\mathbf{x}_0$ becomes $Q\mathbf{x}_1 = QP\mathbf{x}_0$. If our goal was to get rid of the P, we need QP to disappear from the equation. Unfortunately, QP will be a 2×2 matrix and there is no way around that. However, there is a special 2×2 matrix that would be good enough for our purposes.

Definition. The 2×2 *identity matrix* is

$$I = \begin{pmatrix} 1 & 0 \\ 0 & 1 \end{pmatrix}.$$

The $n \times n$ *identity matrix* is a square matrix whose entries are all 0, except for 1's on the main diagonal.

Note that

$$I \begin{pmatrix} x \\ y \end{pmatrix} = \begin{pmatrix} 1 & 0 \\ 0 & 1 \end{pmatrix} \begin{pmatrix} x \\ y \end{pmatrix} = \begin{pmatrix} x \\ y \end{pmatrix},$$

so that I behaves like the number 1 in ordinary algebra with scalars. Multiplying any vector times I leaves the vector unchanged. You should check that $AI = A$ and $IA = A$ for any matrix A as well.

Returning to our attempt to project a population backward in time, we had $Q\mathbf{x}_1 = QP\mathbf{x}_0$ so that if we can just pick Q so that $QP = I$, the equation becomes

$$Q\mathbf{x}_1 = I\mathbf{x}_0 = \mathbf{x}_0.$$

In other words, we will have managed to solve for $\mathbf{x}_0$ by calculating $Q\mathbf{x}_1$.

Definition. If P and Q are both $n \times n$ square matrices with $QP = I$, then we say that Q is the *inverse* of P. We then use the notation $Q = P^{-1}$.

Although we will not prove it here, it is possible to show that, for square matrices, if $QP = I$, then $PQ = I$. So, if Q is the inverse of P, then P is the inverse of Q.

Before we try to calculate the inverse of a matrix, we should ask ourselves if one really has to exist. For instance

$$\begin{pmatrix} -2 & 1 \\ 1.5 & -.5 \end{pmatrix} \begin{pmatrix} 1 & 2 \\ 3 & 4 \end{pmatrix} = \begin{pmatrix} 1 & 0 \\ 0 & 1 \end{pmatrix},$$

so

$$\begin{pmatrix} 1 & 2 \\ 3 & 4 \end{pmatrix}^{-1} = \begin{pmatrix} -2 & 1 \\ 1.5 & -.5 \end{pmatrix}.$$

On the other hand, if $A = \begin{pmatrix} 0 & -2 \\ 0 & -3 \end{pmatrix}$, then A does not have an inverse. To see this, think about

$$\begin{pmatrix} * & * \\ * & * \end{pmatrix} \begin{pmatrix} 0 & -2 \\ 0 & -3 \end{pmatrix} = \begin{pmatrix} 1 & 0 \\ 0 & 1 \end{pmatrix}.$$

You simply cannot fill in the entries in the top row of the matrix on the left so that the upper left entry in the product is 1. Because of the column of 0's in A, the upper left entry in the product will always be 0.

Although this example has shown that are some matrices without inverses, trying to find the inverse of a generic 2×2 matrix $\begin{pmatrix} a & b \\ c & d \end{pmatrix}$ will give us more insight into the problem. We will make guesses as to how to fill out the unknown matrix in the equation

$$\begin{pmatrix} * & * \\ * & * \end{pmatrix} \begin{pmatrix} a & b \\ c & d \end{pmatrix} = \begin{pmatrix} 1 & 0 \\ 0 & 1 \end{pmatrix}.$$

Focusing on the upper right entry in the product first, we can easily get a zero there by putting d and $-b$ in the top row of the empty matrix. To get a zero in the bottom left entry of the product, we can put $-c$ and a in the bottom row. This leaves us with

$$\begin{pmatrix} d & -b \\ -c & a \end{pmatrix} \begin{pmatrix} a & b \\ c & d \end{pmatrix} = \begin{pmatrix} ad - bc & 0 \\ 0 & ad - bc \end{pmatrix}.$$

To make sure we get 1's on the diagonal, we just need to divide every entry in the matrix on the left by $ad - bc$, so

$$\begin{pmatrix} \frac{d}{ad-bc} & \frac{-b}{ad-bc} \\ \frac{-c}{ad-bc} & \frac{a}{ad-bc} \end{pmatrix} \begin{pmatrix} a & b \\ c & d \end{pmatrix} = \begin{pmatrix} 1 & 0 \\ 0 & 1 \end{pmatrix}.$$

The number $ad - bc$ is given a special name:

Definition. The *determinant* of a 2×2 matrix $A = \begin{pmatrix} a & b \\ c & d \end{pmatrix}$ is the scalar $ad - bc$. It is denoted by $\det A$ or $|A|$.

Our formula for the inverse of a 2×2 matrix becomes:

$$\text{If } A = \begin{pmatrix} a & b \\ c & d \end{pmatrix}, \quad \text{then} \quad A^{-1} = \frac{1}{\det A} \begin{pmatrix} d & -b \\ -c & a \end{pmatrix}.$$

Example.

$$\begin{pmatrix} 3 & -1 \\ 2 & 1 \end{pmatrix}^{-1} = \frac{1}{3 \cdot 1 - (-1) \cdot 2} \begin{pmatrix} 1 & 1 \\ -2 & 3 \end{pmatrix} = \begin{pmatrix} .2 & .2 \\ -.4 & .6 \end{pmatrix}.$$

Because not every matrix has an inverse, we cannot have really found a formula for the inverse of all 2×2 matrices. Something must go wrong occasionally. Looking at the formula, we see that it does not make sense if $\det A = 0$. In fact, although we will not prove it, if $\det A = 0$ then A has no inverse. In other words, to find the inverse of a 2×2 matrix, we can just try to use the formula. If the formula does not make sense, then the matrix has no inverse. We summarize this with

Theorem. *A square matrix has an inverse if, and only if, its determinant is nonzero.*

Example. $\begin{pmatrix} 1 & -2 \\ -2 & 4 \end{pmatrix}$ has no inverse, because its determinant is $1 \cdot 4 - (-2)(-2) = 0$.

For a matrix that is 3×3 or larger, calculating the determinant or inverse (if it exists) is harder. Although there are formulas for the determinant and inverse of any square matrix, they are too complicated to be very useful. Inverses are usually calculated through a different approach, called the Gauss-Jordan method, which is taught in linear algebra courses. In this text, for matrices larger than 2×2, we rely on software such as MATLAB to do the calculations for us.

It is important to remember, though, that not every matrix will have an inverse. If you attempt to calculate one when none exists, MATLAB will let you know. Fortunately, most square matrices do in fact have inverses, for any reasonable interpretation of the word "most." For this reason, matrices without inverses are said to be *singular*.

Let's return to our original motivation for developing the matrix inverse.

Example. For the forest model of Section 2.1, suppose at time 1 the populations were $x_1 = (500, 500)$. What must they have been at time 0?

To answer this, because $x_1 = Px_0$, we multiple by P^{-1} to find

$$
\begin{aligned}
x_0 &= P^{-1}x_1 \\
&= \begin{pmatrix} .9925 & .0125 \\ .0075 & .9875 \end{pmatrix}^{-1} \begin{pmatrix} 500 \\ 500 \end{pmatrix} \\
&= \frac{1}{(.9925)(.9875) - (.0075)(.0125)} \begin{pmatrix} .9875 & -.0125 \\ -.0075 & .9925 \end{pmatrix} \begin{pmatrix} 500 \\ 500 \end{pmatrix} \\
&= \frac{1}{.98} \begin{pmatrix} 487.5 \\ 492.5 \end{pmatrix} \approx \begin{pmatrix} 497.449 \\ 502.551 \end{pmatrix}.
\end{aligned}
$$

Problems

2.2.1. The first section of this chapter began with two examples of insect population models. Is either of these a Leslie model? Is either of these an Usher model? Explain why by describing the form of the projection matrices for them.

2.2.2. In MATLAB, create the Leslie matrix for the 1964 U.S. population model of (Keyfitz and Murphy, 1967) described in the text with the commands

```
sd=[.9966,.9983,.9979,.9968,.9961,...
      .9947,.9923,.9987,.9831]
P=diag(sd,-1)
P(1,:)=[.0000,.0010,.0878,.3487,.4761,...
      .3377,.1833,.0761,.0174,.0010]
```

For several choices of initial populations, produce graphs of the population over the next 10 time steps. Describe your observations.

2.2.3. Without using a computer, find the determinants and inverses of

$$
\begin{pmatrix} 1 & 2 \\ 2 & 3 \end{pmatrix}, \quad \begin{pmatrix} 2 & -1 \\ 2 & 3 \end{pmatrix}, \quad \begin{pmatrix} .7 & .3 \\ -1.4 & -.6 \end{pmatrix},
$$

provided they exist. Then check your answers with a computer. (The

MATLAB commands to find the inverse and determinant of a matrix A are inv(A) and det(A).)

2.2.4. Use a computer to find the determinants and inverses of the matrices

$$\begin{pmatrix} 1 & 0 & -1 \\ 2 & 1 & 0 \\ -1 & 1 & -2 \end{pmatrix}, \quad \begin{pmatrix} 3 & 2 & -1 \\ -2 & 0 & 2 \\ 0 & -1 & 1 \end{pmatrix}, \quad \begin{pmatrix} 1 & 0 & 2 \\ -2 & 1 & 1 \\ 3 & -1 & 1 \end{pmatrix},$$

provided they exist. Check to see that the computed inverse times the original matrix really gives the identity matrix.

2.2.5. A simple Usher model of a certain organism tracks immature and mature classes, and is given by the matrix $P = \begin{pmatrix} .2 & 3 \\ .3 & .5 \end{pmatrix}$.

   a. On average, how many births are attributed to each adult in a time step?
   b. What percentage of adults die in each time step?
   c. Assuming no immature individuals are able to reproduce in a time step, what is the meaning of the upper left entry in P?
   d. What is the meaning of the lower left entry in P?

2.2.6. For the model of the last problem,
   a. Find P^{-1}.
   b. If $\mathbf{x}_1 = (1100, 450)$, find $\mathbf{x}_0$ and $\mathbf{x}_2$.

2.2.7. Suppose a structured population model has projection matrix A, which has an inverse.
   a. What is the meaning of the matrix A^{100}? If a population vector is multiplied by it, what is produced? If a population vector is multiplied by $(A^{100})^{-1}$, what is produced?
   b. What is the meaning of the matrix $(A^{-1})^{100}$? If a population vector is multiplied by it, what is produced?
   c. Based on your answers to parts (a) and (b), explain why $(A^n)^{-1} = (A^{-1})^n$ for any positive integer n. This matrix is often denoted by A^{-n}.

2.2.8. A model given in (Cullen, 1985), based on data collected in (Nellis and Keith, 1976), describes a certain coyote population. Three stage classes – pup, yearling, and adult – are used while the matrix

$$P = \begin{pmatrix} .11 & .15 & .15 \\ .3 & 0 & 0 \\ 0 & .6 & .6 \end{pmatrix}$$

describes changes over a time step of 1 year. Explain what each entry in this matrix is saying about the population. Be careful in trying to explain the meaning of the .11 in the upper left corner.

2.2.9. a. Show that $A\mathbf{x} = A\mathbf{y}$ does not necessarily mean $\mathbf{x} = \mathbf{y}$ by calculating $A\mathbf{x}$ and $A\mathbf{y}$ for $A = \begin{pmatrix} 2 & 1 \\ 6 & 3 \end{pmatrix}$, $\mathbf{x} = \begin{pmatrix} 5 \\ 7 \end{pmatrix}$, and $\mathbf{y} = \begin{pmatrix} 6 \\ 5 \end{pmatrix}$.

　　b. Explain why if $A\mathbf{x} = A\mathbf{y}$ and A^{-1} exists, then $\mathbf{x} = \mathbf{y}$

2.2.10. Unlike scalars, for matrices usually $(AB)^{-1} \neq A^{-1}B^{-1}$. Instead, as long as the inverses exist, $(AB)^{-1} = B^{-1}A^{-1}$.

　　a. For $A = \begin{pmatrix} 2 & 1 \\ 1 & 1 \end{pmatrix}$ and $B = \begin{pmatrix} 1 & 2 \\ 3 & 5 \end{pmatrix}$, without using a computer compute $(AB)^{-1}$, $A^{-1}B^{-1}$, and $B^{-1}A^{-1}$ to verify these statements.

　　b. Pick any two other invertible 2×2 matrices C and D and verify that $(CD)^{-1} = D^{-1}C^{-1}$.

　　c. Pick two invertible 3×3 matrices E and F and use a computer to verify that $(EF)^{-1} = F^{-1}E^{-1}$.

2.2.11. The formula $(AB)^{-1} = B^{-1}A^{-1}$ can be explained several ways.

　　a. Explain why $(B^{-1}A^{-1})(AB) = I$. Why does this show $(AB)^{-1} = B^{-1}A^{-1}$?

　　b. Suppose, as in the first section of this chapter, that D is a projection matrix for a forest population in a dry year, and W is a projection matrix for a wet year. Then, if the first year is dry and the second wet, $\mathbf{x}_2 = WD\mathbf{x}_0$. How could you find $\mathbf{x}_1$ from $\mathbf{x}_2$? How could you find $\mathbf{x}_0$ from $\mathbf{x}_1$? Combine these to explain how you could find $\mathbf{x}_0$ from $\mathbf{x}_2$. How does this show $(WD)^{-1} = D^{-1}W^{-1}$?

2.2.12. A forest is composed of two species of trees, A and B. Each year $\frac{1}{3}$ of the trees of species A are replaced by trees of species B, while $\frac{1}{4}$ of the trees of species B are replaced by trees of species A. The remaining trees either survive or are replaced by trees of their own species.

　　a. Letting A_t and B_t denote the number of trees of each type in year t, give equations for A_{t+1} and B_{t+1} in terms of A_t and B_t.

　　b. Write the equations of part (a) as a single matrix equation.

　　c. Use part (b) to get a formula for A_{t+2} and B_{t+2} in terms of A_t and B_t.

　　d. Use part (b) to get a formula for A_{t-1} and B_{t-1} in terms of A_t and B_t.

　　e. Suppose $A_0 = 100$ and $B_0 = 100$. By hand, calculate A_t and B_t for $t = 1, 2$, and 3. Use MATLAB to check your work and extend the calculation through $t = 10$. What is happening to the populations?

f. Choose several different values of A_0 and B_0 and use MATLAB to track how the populations change over time. How do your results compare to those of part (e)?

2.3. Eigenvectors and Eigenvalues

Let's return to the forest model introduced in Section 2.1 of this chapter. Recall that we tracked two types of trees in a forest by

$$\mathbf{x}_{t+1} = P\mathbf{x}_t, \quad \text{with } P = \begin{pmatrix} .9925 & .0125 \\ .0075 & .9875 \end{pmatrix}.$$

The vector $\mathbf{v}_1 = (625, \ 375)$, which gave the population values that the forest approached in our numerical investigation, has the significant property that $P\mathbf{v}_1 = \mathbf{v}_1$. (Make sure you check this.) Using the language of Chapter 1, we might call $\mathbf{v}_1$ an equilibrium vector for our model.

Actually, there is another vector that is almost as well behaved as $\mathbf{v}_1$ for this particular model. If $\mathbf{v}_2 = (1, -1)$, then $P\mathbf{v}_2 = .98\mathbf{v}_2$. (Check this, too.) Although $\mathbf{v}_2$ is not an equilibrium, it does exhibit rather simple behavior when multiplied by P – the effect of multiplying $\mathbf{v}_2$ by P is exactly the same as multiplying it by the scalar .98.

Definition. If A is an $n \times n$ matrix, $\mathbf{v}$ a nonzero vector in $\mathbb{R}^n$, and λ a scalar such that $A\mathbf{v} = \lambda\mathbf{v}$, then we say that $\mathbf{v}$ is an *eigenvector* of A with *eigenvalue* λ.

We require that eigenvectors not be the zero vector, because $A\mathbf{0} = \mathbf{0} = \lambda\mathbf{0}$ for *all* real numbers λ. As long as an eigenvector $\mathbf{v} \neq \mathbf{0}$, there can be only one eigenvalue associated to it.

Using this terminology, the matrix P above has eigenvector $(625, 375)$ with eigenvalue 1, and eigenvector $(1, -1)$ with eigenvalue .98.

Notice, however, that like $(625, 375)$, the vectors $(5, 3)$, $(-10, -6)$, and $(15, 9)$ are also eigenvectors of P with eigenvalue 1. However, because all of these vectors are scalar multiples of one another, this may not seem too surprising. This is explained by:

Theorem. *If $\mathbf{v}$ is an eigenvector of A with eigenvalue λ, then for any scalar c, $c\mathbf{v}$ is also an eigenvector of A with the same eigenvalue λ.*

Proof. If $A\mathbf{v} = \lambda\mathbf{v}$, then $A(c\mathbf{v}) = c(A\mathbf{v}) = c(\lambda\mathbf{v}) = \lambda(c\mathbf{v})$.

Table 2.2. *Linear
Model Simulation
with Eigenvector as
Initial Values*

t	$\mathbf{x}_t$
0	$\mathbf{v}$
1	$A\mathbf{v} = \lambda\mathbf{v}$
2	$A\lambda\mathbf{v} = \lambda^2\mathbf{v}$
3	$A\lambda^2\mathbf{v} = \lambda^3\mathbf{v}$
$\vdots$	$\vdots$

The practical consequence of this is that although people might speak of $(5, 3)$ as "the" eigenvector of P with eigenvalue 1, for instance, they do not really mean there is only one such eigenvector. Any nonzero scalar multiple of $(5, 3)$ is also an eigenvector.

Understanding eigenvectors is crucial to understanding linear models. As a first step to seeing why this is so, consider what happens when the initial values of a linear model are given by an eigenvector. Consider a model $\mathbf{x}_{t+1} = A\mathbf{x}_t$, where we know that $A\mathbf{v} = \lambda\mathbf{v}$. Then, if $\mathbf{x}_0 = \mathbf{v}$, we produce Table 2.2.

The entries in Table 2.2 lead to the formula $\mathbf{x}_t = \lambda^t\mathbf{v}$. This means that, when the initial vector is an eigenvector, we can give a simple formula for all future values. Note that this formula involves a scalar exponential, just like the corresponding formula for the linear model of Chapter 1. The only difference is that this exponential multiplies the eigenvector of initial population values, rather then the single initial population value used in Chapter 1.

Example. If the forest model with $P = \begin{pmatrix} .9925 & .0125 \\ .0075 & .9875 \end{pmatrix}$ has initial vector $\mathbf{x}_0 = (1, -1)$, then $\mathbf{x}_t = .98^t(1, -1) = (.98^t(1), .98^t(-1))$. Thus, as time increases, the entries of $\mathbf{x}_t$ will both decay (rather slowly) to 0.

There are at least two questions that you might be wondering about: 1) Since populations cannot be negative, why is an eigenvector with a negative entry relevant to understanding this biological model? 2) How was the eigenvector $(1, -1)$ found? We address the first of these questions next, and defer the second to the next section.

The use of eigenvectors. For the forest model $P = \begin{pmatrix} .9925 & .0125 \\ .0075 & .9875 \end{pmatrix}$, we have the two eigenvector equations

$$P \begin{pmatrix} 5 \\ 3 \end{pmatrix} = 1 \begin{pmatrix} 5 \\ 3 \end{pmatrix}, \quad P \begin{pmatrix} 1 \\ -1 \end{pmatrix} = .98 \begin{pmatrix} 1 \\ -1 \end{pmatrix}.$$

If we begin with an initial population that is not one of these eigenvectors, how can we use the eigenvectors to understand what will happen?

The key idea is to try to write our initial population vector in terms of eigenvectors. Specifically, given an initial population vector $x_0 = (A_0, B_0)$, we look for two scalars, c_1, and c_2, with

$$\begin{pmatrix} A_0 \\ B_0 \end{pmatrix} = c_1 \begin{pmatrix} 5 \\ 3 \end{pmatrix} + c_2 \begin{pmatrix} 1 \\ -1 \end{pmatrix}.$$

Equivalently, we need to solve

$$\begin{pmatrix} A_0 \\ B_0 \end{pmatrix} = \begin{pmatrix} 5 & 1 \\ 3 & -1 \end{pmatrix} \begin{pmatrix} c_1 \\ c_2 \end{pmatrix}.$$

Notice that the matrix appearing in this equation has the eigenvectors of A as its columns. Now this equation can be solved provided that the matrix has an inverse. We have shown the 2×2 version of the following theorem.

Theorem. *Suppose A is an $n \times n$ matrix with n eigenvectors that form the columns of a matrix S. If S has an inverse, then any vector can be written as a sum of eigenvectors.*

Example. When we investigated the forest model numerically, we used the initial population vector $x_0 = (10, 990)$. The eigenvector matrix is $S = \begin{pmatrix} 5 & 1 \\ 3 & -1 \end{pmatrix}$. To solve $\begin{pmatrix} 10 \\ 990 \end{pmatrix} = \begin{pmatrix} 5 & 1 \\ 3 & -1 \end{pmatrix} \begin{pmatrix} c_1 \\ c_2 \end{pmatrix}$, we compute $S^{-1} = \frac{1}{-8} \begin{pmatrix} -1 & -1 \\ -3 & 5 \end{pmatrix}$, so

$$\begin{pmatrix} c_1 \\ c_2 \end{pmatrix} = \frac{1}{-8} \begin{pmatrix} -1 & -1 \\ -3 & 5 \end{pmatrix} \begin{pmatrix} 10 \\ 990 \end{pmatrix} = \begin{pmatrix} 125 \\ -615 \end{pmatrix}.$$

Thus

$$\begin{pmatrix} 10 \\ 990 \end{pmatrix} = 125 \begin{pmatrix} 5 \\ 3 \end{pmatrix} - 615 \begin{pmatrix} 1 \\ -1 \end{pmatrix}.$$

Technical remark: Not every matrix has eigenvectors that can be used as columns to form an invertible matrix. However, it is possible to prove that if a matrix does not have this property, then by changing the entries an "arbitrarily small" amount, you can get a matrix that does. Moreover, "most" matrices do have this property – if you pick a matrix at random, it is essentially guaranteed to have the property. The consequences of these facts for applying the theory of eigenvectors to biological models is that there is no need to really worry about not having nice enough eigenvectors.

Now that we understand how to express initial values in terms of eigenvectors, how do we use that expression? Let's suppose A is $n \times n$, with n eigenvectors $\mathbf{v}_1, \mathbf{v}_2, \ldots, \mathbf{v}_n$, whose corresponding eigenvalues are $\lambda_1, \lambda_2, \ldots, \lambda_n$. We express our initial vector $\mathbf{x}_0$ as

$$\mathbf{x}_0 = c_1\mathbf{v}_1 + c_2\mathbf{v}_2 + \cdots + c_n\mathbf{v}_n,$$

so then

$$\begin{aligned}
\mathbf{x}_1 = A\mathbf{x}_0 &= A(c_1\mathbf{v}_1 + c_2\mathbf{v}_2 + \cdots + c_n\mathbf{v}_n) \\
&= Ac_1\mathbf{v}_1 + Ac_2\mathbf{v}_2 + \cdots + Ac_n\mathbf{v}_n.
\end{aligned}$$

But each term in this last expression is simply A applied to an eigenvector, so we see

$$\mathbf{x}_1 = c_1\lambda_1\mathbf{v}_1 + c_2\lambda_2\mathbf{v}_2 + \cdots + c_n\lambda_n\mathbf{v}_n.$$

But then

$$\begin{aligned}
\mathbf{x}_2 = A\mathbf{x}_1 &= A(c_1\lambda_1\mathbf{v}_1 + c_2\lambda_2\mathbf{v}_2 + \cdots + c_n\lambda_n\mathbf{v}_n) \\
&= Ac_1\lambda_1\mathbf{v}_1 + Ac_2\lambda_2\mathbf{v}_2 + \cdots + Ac_n\lambda_n\mathbf{v}_n,
\end{aligned}$$

and because each term is again A times an eigenvector,

$$\mathbf{x}_2 = c_1\lambda_1^2\mathbf{v}_1 + c_2\lambda_2^2\mathbf{v}_2 + \cdots + c_n\lambda_n^2\mathbf{v}_n.$$

Continuing to apply A, we obtain the formula

$$\mathbf{x}_t = c_1\lambda_1^t\mathbf{v}_1 + c_2\lambda_2^t\mathbf{v}_2 + \cdots + c_n\lambda_n^t\mathbf{v}_n.$$

Understanding the eigenvectors has allowed us to find a formula for the values of $\mathbf{x}_t$ at any time. Notice the similarity of this formula to the corresponding one for the Malthusian model of Chapter 1. Although there are a number of terms added together, each one has a simple exponential form that is already familiar.

Example. For the populations used in the numerical investigation of the forest model, we have already seen $\mathbf{x}_0 = \begin{pmatrix} 10 \\ 990 \end{pmatrix} = 125 \begin{pmatrix} 5 \\ 3 \end{pmatrix} - 615 \begin{pmatrix} 1 \\ -1 \end{pmatrix}$.
This means

$$\mathbf{x}_t = 1^t (125) \begin{pmatrix} 5 \\ 3 \end{pmatrix} + .98^t (-615) \begin{pmatrix} 1 \\ -1 \end{pmatrix}$$

$$= \begin{pmatrix} 1^t (125)(5) + .98^t (-615)(1) \\ 1^t (125)(3) + .98^t (-615)(-1) \end{pmatrix} = \begin{pmatrix} 625 - (615).98^t \\ 375 + (615).98^t \end{pmatrix}.$$

We have thus found a formula giving all the entries in Table 2.1 that were originally produced by numerical investigation. Try picking a few values of t and seeing that you get the same values that appear in the table. Note also that the formulas make clear that the populations will approach $(625, 375)$ as t grows.

Why does this all work? As far as an eigenvector is concerned, multiplication by the matrix is the same as multiplying by a scalar (the eigenvalue). Thus, initial values given by eigenvectors will produce fully understandable behavior (exponential growth or decay). If we decompose any initial vector into eigenvectors, we can understand the model's effect on the initial vector through its effect on the eigenvectors.

Asymptotic behavior. Given a linear model $\mathbf{x}_{t+1} = A\mathbf{x}_t$ with initial vector $\mathbf{x}_0$, we now know how to find an explicit formula for $\mathbf{x}_t$: If $\lambda_1, \lambda_2, \ldots, \lambda_n$ are eigenvalues of A with $\mathbf{v}_1, \mathbf{v}_2, \ldots, \mathbf{v}_n$ the corresponding eigenvectors, then writing $\mathbf{x}_0$ in terms of the eigenvectors

$$\mathbf{x}_0 = c_1 \mathbf{v}_1 + c_2 \mathbf{v}_2 + \cdots + c_n \mathbf{v}_n$$

means

$$\mathbf{x}_t = c_1 \lambda_1^t \mathbf{v}_1 + c_2 \lambda_2^t \mathbf{v}_2 + \cdots + c_n \lambda_n^t \mathbf{v}_n. \tag{2.5}$$

This formula for $\mathbf{x}_t$ immediately gives us qualitative information on the model. Suppose, for example, that all the λ_i satisfy $|\lambda_i| < 1$; then, as higher powers of the λ_i will shrink to 0, the populations $\mathbf{x}_t$ will also decline to $\mathbf{0}$ as t increases. On the other hand, if for at least one i we have $\lambda_i > 1$ (and the corresponding $c_i \neq 0$), then $\mathbf{x}_t$ should have a component of exponential growth. We also see that a negative value for λ_i should produce some form of oscillatory behavior, because its powers alternate in sign. Viewing the formula this way shows that the eigenvalues are really the key to the qualitative behavior of the model.

Definition. An eigenvalue of A that is largest in absolute value is called a *dominant eigenvalue* of A. An eigenvector corresponding to it is called a *dominant eigenvector*.

Notice that we did not say "the" dominant eigenvalue in the definition, because several eigenvalues may have the same absolute value. If there is an eigenvalue whose absolute value is strictly larger than all the others (e.g., $|\lambda_1| > |\lambda_i|$ for $i = 2, 3, \ldots, n$), we say it is *strictly* dominant.

Numbering the eigenvalues so that λ_1 is a dominant one, then Eq. (2.5) can be rewritten as

$$\mathbf{x}_t = \lambda_1^t \left(c_1 \mathbf{v}_1 + c_2 \left(\frac{\lambda_2}{\lambda_1}\right)^t \mathbf{v}_2 + \cdots + c_n \left(\frac{\lambda_n}{\lambda_1}\right)^t \mathbf{v}_n \right). \qquad (2.6)$$

Assuming λ_1 is strictly dominant, then $\left|\frac{\lambda_i}{\lambda_1}\right| < 1$ for $i = 2, 3, \ldots, n$, so as t increases all the terms in the parentheses will decay exponentially, except for the first. Discarding the decaying terms shows the behavior of $\mathbf{x}_t$ is approximated by

$$\mathbf{x}_t \approx \lambda_1^t c_1 \mathbf{v}_1.$$

Overall, then, the model displays roughly exponential growth or decay, depending on the value of λ_1. For example, the model producing Figure 2.2. must have had a dominant eigenvalue that was larger than 1, because the graph shows exponential growth.

The dominant eigenvalue describes the main component of the model's behavior. For a linear population model, the dominant eigenvalue is often called the *intrinsic growth rate* of the population, and it is the single most important number describing how the population changes over time. It is an example of a summary statistic, because it extracts the most important feature from all the entries in the matrix.

Equation (2.6) can tell us more, though. Dividing each side by λ_1^t, it becomes

$$\frac{1}{\lambda_1^t} \mathbf{x}_t = c_1 \mathbf{v}_1 + c_2 \left(\frac{\lambda_2}{\lambda_1}\right)^t \mathbf{v}_2 + \cdots + c_n \left(\frac{\lambda_n}{\lambda_1}\right)^t \mathbf{v}_n.$$

As $t \to \infty$, we see $\frac{1}{\lambda_1^t} \mathbf{x}_t \to c_1 \mathbf{v}_1$. In words, once we counteract the growth the model predicts for $\mathbf{x}_t$, the vector will simply approach a multiple of the dominant eigenvector. Therefore, for large t, the entries of $\mathbf{x}_t$ should be in roughly the same proportions to one another as the entries of $\mathbf{v}_1$. We see this in Figure 2.2., after the first few time steps have passed.

For a population model, the dominant eigenvector is hence often referred to as the *stable age distribution* or *stable stage distribution*, because it gives us the proportions of the population that should appear in each age or stage class, once we account for the growth trend.

Up to this point, we have avoided commenting on the significance of the coefficients c_i in Eq. (2.5) and (2.6). Recall that they were found by letting $\mathbf{c} = (c_1, c_2, \ldots, c_n)$ and solving $\mathbf{x}_0 = S\mathbf{c}$, where S is a matrix with the eigenvectors as its columns. This means that if we change $\mathbf{x}_0$, we change the values of the c_i's. It is only through the c_i's that the initial vector $\mathbf{x}_0$ enters into formulas (2.5) and (2.6).

Even though it was not pointed out previously, the discussion of the growth rate and stable distribution actually required an assumption that $c_1 \neq 0$. If we slough over this point, we reach the rather significant conclusion that the main features of the qualitative behavior of the model – the intrinsic growth rate and the stable age distribution – are independent of the initial vector. The dominant eigenvector and eigenvalue alone tell us the most important features of the model. This result is sometimes called the *Strong Ergodic Theorem* for linear models, or, in the context of population models, the *Fundamental Theorem of Demography*.

Although certain choices of $\mathbf{x}_0$ might cause $c_1 = 0$, that happens rarely; for most choices of $\mathbf{x}_0$, we expect $c_1 \neq 0$. For many types of models, it can even be proved $c_1 \neq 0$ for all biologically meaningful choices of $\mathbf{x}_0$.

Example. Consider an Usher model for a population with two stage classes given by the matrix

$$P = \begin{pmatrix} 0 & 2 \\ .5 & .1 \end{pmatrix}.$$

Because we have only two classes, we can make some reasonable guesses as to how the population should change. Note each adult produces two offspring, but only half of these make it to adulthood. If the lower right-hand entry were not .1, we might expect a stable population size, but the small fraction of adults surviving for more than one time step, and therefore reproducing again, should result in a growing population. Because the fraction of adults surviving for an additional time step is small, the population will probably grow slowly.

Now using a computer to calculate eigenvectors and eigenvalues gives us

$$P \begin{pmatrix} .8852 \\ .4653 \end{pmatrix} = 1.0512 \begin{pmatrix} .8852 \\ .4653 \end{pmatrix}, \quad P \begin{pmatrix} .9031 \\ -.4295 \end{pmatrix} = -.9512 \begin{pmatrix} .9031 \\ -.4295 \end{pmatrix}.$$

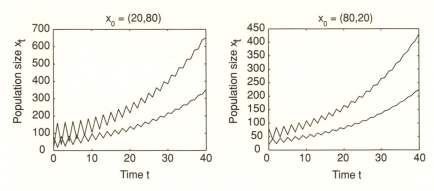

Figure 2.3. Two simulations of a linear model; note similar qualitative features despite different initial values.

This means that if we write our initial population (which has not been given here!) as

$$\mathbf{x}_0 = c_1 \begin{pmatrix} .8852 \\ .4653 \end{pmatrix} + c_2 \begin{pmatrix} .9031 \\ -.4295 \end{pmatrix},$$

for some numbers c_1 and c_2, then future populations are given by

$$\mathbf{x}_t = c_1 (1.0512)^t \begin{pmatrix} .8852 \\ .4653 \end{pmatrix} + c_2 (-.9512)^t \begin{pmatrix} .9031 \\ -.4295 \end{pmatrix}.$$

The first term here will produce slow growth, while the second term will decline in size. Note that the sign of the eigenvalue in the second term will cause the numbers in that term to oscillate between negative and positive values as they approach zero. This means that if we pick any initial population, calculate future populations, and graph them, we should expect a slow exponential growth trend, with a decaying oscillation superimposed on it. You can see this for two choices of initial population vectors in Figure 2.3.

The stable stage distribution for the model is given by $\mathbf{v}_1 = (.8852, .4653)$. Even though the population continues to grow, after enough time has elapsed we should see the populations in the two classes in proportion approximately $\frac{.8852}{.4653} = 1.9024$. That is, for every adult, there will be about 1.9 immature individuals.

Many theorems have been proved about the particular types of matrices appearing in Leslie and Usher models. One of these is:

Theorem. *A Leslie model in which two consecutive age classes are fertile (i.e., both $f_i > 0$ and $f_{i+1} > 0$) will have a positive real strictly dominant eigenvalue, and hence a stable age distribution.*

Although such theorems are useful for making general statements about the way populations must behave, when it comes to any particular model, it is always necessary to actually find the eigenvectors and eigenvalues.

Complex numbers. As you will see when you compute eigenvectors and eigenvalues, the above examples have been a little misleading, since eigenvectors and eigenvalues often involve complex numbers. Despite this, the discussion of asymptotic behavior is still valid, provided we explain how to measure the size of complex numbers.

Definition. The *absolute value* of a complex number $a + bi$ is $|a + bi| = \sqrt{a^2 + b^2}$.

Note that if $b = 0$, then $|a + 0i| = \sqrt{a^2} = |a|$ is the usual meaning of absolute value for real numbers. Also, $|a + bi| \geq 0$, and $|a + bi| = 0$ only when $a + bi = 0$, as we would like for something that purports to measure the size of a number. Less obvious properties are:

Theorem. *For real numbers a, b, c, d,*

   a) $|(a + bi)(c + di)| = |a + bi||c + di|$

   b) $|(a + bi)^n| = |a + bi|^n$

   c) $\left| \dfrac{a + bi}{c + di} \right| = \dfrac{|a + bi|}{|c + di|}.$

Notice that all three of these statements are familiar to you in the special case when $b = 0$ and $d = 0$, when the absolute value simply means the one you are familiar with for real numbers.

The proof of statement (a) appears as an exercise and just requires multiplying out each side. Statement (b) is shown by just applying (a) repeatedly, since $(a + bi)^n = (a + bi)^{n-1}(a + bi)$. Statement (c) also follows from (a), if you first multiply the equation in (c) through by $|c + di|$ to clear the denominator.

To see how the discussion of the asymptotic behavior of a linear model is affected by complex eigenvalues, look back at Eq. (2.6). Even if some of the eigenvalues λ_i are complex, if λ_1 is strictly dominant so $|\lambda_1| > |\lambda_i|$ for $i = 2, 3, \ldots, n$, then by part (c) of the theorem, $\left|\frac{\lambda_i}{\lambda_1}\right| < 1$ as before, and so $\left|\frac{\lambda_i}{\lambda_1}\right|^t$ approaches 0 as t increases. By part (b) of the theorem, this would mean $\left|\left(\frac{\lambda_i}{\lambda_1}\right)^t\right|$ approaches 0, and so we must have that $\left(\frac{\lambda_i}{\lambda_1}\right)^t$ approaches 0. Just as before, we see that all the terms inside the parentheses in Eq.(2.6), except for the first, vanish as t increases. Our earlier argument is still valid even if some eigenvalues are complex.

Although the appearance of complex eigenvalues can be confusing at first, once you understand how to measure their size with the absolute value, they do not create any difficulties for analyzing a model. Their presence will usually result in irregular-looking oscillations as part of the model's behavior, just as negative eigenvalues cause oscillations. For population models, a strictly dominant eigenvalue will always turn out to be real.

Problems

2.3.1. Use MATLAB to investigate the model $P = \begin{pmatrix} 0 & 2 \\ .5 & .1 \end{pmatrix}$ discussed in the text. Show that, for a variety of choices of initial populations, the model behaves exactly as one would predict from knowing only the two eigenvalues.

2.3.2. The MATLAB command [S D]=eig(A) computes the eigenvectors and eigenvalues of a matrix A. The columns of S will be the eigenvectors and the corresponding diagonal entries of D their eigenvalues.

   Use MATLAB to compute the eigenvectors and eigenvalues for the matrix P in the text for the forest succession model. Are they the "same" ones given in the text? Explain.

2.3.3. Use MATLAB to compute the eigenvalues of the matrix given in Eq. (2.4) of Section 2.2 describing a plant model. Explain how the eigenvalues are related to the graph in Figure 2.2..

2.3.4. Consider the plant model of Eq. (2.4) of Section 2.2, as well as another plant model obtained by replacing all entries in the first row and column of that matrix with 0's.

   a. In biological terms, what is the meaning of replacing the specified entries with 0's?

b. Compute the dominant eigenvalue for each model. Is there much difference in the intrinsic growth rate? Did the intrinsic growth rate change the way you thought it would? Explain.

c. If the ungerminated seeds have little affect on the intrinsic growth rate of this plant, why might they still be biologically advantageous to the species?

2.3.5. Consider the Leslie model with $P = \begin{pmatrix} .3 & 2 \\ .4 & 0 \end{pmatrix}$.

a. By thinking about the biological meaning of each entry in this matrix, do you think it describes a growing or declining population? Would you guess the population size would change rapidly or slowly?

b. Compute eigenvectors and eigenvalues of the model with MAT-LAB.

c. What is the intrinsic growth rate? the stable stage distribution?

d. Express the initial vector $x_0 = (5, 5)$ as a sum of the eigenvectors.

e. Use your answer in part (d) to give a formula for the population vector x_t.

2.3.6. Repeat the last problem for the Usher model $P = \begin{pmatrix} 0 & 0 & 73 \\ .04 & 0 & 0 \\ 0 & .39 & .65 \end{pmatrix}$

with $x_0 = (100, 10, 1)$.

2.3.7. Find the growth rate and stable stage distribution of the coyote model whose matrix is

$$P = \begin{pmatrix} .11 & .15 & .15 \\ .3 & 0 & 0 \\ 0 & .6 & .6 \end{pmatrix}.$$

Will the population grow or decline? Quickly or slowly?

2.3.8. Find the intrinsic growth rate and stable age distribution for the U.S. population model described in the text and in Problem 2.2.2. Recalling that the time step for this model was 5 years, how would you express the intrinsic growth rate on a yearly basis?

2.3.9. Suppose a simple population is broken up into immature and mature developmental classes. Only one-sixth of the immature individuals make it to maturity at each time step (with the rest dying). A typical mature individual gives birth to five young at each time step. Finally, three-quarters of the adults die (after producing young) at each time step, while the rest survive.

a. Model this situation using a matrix. Is this a Leslie or an Usher model, or neither?

b. Compute eigenvectors and eigenvalues of the projection matrix using MATLAB.

c. What is the intrinsic growth rate? The stable stage distribution?

2.3.10. Show that the absolute value for complex numbers satisfies

$$|(a + bi)(c + di)| = |a + bi||c + di|.$$

Projects

1. Consider a specific Leslie model with two age groups. After interpreting each matrix entry in biological terms, investigate the behavior of your model numerically using MATLAB for a variety of initial populations, including the eigenvectors of the matrix. Explain how the eigenvalues and eigenvectors are reflected in the behaviors you see when you plot the populations over time. Repeat for several other matrices.

Suggestions
- Begin with the Leslie model

$$\begin{pmatrix} \frac{1}{8} & 6 \\ \frac{1}{5} & 0 \end{pmatrix},$$

using a MATLAB command sequence like:

```
P=[1/8  6; 1/5 0]
x=[10; 990]
xhistory=x
x=P*x, xhistory=[xhistory x]
x=P*x, xhistory=[xhistory x]
x=P*x, xhistory=[xhistory x]
    ⋮
plot(xhistory')
```

- For a variety of choices of the initial populations, describe what appears to be happening to the populations over time. Do the number of individuals in each group get bigger or smaller? Do they oscillate? Compute the ratio of immature individuals to adults at various times. How does this ratio change? Repeat this work with several different

choices of an initial vector. Qualitatively describe all the behaviors you
see.
- Compute the eigenvectors and eigenvalues of A by entering [S,D]=
  eig(A). Use the first eigenvector as your initial vector by letting
  x=S(:,1) and repeat the work above, including producing a plot.
  Do this again using the second eigenvector with x=(:,2). Describe
  the behavior of the model with these choices on initial vectors. How
  is the behavior different? How is it the same? How are the eigenvalues
  responsible for these behaviors?
- How are the behaviors you see for the eigenvectors reflected in the
  behavior you saw for other initial vectors?
- Repeat all of the above for a few other models such as:

$$\begin{pmatrix} 0 & 6 \\ \frac{1}{5} & \frac{1}{4} \end{pmatrix}, \begin{pmatrix} 0 & 6 \\ \frac{1}{6} & 0 \end{pmatrix}, \begin{pmatrix} 0 & 6 \\ \frac{1}{7} & 0 \end{pmatrix}, \begin{pmatrix} 0 & 6 \\ \frac{1}{6} & \frac{1}{4} \end{pmatrix}, \begin{pmatrix} 0 & 6 \\ \frac{1}{12} & \frac{1}{4} \end{pmatrix}.$$

Explain in biological terms why each of these models produces the
behavior it does. Then explain in terms of the eigenvalues and eigenvectors of the matrix why the behavior occurs.
- Characterize the possible behavior of these 2×2 matrix models in
  terms of the *sign* and *size* of the eigenvalues.

2. Leslie and Usher models can be used to help design intervention strategies
to help declining populations recover. A well-known example of this
was the study in (Crouse *et al.*, 1987) on sea turtle populations that
supported the U.S. government-mandated use of turtle exclusion devices
in shrimpers' nets.

   An intervention might be designed to affect any one of the entries in the
Leslie matrix modeling a population. Because the dominant eigenvalue of
the matrix determines the overall growth rate, it is necessary to study how
changing each matrix entry affects the dominant eigenvalue. Determining
the effect of small changes in each of the entries is sometimes called
a *sensitivity analysis*. Imagine an endangered population grouped into
immature and mature subpopulations and modeled by the Usher model
with matrix

$$\begin{pmatrix} 0 & 1.7 \\ .3 & .1 \end{pmatrix}.$$

Analyze the effect of small changes in each of the nonzero entries in the
matrix on the population's future.

Suggestions

- What is the dominant eigenvalue of the model? How fast will the population increase or decline if no changes are made?
- For the matrix

$$\begin{pmatrix} 0 & 1.7 \\ c & .1 \end{pmatrix},$$

what values of c give a biologically meaningful model? For a variety of values of c in that range, compute the dominant eigenvalue λ_1. Present the results of your computations in a table and as a graph of c vs. λ_1. Useful MATLAB commands are

```
lambda1vec=[]
cvec=[0:.1:1]
for c=cvec
     A=[ 0 1.7;c .1]
     lambda1=max(eig(A))
     lambda1vec=[lambda1vec, lambda1]
end
plot(cvec, lambda1vec)
```

- If you have read ahead to the next section of this text, find a formula for λ_1 as a function of c? Does its graph agree with the one you produced?
- If an intervention strategy attempts to change c in this matrix, describe in biological terms what its focus might be. What value of c must be achieved so that the population will recover?
- Repeat the same sort of analysis to understand the affect of changing the other nonzero entries of the matrix.
- Without regard to the cost of implementing any recovery plan, which entry do you think would be most effective to try to change? What biological issues might you need to understand better about the population to answer this question adequately?
- Why might a plan to change a fertility rate by a small amount have different costs than a plan to change a survival rate?
- Perform a sensitivity analysis on a Leslie or Usher model described by a larger matrix of your choice.

2.4. Computing Eigenvectors and Eigenvalues

We first show how eigenvectors and eigenvalues can be computed by hand for 2×2 matrices.

Given a matrix A, the eigenvector equation we wish to solve is $A\mathbf{x} = \lambda\mathbf{x}$, where both the vector $\mathbf{x}$ and the scalar λ are unknown. This equation can be rewritten as

$$A\mathbf{x} - \lambda\mathbf{x} = \mathbf{0},$$
$$A\mathbf{x} - \lambda I\mathbf{x} = \mathbf{0},$$
$$(A - \lambda I)\mathbf{x} = \mathbf{0}.$$

Notice that we had to stick the identity matrix into the middle equation so that factoring the $\mathbf{x}$ out of each term would be valid. Without the identity, we would have had $A - \lambda$, which makes no sense since we never defined subtraction of a scalar from a matrix.

Now, if $\mathbf{x}$ and λ are really an eigenvector and its eigenvalue, this last equation shows that the matrix $(A - \lambda I)$ *cannot* have an inverse. For if it did, then we could multiply each side of $(A - \lambda I)\mathbf{x} = \mathbf{0}$ by the inverse to get $\mathbf{x} = (A - \lambda I)^{-1}\mathbf{0}$. But even without knowing what $(A - \lambda I)^{-1}$ is explicitly, we can tell that $(A - \lambda I)^{-1}\mathbf{0} = \mathbf{0}$, and this would mean $\mathbf{x} = \mathbf{0}$. But our definition of eigenvectors requires that they be nonzero, so we know this cannot happen.

Now, if $(A - \lambda I)$ does not have an inverse, then $\det(A - \lambda I)$ must be 0. We have shown then that if λ is any eigenvalue of a matrix A, it must satisfy the equation

$$\det(A - \lambda I) = 0.$$

Example. For the forest model matrix

$$P = \begin{pmatrix} .9925 & .0125 \\ .0075 & .9875 \end{pmatrix}, \quad P - \lambda I = \begin{pmatrix} .9925 - \lambda & .0125 \\ .0075 & .9875 - \lambda \end{pmatrix},$$

and so $\det(P - \lambda I) = 0$ becomes

$$(.9925 - \lambda)(.9875 - \lambda) - (.0125)(.0075) = 0$$
$$\lambda^2 - 1.98\lambda + .98 = 0$$
$$(\lambda - 1)(\lambda - .98) = 0.$$

This means the only possible eigenvalues for P are the numbers 1 and .98.

The equation $\det(A - \lambda I) = 0$ is called the *characteristic equation* of A. For a 2×2 matrix, it will always be a quadratic equation, so solving it will be no harder than factoring or using the quadratic formula to find the roots.

While our argument above also applies to larger matrices (provided you learn to compute determinants of larger matrices in some other course), solving the characteristic equation will be much harder because for an $n \times n$ matrix, it involves an nth degree polynomial. Unless n is quite small, or you are very lucky, this is simply not practical to do by hand. Nonetheless, we can see that there can be at most n eigenvalues, since the characteristic equation can have at most n roots. We also see that complex numbers may enter into our calculations since the roots of a polynomial may not be real.

In summary, we have:

Theorem. *If λ is an eigenvalue for an $n \times n$ matrix A, then it satisfies the nth degree polynomial equation* $\det(A - \lambda I) = 0$. *Thus, there are at most n eigenvalues for A.*

Once we have determined possible eigenvalues for a matrix, we need to find corresponding eigenvectors. We illustrate the process for $P = \begin{pmatrix} .9925 & .0125 \\ .0075 & .9875 \end{pmatrix}$ and $\lambda = .98$. We need to find a vector $\mathbf{v}$ such that $(P - .98I)\mathbf{v} = \mathbf{0}$, so we must solve

$$\begin{pmatrix} .9925 - .98 & .0125 \\ .0075 & .9875 - .98 \end{pmatrix} \begin{pmatrix} x \\ y \end{pmatrix} = \begin{pmatrix} 0 \\ 0 \end{pmatrix}$$

$$\begin{pmatrix} .0125 & .0125 \\ .0075 & .0075 \end{pmatrix} \begin{pmatrix} x \\ y \end{pmatrix} = \begin{pmatrix} 0 \\ 0 \end{pmatrix} .$$

Since we cannot solve this by finding the inverse of the matrix (why not?), we write out the two equations this represents in nonmatrix form.

$$.0125x + .0125y = 0,$$
$$.0075x + .0075y = 0.$$

While guessing a nonzero solution is a perfectly valid way to proceed, we'll be more methodical in our approach. Because one of these equations is a multiple of the other, to solve them we just need to solve $.0125x + .0125y = 0$. With only one equation in two unknowns, we can take one of the unknowns to have any value we like, and let that determine the second. For instance, if we solve for y in terms of x, we find $y = -x$. Thus, any vector of the form $\mathbf{v} = (x, y) = (x, -x) = x(1, -1)$ is an eigenvector with eigenvalue .98. Since we have freedom to choose x as we wish, we'll take it to be 1. Thus, we have found the eigenvector $(1, -1)$ that was used throughout this chapter.

The eigenvector associated with $\lambda_1 = 1$ can be found similarly:

$$P - 1I = \begin{pmatrix} .9925 - 1 & .0125 \\ .0075 & .9875 - 1 \end{pmatrix} = \begin{pmatrix} -.0075 & .0125 \\ .0075 & -.0125 \end{pmatrix}$$

so we must solve

$$-.0075x + .0125y = 0$$
$$.0075x - .0125y = 0.$$

Because the equations are multiples of each other, we solve $-.0075x + .0125y = 0$ to get $y = \frac{.0075}{.0125}x = \frac{3}{5}x$, so $\mathbf{v} = \left(x, \frac{3}{5}x\right) = x\left(1, \frac{3}{5}\right)$. Choosing $x = 5$ (since it makes the vector have the simplest form), we find $\mathbf{v} = (5, 3)$.

Although this was only one example of calculating an eigenvector for a particular matrix P, for any 2×2 matrix the procedure works the same way. Although we will not prove it here, you will always find one of the equations is a multiple of the other and then be able to solve for y in terms of x (or x in terms of y) to find all the eigenvectors.

As with eigenvalues, calculating eigenvectors for 3×3 or larger matrices is done analogously to the 2×2 case, although some additional complications come in. We'll leave a discussion of those to a full course in linear algebra and instead suggest that you let MATLAB do the computations for you.

Computer methods of calculation. Actually, MATLAB and other computer packages do not really calculate eigenvectors and eigenvalues in the way described previously. Because the computation of these is so important, not only for biological models but for a host of problems throughout science and engineering, quite clever and sophisticated methods have been developed and incorporated into many standard software packages.

Although we will not really explain any methods these packages use, we will give a hint at one type of approach by discussing the *power method*.

Given A, pick any initial vector $\mathbf{x}_0$ and compute $\mathbf{x}_1 = A\mathbf{x}_0$. According to the Strong Ergodic Theorem, if λ_1 is the dominant eigenvalue of A with corresponding eigenvector $\mathbf{v}_1$, then we should expect $\frac{1}{\lambda_1}\mathbf{x}_1$ to be closer to $\mathbf{v}_1$ than $\mathbf{x}_0$ was. Because we do not yet know what λ_1 is, we have to somehow adjust $\mathbf{x}_1$ to account for the growth factor. One way of doing this is to simply divide each entry of $\mathbf{x}_1$ by its largest entry to get a new vector we call $\mathbf{x}_1'$. This means $\mathbf{x}_1'$ will have one entry that is a 1 and will be "closer" to being an eigenvector than $\mathbf{x}_0$ was.

We can then repeat the process using $\mathbf{x}_1'$ in place of $\mathbf{x}_0$ to get an even better approximate eigenvector. Of course, we should then repeat the process again, and again, until we find our approximate eigenvectors are not changing by much.

Example. Suppose $A = \begin{pmatrix} 1 & 3 \\ 7 & 5 \end{pmatrix}$ and we choose $\mathbf{x}_0 = \begin{pmatrix} 1 \\ 1 \end{pmatrix}$. Then

$$\mathbf{x}_1 = \begin{pmatrix} 1 & 3 \\ 7 & 5 \end{pmatrix} \begin{pmatrix} 1 \\ 1 \end{pmatrix} = \begin{pmatrix} 4 \\ 12 \end{pmatrix}$$

$$\mathbf{x}_1' = \frac{1}{12} \begin{pmatrix} 4 \\ 12 \end{pmatrix} \approx \begin{pmatrix} .33 \\ 1.00 \end{pmatrix}$$

$$\mathbf{x}_2 = \begin{pmatrix} 1 & 3 \\ 7 & 5 \end{pmatrix} \begin{pmatrix} .33 \\ 1.00 \end{pmatrix} = \begin{pmatrix} 3.33 \\ 7.33 \end{pmatrix}$$

$$\mathbf{x}_2' = \frac{1}{7.33} \begin{pmatrix} 3.33 \\ 7.33 \end{pmatrix} \approx \begin{pmatrix} .45 \\ 1.00 \end{pmatrix}$$

$$\mathbf{x}_3 = \begin{pmatrix} 1 & 3 \\ 7 & 5 \end{pmatrix} \begin{pmatrix} .45 \\ 1.00 \end{pmatrix} = \begin{pmatrix} 3.45 \\ 8.18 \end{pmatrix}$$

$$\mathbf{x}_3' = \frac{1}{8.18} \begin{pmatrix} 3.45 \\ 8.18 \end{pmatrix} \approx \begin{pmatrix} .42 \\ 1.00 \end{pmatrix}$$

$$\mathbf{x}_4 = \begin{pmatrix} 1 & 3 \\ 7 & 5 \end{pmatrix} \begin{pmatrix} .42 \\ 1.00 \end{pmatrix} = \begin{pmatrix} 3.42 \\ 7.96 \end{pmatrix}$$

$$\mathbf{x}_4' = \frac{1}{7.96} \begin{pmatrix} 3.42 \\ 7.96 \end{pmatrix} \approx \begin{pmatrix} .43 \\ 1.00 \end{pmatrix}$$

$$\mathbf{x}_5 = \begin{pmatrix} 1 & 3 \\ 7 & 5 \end{pmatrix} \begin{pmatrix} .43 \\ 1.00 \end{pmatrix} = \begin{pmatrix} 3.43 \\ 8.01 \end{pmatrix}$$

$$\mathbf{x}_5' = \frac{1}{8.01} \begin{pmatrix} 3.43 \\ 8.01 \end{pmatrix} \approx \begin{pmatrix} .43 \\ 1.00 \end{pmatrix}.$$

Thus, $\mathbf{v}_1 \approx \begin{pmatrix} .43 \\ 1.00 \end{pmatrix}$ is the dominant eigenvector, at least to within a digit or so of accuracy. To get the dominant eigenvalue λ_1, we just note

$$\begin{pmatrix} 1 & 3 \\ 7 & 5 \end{pmatrix} \begin{pmatrix} .43 \\ 1.00 \end{pmatrix} = \begin{pmatrix} 3.43 \\ 8.01 \end{pmatrix} = 8.01 \begin{pmatrix} .43 \\ 1.00 \end{pmatrix},$$

so $\lambda_1 \approx 8.01$.

▶ Compute the eigenvalues and eigenvectors for this matrix *exactly* and see if they agree with this result.

While this technique has only given us the dominant eigenvector and eigen-value, variations on the idea can find others.

Problems

2.4.1. For $A = \begin{pmatrix} .9 & .3 \\ .1 & .7 \end{pmatrix}$, $B = \begin{pmatrix} 1 & 4 \\ 2 & 3 \end{pmatrix}$, and $C = \begin{pmatrix} -1 & 3 \\ 2 & 0 \end{pmatrix}$,

    a. Find the eigenvalues by solving the characteristic equations.

    b. Find an eigenvector for each eigenvalue.

2.4.2. For the matrices in the preceeding problem, use the power method to find the dominant eigenvectors and eigenvalues. Do your answers agree with those you found before?

2.4.3. Use the power method to find the dominant eigenvector and eigenvalue of the Leslie matrix

$$\begin{pmatrix} 0 & 0 & 73 \\ .04 & 0 & 0 \\ 0 & .39 & .65 \end{pmatrix}.$$

Check your answer by asking MATLAB for the eigenvectors and eigen-values.

2.4.4. Actually, things can be more complicated than the text may have led you to believe:

    a. For $A = \begin{pmatrix} 2 & 0 \\ 0 & 2 \end{pmatrix}$, find two eigenvalues and two eigenvectors. (They really do exist.)

    b. For $B = \begin{pmatrix} 2 & 1 \\ 0 & 2 \end{pmatrix}$, find two eigenvalues and *try* to find two eigenvectors. What goes wrong?

2.4.5. Explain the connection between the power method and the Strong Ergodic Theorem for linear models.

3

Nonlinear Models of Interactions

Our attention so far has been focused on modeling single populations. Although we have broken a single population into subgroups, such as by age or developmental stage, we have still treated it as if it is unaffected by the other species or populations with which it might share an environment. Although these models have provided valuable insights into how population sizes can change, we now move our attentions to interactions between species or populations.

Most living things interact with many coinhabitants of their environment. Preying on other species, whether plant or animal, is a common way of taking in energy; and most organisms are at risk of being preyed on themselves. But not all important interactions between species are so obvious. Species may find themselves in competition for limited resources, whether food or space, so that growth in one population is detrimental to another. Mutualism, where several species interact in a way that benefits both, also occurs in nature. A real ecosystem may have hundreds or thousands of interacting populations, with all sorts of direct and indirect interactions between them. How can we understand the effects of these interactions without being lost in the complexity of it all?

To begin to understand the dynamics of interacting populations in such systems, we start by focusing on only two populations and a single type of interaction. The questions that will guide us are modifications of those we have already asked when modeling a single population. For instance, what mathematical formulas might model an interaction such as that of a predator and its prey? What behavior does a computer simulation of such a model show? Does one species disappear, and if so why? Do the populations reach some equilibrium, oscillate, or jump wildly? Can such a system of interacting populations show stability, and if so, under what circumstances?

3.1. A Simple Predator–Prey Model

Imagine two species, one of which, the predator, preys on the other, the prey. To keep things simple, we imagine that the predator–prey interaction between these species is the most important one for determining population sizes. An example of this might be arctic hares and foxes confined to an island. The hares are the primary food source for the foxes, and the foxes provide the primary limitation to the unchecked growth of the hare population.

▶   For what other pairs of species might a predator–prey interaction be the dominant one in determining population size?

Even for species experiencing many other interactions, the model we develop can be viewed as a first step to understanding one contributing factor in population dynamics.

Letting P_t denote the size of the prey population and Q_t the size of the predator population at time t, we need to choose equations

$$\Delta P = F(P, Q)$$
$$\Delta Q = G(P, Q)$$

that might give the changes in each of the populations over one time step in terms of *both* of the populations. But what are appropriate expressions for the right-hand sides?

First, it is helpful to think of how the two populations would change in the absence of one another. For instance, a reasonable assumption for the prey is that, if no predators are around, the population would be described by the discrete logistic model:

$$\Delta P = rP(1 - P/K) \qquad \text{in the absence of predators.}$$

If we assume the predator's primary source of food is the prey, then we would expect the population of predators to decline in the absence of prey:

$$\Delta Q = -uQ \qquad \text{in the absence of prey.}$$

Here, u should be a positive constant that is at most 1, since $-u$ gives a per-capita death rate.

To introduce the interaction between the species, we include terms involving the product PQ. This product has several qualitative features that make it a good candidate for describing interactions. If both populations P and Q are small, so that we would expect little effect from the interaction, then PQ is small. If both P and Q are large, so that we would expect major effects from the interaction, then PQ is large. If one of P and Q is small, and the other is large, then (at least for some values of P and Q) the product PQ will be midsize. Most importantly, if *either* P or Q is increased, so that we would expect the interaction to be greater, then PQ increases. The product PQ, then, behaves roughly as we would want to give a good description of the amount of interaction we might expect between the populations.

We model the two populations by:

$$\Delta P = rP(1 - P/K) - sPQ,$$
$$\Delta Q = -uQ + vPQ.$$

Here, s and v both denote positive constants. The term "$-sPQ$" describes a detrimental effect of the predator–prey interaction on the prey, and the term "vPQ" describes a beneficial effect of the interaction on the predator. There is no reason to expect that the values of s and v need to be of the same size, since the predator may well benefit more than the prey is harmed, or the prey may be harmed more than the predator benefits.

The use of a term such as kPQ to model population interactions is sometimes called a law of *mass action*. One way to justify it is to imagine individuals in two populations of size P and Q moving around at random and mixing homogeneously. Then, over a certain time interval, we might expect the number of chance meetings between individuals in the different populations to be PQ. A fraction of these meetings will be significant enough to result in kPQ predation interactions during a time step. Note that a mass action term in a model means the equations are nonlinear; even though this is a very simple interaction term, we should perhaps expect complicated dynamics.

Rewriting our simple predator–prey model in terms of populations, rather than changes in populations, gives:

$$P_{t+1} = P_t(1 + r(1 - P_t/K)) - sP_tQ_t,$$
$$Q_{t+1} = (1 - u)Q_t + vP_tQ_t.$$

with r, s, u, v, and K all positive constants, and $u < 1$.

With any model, getting accurate ideas of parameter values from data collected on experimental study populations can be quite difficult. The appropriate values of the constants appearing in the species–interaction terms are perhaps even harder to have an intuitive feel for than the r and K of the logistic model. For the time being, it's enough to know that, the larger s and v are, the stronger the effect of the predator–prey interaction.

Table 3.1. *Predator–Prey Model Population Values*

t	0	1	2	3	4	5	6	7	8	9	10
P_t	1.10	0.74	0.68	0.55	0.40	0.31	0.30	0.34	0.44	0.54	0.60
Q_t	0.40	0.83	1.22	1.71	2.00	1.89	1.51	1.17	0.99	0.99	1.15

t	11	12	13	14	15	16	17	18	19	20	...
P_t	0.57	0.48	0.39	0.34	0.34	0.39	0.46	0.52	0.54	0.50	...
Q_t	1.44	1.74	1.85	1.70	1.43	1.21	1.12	1.16	1.33	1.56	...

The phase plane. To be concrete, consider the parameter values

$$K = 1, \; r = 1.3, \; s = .5, \; u = .7, \quad \text{and} \quad v = 1.6,$$

so that our model becomes

$$P_{t+1} = P_t(1 + 1.3(1 - P_t)) - .5P_tQ_t,$$
$$Q_{t+1} = .3Q_t + 1.6P_tQ_t.$$

What are appropriate means for studying the behavior of this basic predator–prey model?

The first thing to do is to simply pick some initial values (P_0, Q_0), compute a long list of future populations, and plot the populations over time. For instance, with $(P_0, Q_0) = (1.10, 0.40)$, we produce Table 3.1.

▶ Look carefully at the population values in Table 3.1. Do they seem to be approaching any limiting values? Do they oscillate or change monotonically?

Plotting the populations as functions of time produces Figure 3.1. There, we see the populations seem to oscillate, but the oscillations also appear to be decreasing in amplitude.

We could redo this numerical experiment with more time steps and a variety of different initial population choices and see that most reasonable choices seem to lead to the populations ultimately settling down to around $(P^*, Q^*) \approx (.4, 1.5)$.

Next, we might want to produce some sort of a cobweb diagram, as we did for single populations in Chapter 1. After all, the cobweb was very helpful in understanding the simple nonlinear models, such as the discrete logistic model. It led to direct insight into equilibria and stability. Unfortunately, with two population values, P_t and Q_t, giving us two new population values, P_{t+1} and Q_{t+1}, we would need four dimensions to draw such a graph.

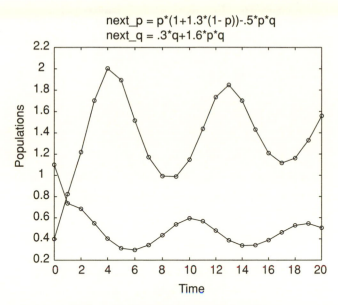

Figure 3.1. Predator–prey model time plot.

Instead, we introduce a new type of plot called a *phase plane*. Rather than plot both P_t and Q_t as functions of t as in Figure 3.1, we label our axes P and Q, and put a dot at the point representing (P_0, Q_0). We then put another dot at (P_1, Q_1) and draw a line from our first dot to it. Next, we plot (P_2, Q_2) and connect it to its predecessor, and continue on connecting each consecutive pair of points representing the two population sizes. In the case of the model data in Table 3.1, this gives Figure 3.2.

The succession of points (P_0, Q_0), (P_1, Q_1), (P_2, Q_2), ... is called the population *orbit*. Although we can only draw a finite number of points in an orbit, it really continues forever. (It might, however, hit an equilibrium point, in which case we would need to plot the same point repeatedly.)

Notice there may be some loss of information when we produce a phase plane plot. For instance, unless we label the points, we no longer know the value of t that corresponds to each point (P_t, Q_t) we have plotted. Of course, we could follow the lines connecting the points in an orbit, counting time steps, to figure out a value for t, but we would still need to know where to start. It's a good idea to at least label the point where $t = 0$, so that you know the correct direction to follow in the orbit.

▶   Does the orbit in Figure 3.2 spiral inward or outward? How does knowing $(P_0, Q_0) = (1.10, 0.40)$ let you decide?

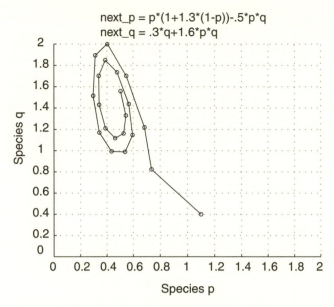

Figure 3.2. Predator–prey model phase plane plot; single orbit.

By drawing many different orbits on the same phase plane, you can get a good feel for the behavior of the model. For instance, for the model above, we can produce the phase plane diagram in Figure 3.3.

Most of the orbits in this plot spiral counterclockwise inward toward the vicinity of (0.45, 1.5). A few orbits, such as one starting around (1.56, 0.66), fly off the graph. A few others, such as one starting around (1.35, 0.42), take a step in the clockwise direction before falling into the general pattern of counterclockwise progression. Despite the few exceptions, this plot illustrates a remarkable qualitative regularity in the behavior of population sizes over time, regardless of initial values. We would certainly be tempted to say that there is a stable equilibrium somewhere around (0.45, 1.5), since most population values seem to be drawn in toward that point.

▸   What is the biological meaning of the orbit beginning around (1.56, 0.66) leaving the phase plane plot? What must be happening to the populations?

The fact that the predator–prey model often shows repeated oscillations is intriguing. Might it be possible for a predator–prey interaction to lead to stable oscillations that, unlike those in the figures here, do not "damp out"? Could two populations in nature endlessly cycle in size?

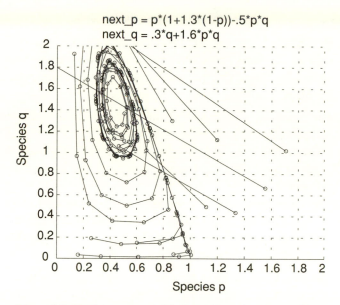

Figure 3.3. Predator–prey model phase plane plot; many orbits.

But before getting too drawn into speculation, we should reflect on the biological situation we are studying to make sure we are not being led astray. Are there real predator and prey populations that behave in an oscillatory fashion, even for a short time?

Certainly, there is widespread evidence of a large prey population being followed by a growth in predators, which then is followed by a decline in prey, and then a decline in predators. One obvious example is in human hunting societies over-exploiting their food sources. Bird and insects, as well as mammal predator–prey pairs, also exhibit such growth and decline. However, these cycles usually "damp out" to an approximate equilibrium, or lead to the extermination of one of the populations. There is little unequivocal evidence from nature of long-term repeated oscillations, although carefully controlled laboratory experiments have shown oscillations.

Sustained oscillations of constant size would require a very delicate balance that is unlikely in real populations. Mathematically, we say such systems are *structurally unstable,* in that slight changes to the model cause the oscillations to either shrink or grow. Because of the complexity of biological systems, numerous factors left out of our model are likely to prevent regular, constant-size oscillations to continue for any significant length of time.

Problems

3.1.1. In Figure 3.1, you see oscillations of both the predator and the prey populations. Which population peaks first, and which one lags slightly behind the other? Can you explain why this should happen through biological intuition?

3.1.2. The oscillations of P_t and Q_t in Figure 3.1 lead to the clockwise pattern of the orbits in Figure 3.2.

    a. Graph $P_t = \cos t$ and $Q_t = \sin t$. Do the peaks on these oscillatory graphs have the same relative location to one another as those in Figure 3.1?

    b. In a phase plane, plot the points $(P_t, Q_t) = (\cos t, \sin t)$ for various $t = 0, .1, .2, \ldots$. As you increase t, how do the points move?

    c. What features of the plots in Figure 3.1 are responsible for the inwardness of the spiral in Figure 3.2?

3.1.3. Use the MATLAB program `twopop` to investigate how the interaction parameters s and v affect the long-term behavior of the predator–prey model presented in the text. Holding all other parameters fixed, vary s and describe how the location of the equilibrium point changes. Then vary v. Give an intuitive explanation of why these behaviors are reasonable.

3.1.4. If you begin with the particular parameter values used in the text for the predator–prey model, and gradually increase s and v, what qualitative changes do you think you will see in the orbits? Check your conjecture by experimenting with `twopop`. Increase s and v until orbits leave the phase plane plot. How should you interpret this behavior biologically?

3.1.5. In this section, we have observed that the model

$$P_{t+1} = P_t(1 + 1.3(1 - P_t)) - .5P_tQ_t,$$
$$Q_{t+1} = .3Q_t + 1.6P_tQ_t$$

appears to have a stable equilibrium that is approached through oscillations. Because the discrete logistic model $P_{t+1} = P_t(1 + 1.3(1 - P_t))$ on which it is based has $r = 1.3$, we know that it alone would produce damped oscillations toward the carrying capacity. We might think, then, that the observed oscillations were due to this value of r.

    By using the MATLAB program `twopop` with a number of values of r smaller than 1.3 in the predator–prey model, investigate whether that is indeed the sole reason for the oscillations that were observed.

Can you find a value of r that would result in no oscillations in the one-population discrete logistic model, but still yields oscillations in the predator–prey model? If you can, the predator–prey interaction must be contributing the oscillatory behavior.

3.1.6. The one-population discrete logistic model has only unstable equilibria for $r > 2$. By experimenting with MATLAB and twopop, investigate whether with such an r, the presence of a predator might produce a stable equilibrium. Starting with the specific parameter values of the model in the text, try to find a value of $r > 2$, which gives the predator–prey model a stable equilibrium. You might also vary the coefficients of the interaction terms to try to achieve this. What biological lessons can you draw from this experimentation?

3.1.7. Imagine a predator–prey interaction in which a certain number of the prey population cannot be eaten because of a refuge in their environment that the predator cannot enter. Why might interaction terms like $-s(P - w)Q$ and $v(P - w)Q$ be reasonable in the modeling equations? What is the meaning of w? If $P < w$, is this reasonable?

3.1.8. Investigate the behavior of the model of the last problem with twopop. Describe your observations.

3.1.9. The mass action term PQ is not the only possibility for describing interaction between populations. Which of the following have the same qualitative feature as the mass action term, in that they increase if *either* P or Q is increased?

$$P(1 - e^{-vQ}), \quad \frac{P}{Q}, \quad \sqrt{PQ}, \quad P + Q.$$

Which of these might be the basis of a reasonable interaction term? Explain.

3.1.10. Another version of a predator–prey model is

$$P_{t+1} = P_t e^{r(1 - P_t/K) - sQ_t},$$
$$Q_{t+1} = u(1 - e^{-vQ_t})P_t,$$

where $s, u, v > 0$ and r, K are as in the logistic model. Explain why these equations are reasonable. What happens to the prey in the absence of predators? What happens to the predators in the absence of prey? What are the effects of e^{-sQ_t} and $(1 - e^{-vQ_t})$ in the model equations? [You might want to graph e^{-x} and $(1 - e^{-x})$ as part of your explanation.]

3.1.11. Investigate the model of the last problem using twopop. Does it behave qualitatively like the model presented in the text? Are there any important differences you can identify?

3.2. Equilibria of Multipopulation Models

The phase plane is a good tool for exploring how the behavior of a particular model depends on the initial values used. Often, we observe an orbit jumping around a bit before settling into some pattern. Just as for single population models, we call the behavior exhibited in the early steps of an orbit *transient*. Often, what we are more interested in are the *equilibria* of the model.

Definition. For a model of two populations given by $P_{t+1} = F(P_t, Q_t)$ and $Q_{t+1} = G(P_t, Q_t)$, an *equilibrium* is a point (P^*, Q^*) with $P^* = F(P^*, Q^*)$ and $Q^* = G(P^*, Q^*)$. For a model given in the form $\Delta P = f(P, Q)$ and $\Delta Q = g(P, Q)$, it is a point (P^*, Q^*) with $f(P^*, Q^*) = 0$ and $g(P^*, Q^*) = 0$.

Thus, when populations are at an equilibrium, *neither* population will change in future time steps.

Solving for equilibria is not much harder for our predator–prey model than it was for the logistic model alone, although there are now two equations to solve simultaneously:

$$P^* = P^*(1 + 1.3(1 - P^*)) - .5P^*Q^*,$$
$$Q^* = .3Q^* + 1.6P^*Q^*,$$

or

$$0 = P^*1.3(1 - P^*) - .5P^*Q^* = P^*(1.3 - 1.3P^* - .5Q^*),$$
$$0 = -.7Q^* + 1.6P^*Q^* = Q^*(-.7 + 1.6P^*).$$

From the factorization in the second equation, we see

$$\text{either} \quad Q^* = 0 \quad \text{or} \quad P^* = .7/1.6 = .4375.$$

If $Q^* = 0$, then the first equation says either $P^* = 0$ or $P^* = 1$, giving us the two equilibria $(0, 0)$ and $(1, 0)$. If $P^* = .4375$, on the other hand, then the first equation requires that $0 = 1.3 - 1.3(.4375) - .5Q^*$, so that $Q^* = 1.4625$. This means there is a third equilibrium at $(.4375, 1.4625)$.

The first two equilibria are easily understood: Obviously $(0, 0)$ is an equilibrium from biological interpretations alone; if both populations are 0, they will stay that way. The equilibrium $(1, 0)$ would simply correspond to having no predators, so the prey remains at the carrying capacity of 1 in the logistic model. The third equilibrium is perhaps the most interesting and is the one observed from plotting orbits on the phase plane. Interestingly, the presence of the predator seems to have depressed the prey population to a size well below the carrying capacity.

▶ Why did we not see orbits being drawn to the equilibria $(1, 0)$ and $(0, 0)$ in the phase plane plot of Figure 3.3? Is this reasonable biologically?

Choosing particular parameter values for the model has actually limited our insight. Rather than finding the equilibria for a single choice of the parameters, it would be more useful to have formulas for the equilibria in terms of the parameters. To find these, we return to our general predator–prey model (in units where the carrying capacity $K = 1$):

$$P_{t+1} = P_t(1 + r(1 - P_t)) - sP_tQ_t,$$
$$Q_{t+1} = (1 - u)Q_t + vP_tQ_t.$$

The equilibrium equations are:

$$0 = P^*(r(1 - P^*)) - sP^*Q^* = P^*(r(1 - P^*) - sQ^*), \qquad (3.1)$$
$$0 = -uQ^* + vP^*Q^* = Q^*(-u + vP^*).$$

The factorizations of these equations mean equilibria are determined by at least one of the conditions

$$P^* = 0 \quad \text{or} \quad r(1 - P^*) - sQ^* = 0,$$

and at least one of

$$Q^* = 0 \quad \text{or} \quad -u + vP^* = 0.$$

Therefore, if we graph the four lines

$$P = 0, \quad Q = \frac{r}{s}(1 - P), \quad Q = 0, \quad P = \frac{u}{v}$$

in the phase plane, we will find the equilibria at some of their points of intersection. The four lines plotted in Figure 3.4 are called *nullclines* of the model, because they are lines on which either $\Delta Q = 0$ or $\Delta P = 0$.

▶ Explain why only three of the five points of intersection of these four lines are equilibria. Which points of intersection do not satisfy Eqs. (3.1)?

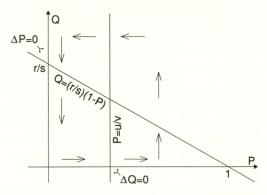

Figure 3.4. Nullclines, $\Delta P = 0$ and $\Delta Q = 0$, for the predator–prey model.

A little algebra to find the points of intersection of the nullclines gives us formulas for the equilibria:

$$(0, 0), \ (1, 0), \quad \text{and} \quad \left(\frac{u}{v}, \frac{r}{s}\left(1 - \frac{u}{v}\right)\right).$$

Although we can use the formula for this last equilibrium to see how it changes if the parameters are varied, it's more interesting to investigate this graphically through Figure 3.4.

If either r or s is changed, the Q-nullcline is unaffected. However, either increasing r or decreasing s causes r/s to increase, causing the downward sloping line in the P-nullcline to intersect the Q-axis higher up. Note the P-nullcline still crosses the P-axis at 1. This means that the equilibrium $(u/v, (r/s)(1 - u/v))$ will simply move higher, so that P^* is unchanged while Q^* increases.

▶   From the model equations, increasing r or decreasing s seems to be likely to benefit the prey population. If the populations are at equilibrium, is that how things turn out?

To think through the implications of this, suppose we imagine our model applies to a certain agricultural crop (the prey) and an insect that eats it (the predator). To get a higher yield, a new crop variety is introduced with a larger value of r. The carrying capacity does not change, but we hope to "outgrow" the predator, so that we end up with a greater crop. But, according to our model, assuming the system settles into the equilibrium, this will cause P^* to remain unchanged while Q^* increases. In other words, the insect predator actually benefits whereas the crop prey gains nothing.

Although this may seem surprising at first, on further reflection you might convince yourself it is reasonable. Even if you find it unreasonable, or not in accord with observation of a real predator–prey interaction, and therefore decide to reject our model as not applicable, you have learned something. By writing our assumptions of the predator–prey interaction mathematically in the form of a model and then analyzing it, we were able to deduce the consequences of our assumptions. If these consequences are not in accord with a real population, then we need to rethink our assumptions and try to see what important features of the real situation our model has overlooked.

Even if our goals are more theoretical, and we are not interested in precisely predicting future populations, the mathematical model is a tool both for expressing our beliefs as to what factors affect population changes, and for deducing the effects of those factors alone. If the deduced effects do not fit with observation, we have discovered a gap in our knowledge. Identifying such a gap could be viewed as progress toward both producing a better model and understanding the real interspecies interaction.

Of course we can analyze the affect of varying u and v on the equilibrium also and think through the biological implications similarly. We'll leave that as an exercise.

Nullclines and the direction of orbits. The nullclines are actually of use not only for determining equilibria, but also for understanding better the dynamics of the model. Consider again the P-nullcline, which consists of a vertical line and a downward sloping line. This nullcline divides the plane into several regions. Inside any of these regions, ΔP must always be positive or always be negative. This is because if it is positive at one point and negative at another point, then at some point lying between the two it would have to be zero. That point would lie on the nullcline, and so the nullcline must separate the points where ΔP has different signs. (We have implicitly assumed that ΔP is a continuous function of P and Q, so that there are no sudden jumps in the value of ΔP.)

To determine whether ΔP is positive or negative on one of the regions, we simply pick one point and evaluate. For instance, below the sloping line, pick a point where P and Q are both quite small, but positive. Then, because $\Delta P = P(r(1 - P) - sQ)$, we can analyze the sign of ΔP as follows: Because P is small, $1 - P$ will be near 1, so $r(1 - P)$ will be near r. Because Q is small, sQ will be near 0. Thus, $(r(1 - P) - sQ) > 0$, and because $P > 0$, then $\Delta P = P(r(1 - P) - sQ) > 0$.

If ΔP is positive on a region in the phase plane, that means that any population values in that region will produce an increase in P over the next time step. In other words, the orbit will move to the right (and possibly up or down as well, depending on the sign of ΔQ). In Figure 3.4 we have placed horizontal arrows pointing to the right on the region where we have shown $\Delta P > 0$.

▶ Why do we mark two regions in Figure 3.4 with arrows pointing to the right, rather than just one?

As an exercise, you can show that $\Delta P < 0$ above the sloping line of the P-nullcline. Thus, in that region, each time step will produce a decrease in the value of P, and so arrows pointing to the left have been drawn there in Figure 3.4. A similar analysis can be done of the sign of ΔQ, which gives the upward and downward pointing arrows in the figure.

Notice that these arrows strongly suggest that orbits will move in a counterclockwise fashion around the interesting equilibrium, as in fact we have seen in numerical experiments. However, from these arrows, we are unable to tell if the orbit will spiral in toward the equilibrium or spiral outward, so we cannot draw conclusions as to the stability of the equilibrium.

We also need to be slightly careful because with a discrete model, the populations at successive time steps may well jump over the nullclines from one time step to the next. This can lead to more erratic behavior, although often it is transient. In fact, this happened in the orbit beginning near $(1.35, 0.42)$ in Figure 3.3 that appears to start moving around the equilibrium in a clockwise direction before falling into the counterclockwise pattern. If you sketch the nullclines in on Figure 3.3, you will see even this first step is in the general direction predicted by the arrows of Figure 3.4.

Problems

3.2.1. The nullclines in Figure 3.4 are drawn assuming $u/v < 1$. What would they look like If $u/v = 1$? If $u/v > 1$? Are there still three equilibria in these situations? Are all equilibria biologically meaningful? Explain.

3.2.2. What is the effect of varying u and v on the equilibrium of the predator–prey model of the text. Explain by describing the effects of these parameters on the locations of the nullclines in Figure 3.4. Do these effects seem reasonable biologically?

3.2.3. For each of the cases $u/v = 1$ and $u/v > 1$, draw the correct version of Figure 3.4 and on it draw arrows indicating where ΔP and ΔQ are positive and negative. Use your diagrams to predict the behavior of some orbits, and then check your predictions using twopop.

3.2.4. Using the specified values of r, s, u, and v, sketch the nullclines in on a copy of Figure 3.3, and mark arrows showing where ΔP and ΔQ are positive and negative.

a. Are all orbit steps in the figure in accord with the arrows you drew?

b. If a point lies exactly on top of the ΔP nullcline, do the arrows indicate the correct direction of the orbit over the next time step?

c. If a point lies exactly on top of the ΔQ nullcline, do the arrows indicate the correct direction of the orbit over the next time step?

3.2.5. For the predator–prey model of the text, complete the work of analyzing the signs of ΔP and ΔQ on each region created by the nullclines, justifying all the arrows shown in Figure 3.4.

a. Show that above the sloped line of the P-nullcline, $\Delta P < 0$.

b. Show that to the left of the vertical line of the Q-nullcline, $\Delta Q < 0$.

c. Show that to the right of the vertical line of the Q-nullcline, $\Delta Q > 0$.

3.2.6. Analyze the predator–prey model given in Problem 3.1.10 of the last section by doing the following:

a. Plot the nullclines and label regions in the phase plane with arrows indicating the signs of ΔP and ΔQ.

b. Find all equilibria in terms of the parameters of the model, if possible.

3.2.7. A simple model for a predator–prey interaction where the predator has a source of food in addition to the prey is:

$$\Delta P = rP(1 - P/K_1) - sPQ,$$
$$\Delta Q = uQ(1 - Q/K_2) + vPQ.$$

a. Explain why these equations model the described situation.

b. Choosing units so that $K_1 = K_2 = 1$, find and plot the nullclines for this model. Draw arrows on your plot indicating the signs of ΔP and ΔQ.

c. Compute all equilibria for the model.

d. What does your analysis of this model lead you to expect as typical behavior of orbits? Check your predictions with a computer.

3.3. Linearization and Stability

An intuitive concept of stability of an equilibrium has already been used in our discussion of the predator–prey model. Because in Figure 3.2 we observed population values close to $(P^*, Q^*) = (.4375, 1.4625)$ appear to move closer to those equilibrium values as we follow their orbit, we said that equilibrium appeared to be stable. Notice, however, that the movement toward equilibrium is not simple. Because the inward spiral of the orbit has an oval shape, the orbit is closer to the equilibrium on the more vertical parts of the spiral than on the more horizontal parts. The orbit gets closer to the equilibrium, then further away, then closer, then further away, and so on. The oscillatory overshoot is even clearer in Figure 3.1. However, the overshoot does damp out, so that over time populations near the equilibrium do seem to approach it, and never again move as far away as they were.

As compelling as a computer simulation can be, it is also desirable to be able to show mathematically the stability of the model. To do this, we use the same basic approach as developed earlier for one-population models, such as the logistic model.

To focus attention at the equilibrium, we let

$$P_t = P^* + p_t,$$
$$Q_t = Q^* + q_t,$$

where p_t and q_t represent small perturbations from the equilibrium. We are interested in seeing how these perturbations change over time. Do they grow or do they shrink?

For the model

$$P_{t+1} = P_t(1 + 1.3(1 - P_t)) - .5P_tQ_t,$$
$$Q_{t+1} = .3Q_t + 1.6P_tQ_t,$$

with equilibrium $(P^*, Q^*) = (.4375, 1.4625)$, substituting in the expressions for populations in terms of perturbations from equilibria, gives

$$.4375 + p_{t+1} = (.4375 + p_t)(1 + 1.3(1 - (.4375 + p_t)))$$
$$- .5(.4375 + p_t)(1.4625 + q_t),$$
$$1.4625 + q_{t+1} = .3(1.4625 + q_t) + 1.6(.4375 + p_t)(1.4625 + q_t).$$

Some rather messy algebra, which is nonetheless worth checking, gives us

$$p_{t+1} = .43125p_t - .21875q_t - 1.3p_t^2 - .5p_tq_t,$$
$$q_{t+1} = 2.34p_t + q_t + 1.6p_tq_t.$$

Now we are imagining population values P_t and Q_t that are close to P^* and Q^*, and so p_t and q_t are near 0. But then all second-order terms p_t^2, p_tq_t, and q_t^2 will be much smaller than the first-order terms p_t and q_t, because the product of two small numbers is much smaller than the original numbers. We can therefore approximate our model by simply discarding the second-order terms, since they are of negligible size. This gives

$$p_{t+1} \approx .43125 p_t - .21875 q_t,$$

$$q_{t+1} \approx 2.34 p_t + q_t.$$

Of course the accuracy of this approximation increases the closer p_t and q_t are to 0. Thus, we have approximated our nonlinear model, at least near the equilibrium, by a linear one.

Now we can address the question of whether the equilibrium is stable or not, by using our understanding of linear models. As was shown in Chapter 2, the eigenvalues of a linear model indicate whether the model predicts long-term growth or decay.

Because

$$\begin{pmatrix} p_{t+1} \\ q_{t+1} \end{pmatrix} \approx \begin{pmatrix} .43125 & -.21875 \\ 2.34 & 1 \end{pmatrix} \begin{pmatrix} p_t \\ q_t \end{pmatrix},$$

we just need to compute the eigenvalues of the matrix. Using MATLAB to do this gives the two complex eigenvalues $\lambda = .7156 \pm .6565i$. Computing the absolute value of both eigenvalues yields $|\lambda| = \sqrt{.7156^2 + .6565^2} = .9711$. Since this number is less than 1, the perturbations from equilibrium must get smaller over time. Thus, the equilibrium really is stable as suspected.

As when we linearized the logistic model to understand its stability, we could reinterpret the linearization process in terms of the calculus concept of derivatives. Here, however, because the model tracks several different variables, we need *partial derivatives*, which are usually taught in multivariable calculus. Therefore, we'll omit that discussion, although one of the exercises does move in that direction.

While the particular model and equilibrium we analyzed above turned out to be stable, how else could the analysis have turned out? What would instability look like? Are there different types of stability and instability?

As long as we look at a model of only two interacting populations, once we find an equilibrium and linearize the model near the equilibrium, we get a model described by a 2×2 matrix. Therefore, when the eigenvalues are computed for this matrix, we will find two of them. Denoting the two eigenvalues by λ_1 and λ_2, we might see several different types of behavior around the equilibrium, depending on the sizes of λ_1 and λ_2.

- If both $|\lambda_1| < 1$ and $|\lambda_2| < 1$, as was the case in our example, then all small perturbations from the equilibrium shrink. In this case, nearby populations values get closer to the equilibrium values, and so the equilibrium is stable.

- If both $|\lambda_1| > 1$ and $|\lambda_2| > 1$, then all small perturbations from the equilibrium grow, and so the equilibrium is unstable.
- If $|\lambda_1| > 1$ and $|\lambda_2| < 1$, or vice versa, then different perturbations behave qualitatively differently. A perturbation that is an eigenvector with eigenvalue λ_1 will grow, whereas one that is an eigenvector with eigenvalue λ_2 will shrink. Most perturbations are some combination of these, and so will exhibit a combined behavior.

An equilibrium that exhibits this last type of behavior is a new sort that is often referred to as a *saddle* equilibrium. To understand the name, imagine a ball placed on a saddle, where a horse rider normally sits. If the ball is pushed slightly forward or backward, so that it is forced to move in an uphill direction, it will tend to move back toward the saddle point, as if it is at a stable equilibrium. On the other hand, if it is pushed to the sides, where the saddle goes downhill, it will tend to roll away from the saddle point, behaving as if the equilibrium is unstable. Actually, most directions in which the ball can be pushed move it in a combination of these directions, and it responds by a combination of these behaviors. The net effect is that, although it may move back toward the saddle point for a bit, it typically then heads downhill. A saddle equilibrium, then, should be thought of as having a special type of instability.

Schematically, these three cases are represented in Figure 3.5.

Figure 3.5 should be taken rather loosely, since we've ignored information about the sign of the eigenvalues. If an eigenvalue is negative (or complex), it will produce oscillatory behavior around the equilibrium as populations move toward or away from the equilibrium. After all, that is what we saw in studying linear models earlier, and our analysis here is based on analyzing a

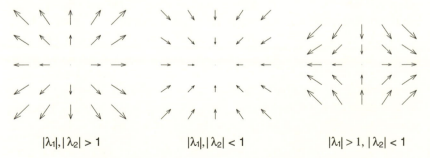

$|\lambda_1|, |\lambda_2| > 1$ $|\lambda_1|, |\lambda_2| < 1$ $|\lambda_1| > 1, |\lambda_2| < 1$

Figure 3.5. Possible behaviors near equilibrium points.

linear approximation to our model. In our predator–prey example, the complex nature of the eigenvalues explains why the approach to the stable equilibrium involved the oscillations of the predator and prey populations, and hence the spiral orbit in the phase plane.

Keep in mind that we have ignored the possibility that $|\lambda_i| = 1$, where we may not be able to draw conclusions as to stability.

Also, it is important to realize that we have only discussed *local stability* of a model – that is, what happens in close proximity to an equilibrium. Because discrete population models can result in large jumps of populations from one time step to the next, it is much harder to analyze *global stability* – that is, whether all initial populations eventually approach an equilibrium or not.

Finally, imagine modeling n interacting populations. Such a model is given by n equations, and on computing equilibria, will have an $n \times n$ matrix model as its linearization at an equilibrium. Thus, there will be n eigenvalues to consider. Only when all eigenvalues are less than one in absolute value can we conclude that an equilibrium is stable.

Problems

3.3.1. In the text, the equilibrium (.4375, 1.4625) of the predator–prey model was investigated for stability by linearizing and computing eigenvalues. Investigate the stability of its other two equilibria the same way. Are your results biologically reasonable? Do they agree with numerical experiments?

3.3.2. For the predator–prey model,

$$P_{t+1} = P_t(1 + .8(1 - P_t)) - 4P_t Q_t,$$
$$Q_{t+1} = .9Q_t + 2P_t Q_t,$$

a. Compute the equilibria.
b. Use MATLAB and `twopop` to make an informed guess as to whether the equilibria are stable or unstable.
c. Linearize the model at each of the equilibria and compute eigenvalues to determine stability.

3.3.3. For the predator–prey model,

$$P_{t+1} = P_t(1 + 1.6(1 - P_t)) - .1P_t Q_t,$$
$$Q_{t+1} = .3Q_t + .6P_t Q_t,$$

    a. Compute the equilibria. Which ones are biologically meaningful?

    b. Use MATLAB and twopop to make an informed guess as to whether the equilibria are stable or unstable.

    c. Linearize the model at each of the equilibria and compute eigenvalues to determine stability.

3.3.4. Rolling a ball on a saddle provided a good image for the third type of equilibrium in Figure 3.5. On what surface would a ball roll to depict the first type? The second type?

3.3.5. Determining the stability of a model only makes sense at an equilibrium. If a point $(P^{\#}, Q^{\#})$ is not an equilibrium of a model, it is meaningless to ask whether it is stable or unstable.

    Suppose, however, you mistakenly thought $(.5, 1.5)$ was an equilibrium of the predator–prey model analyzed in the text. What happens mathematically if you attempt to determine its stability through linearization? What indications are there that you have made a mistake?

3.3.6. For the predator-prey model of this chapter, the equilibrium $(0,0)$ is always a saddle, regardless of parameter values.

    a. Explain why this is so intuitively, by considering initial populations of $(P_0, 0)$ and $(0, Q_0)$, for small P_0 and Q_0.

    b. Show the linearization of the model at $(0,0)$ gives $\begin{pmatrix} 1+r & 0 \\ 0 & 1-u \end{pmatrix}$.

What are the eigenvalues of this matrix? Explain why, for biologically reasonable values of the parameters, this means $(0,0)$ is a saddle.

3.3.7. (Calculus) Stability of equilibria can be determined through derivatives, as in Chapter 1, provided you understand partial derivatives. The *Jacobian matrix* of a model

$$P_{t+1} = F(P_t, Q_t),$$
$$Q_{t+1} = G(P_t, Q_t)$$

is the matrix

$$\begin{pmatrix} \frac{\partial F}{\partial P_t} & \frac{\partial F}{\partial Q_t} \\ \frac{\partial G}{\partial P_t} & \frac{\partial G}{\partial Q_t} \end{pmatrix}.$$

    a. For the predator–prey model,

$$P_{t+1} = P_t(1 + 1.3(1 - P_t)) - .5P_tQ_t,$$
$$Q_{t+1} = .3Q_t + 1.6P_tQ_t$$

compute the Jacobian matrix.

    b. Evaluate the Jacobian matrix at the point $(P^*, Q^*) = (.4375,$ 1.4625)$. Is it the same matrix obtained in the text by linearizing algebraically?

3.3.8. (Calculus) Repeat the last problem more generally, without specifying parameter values.

    a. For the predator–prey model,

$$P_{t+1} = P_t(1 + r(1 - P_t)) - sP_tQ_t,$$
$$Q_{t+1} = uQ_t + vP_tQ_t,$$

compute the Jacobian matrix.

    b. Evaluate the Jacobian matrix at the three equilibria $(P^*, Q^*) =$ $(0, 0)$, $(1, 0)$, and $(\frac{u}{v}, \frac{r}{s}(1 - \frac{u}{v}))$ to get linearizations of the model at those points.

3.4. Positive and Negative Interactions

So far in this chapter, a predator-prey model has been the guiding example for the development and mathematical analysis of a two-population model. Now that the tools have been introduced, they can be applied to models of other interactions as well. By linking the basic elements of the models we have already developed in new ways, a surprisingly wide range of phenomena can be modeled. As illustrations, we present three examples. We'll leave the complete analysis to you, but give some simple examples of models of interesting interactions.

Competition. Competition between two species that fill the same niche in an environment might be modeled by the equations

$$\Delta P = rP(1 - (P + Q)/K),$$
$$\Delta Q = uQ(1 - (P + Q)/K).$$

As written, these equations show the competition occurring solely through the struggle for the resources represented by the carrying capacity. Each population grows according to the logistic model if the other population is absent. The presence of the second population acts to reduce the growth of the first, just as the presence of the first reduces the growth of the second. Note also that the two species may have different intrinsic growth rates.

    A natural question about competing species is whether one always "wins," or if coexistence is possible. Because the only allowed difference between the two populations in the model is their intrinsic growth rates r and u, you might conjecture that whichever has the larger rate will take over. But, then again, since the logistic growth declines as the population increases, maybe the two do reach some constant level where both populations persist. Seeing what actually happens, and analyzing the model, will be left to the exercises.

One drawback of this model for some realistic situations is that it describes no negative interactions between the two populations other than the resource competition. A more general version of a competition model is

$$\Delta P = r P(1 - (P + Q)/K) - s P Q,$$
$$\Delta Q = u Q(1 - (P + Q)/K) - v P Q,$$

where the new terms describe how each population has an additional negative effect on the growth rate of the other, through any mechanism other than resource competition. Notice that, if $s \neq v$, these interactions may have differing effects on each species.

▶ What types of interactions might be described by such a term? Give an example of species for which these might be important, and an example where the first version of the model might be fine.

Of course many variations on this model are possible. Any of the many single population models in Chapter 1 could be used as a basis in place of the logistic terms. Various negative terms, other than the mass action one, could also be used to model the competitive interaction.

Immune system vs. infective agent. When an organism is infected with a disease-causing agent, the immune system and the infecting agent often engage in an interaction that is detrimental to both. Imagine a bacterial infection in a human. Various infection-fighting cells, such as T cells, are produced to wipe out the bacteria, but the cells themselves are destroyed in the process.

A first attempt at modeling an immune system fighting a bacterial disease might proceed as follows. Let P measure the amount of immune response (T cells, etc., that fight the disease), and let Q measure the level of infection (the bacteria) in the body.

A simple model might then be given by:

$$\Delta P = r Q - s P Q,$$
$$\Delta Q = u Q - v P Q.$$

Here, we have a linear term in the formula for ΔP indicating that new immune cells are created in response to the infective agent. The mass action term in that formula indicates the negative effect on the immune system through fighting the infection.

The formula for ΔQ indicates by its linear term that the bacteria reproduce in proportion to their presence. The mass action term shows the, perhaps limited, success of the immune system in fighting the infection.

▶ This model is not built on logistic one-population models, so it does not model either P or Q experiencing resource-limited growth. Why is this reasonable for this situation?

Naturally, much more elaborate models of this sort are necessary for capturing the finer details of an infection's course. This one has been built using only the crudest guesses as to how an infection proceeds and cannot be expected to be particularly helpful for deeply understanding any given disease. Incorporating more detailed knowledge of the body's defenses, and the particulars of the infective agent, however, can greatly improve the model. The immune response could be broken down into various actors and much more attention paid to their mode of interaction with the infective agent. Collaboration with experimentalists is necessary in model development to ensure that the important dynamical features observed in living organisms are captured. Work of this sort led to quite useful models of human immunodeficiency virus (HIV) infection, which have played a role in understanding and developing effective treatment strategies for acquired immune deficiency syndrome (AIDS).

Mutualism. In the competition model, two populations have negative effects on one another; and, in the predator–prey model, one population experiences a positive effect and one a negative effect from their interaction. It is also possible that two populations both experience a positive effect from an interaction.

Leaf-cutter ants and the fungi they cultivate are a classic example of such a relationship, but others abound. Some plants produce fruit attractive to birds, so that the plants seeds are dispersed more effectively while the birds have a food source. Both benefit from the other's presence.

To describe such a mutualistic relationship, a simple model one could propose is:

$$\Delta P = rP + sPQ,$$
$$\Delta Q = uQ + vPQ.$$

These equations result in exponential growth of each population in the absence of the other. Moreover, an extra boost in growth occurs if both populations are present. The greater the size of either population, the greater the boost. It's not too surprising that these equations lead to populations quickly growing to astronomical size.

With such unrealistic behavior coming from our first attempt at modeling mutualism, we should look for factors that might prevent real populations from behaving similarly. Since resource limitations, for one or both species, provide a mechanism preventing unchecked growth, we reformulate the model to include such density-dependent effects.

For instance, if we image that, in the absence of species Q, species P follows a logistic model, then we might consider:

$$\Delta P = rP(1 - P/K) + sPQ,$$
$$\Delta Q = -uQ + vPQ.$$

Note we have made another change, on the sign in front of the u, to signify that the population Q will die out in the absence of its partner. This model is one that the presence of species P is necessary for species Q's long-term persistence, although P is not so dependent on Q.

Of course, until this model has been investigated through simulation, mathematical analysis, and comparison to real populations, it is just speculation that it describes mutualism. Using computer experimentation and the mathematical tools developed in this section, we can begin to understand how the model behaves and why. Then, depending on how well the behavior compares with that observed in a biological system, we will be able to evaluate whether our model captures the basic interactions we are attempting to describe. A preliminary modeling failure at least provides a basis for thoughtful reexamination of our understanding, and that may lead to a better model.

Problems

3.4.1. In creating a mutualism model for certain species, we might suspect that there is a maximum benefit that each individual in one population could get from the other. If that is the case, explain why it might be reasonable to replace the mass action term sPQ used above with a term such as

$$sP(1 - e^{-wQ}).$$

Explain why this term has the appropriate qualitative features for this situation.

3.4.2. Cannibalism is rather common among certain insects and fish. A model studied in (Cushing *et al.*, 2001) for *tribolium* (flour beetles) uses three stages of development (L = larvae, P = pupae, and A = adults), and the equations

$$L_{t+1} = fA_t e^{-c_{EL}L_t - c_{EA}A_t},$$
$$P_{t+1} = \tau_{1,2}L_t,$$
$$A_{t+1} = P_t e^{-c_{PA}A_t} + \tau_{3,3}A_t.$$

(Although the beetles also go through an egg stage before becoming larvae, the model does not explicitly track eggs.) All parameters are ≥ 0.

a. If $c_{EL} = 0$, $c_{EA} = 0$, and $c_{PA} = 0$, this becomes an Usher model. Give the corresponding Usher matrix.

b. Because $0 < e^{-a} < 1$ for any positive number a, and e^{-a} decreases as a increases, the presence of the $e^{-c_{EL}L_t}$ in the first equation means the larger L_t is, the smaller the fraction of the fA_t eggs that are laid make it to the larval stage. This is how cannibalism of eggs by larvae enters the model and so c_{EL} is referred to as the egg-larva *cannibalism coefficient*.

Explain the other two cannibalism coefficients in the model. Which stages prey on which other stages?

3.4.3. Investigate the behavior of the infection model of the text using MAT-LAB and twopop first for $r = .05$, $s = .01$, $u = .05$, and $v = .02$, and then for $r = .05$, $s = .02$, $u = .05$, and $v = .01$. Does the model behave as you would expect? Explain. If this were your immune system you were modeling, which set of parameters would you prefer?

3.4.4. For the infection model, compute and draw the nullclines, mark the equilibria, and determine directions of change for each resulting region in the phase plane. (You will need to consider the cases $\frac{r}{s} > \frac{u}{v}$ and $\frac{r}{s} < \frac{u}{v}$ separately.) Explain your results.

3.4.5. What can you say about the stability of the equilibria of the infection model? Is "equilibrium" an important concept in this modeling situation?

Projects

1. Investigate the competition model, for two species that fill the same niche in an environment, given by the equations

$$\Delta P = rP(1 - (P + Q)/K) - sPQ,$$
$$\Delta Q = uQ(1 - (P + Q)/K) - vPQ.$$

Suggestions

- Explain why these two equations model this situation by explaining each term in the modeling equations. Why should K be viewed as the combined carrying capacity for the two species? What sorts of biologically reasonable interactions might be captured by the sPQ and vPQ terms?
- Using $r = s = u = v = 1$ for parameter values, investigate the model using MATLAB and `twopop`. What happens when $r = u = 1$ and $s = v = 0$? Vary the parameter values to get a sense of the effect of each on the model behavior.
- Taking $K = 1$ for convenience, find formulas for the nullclines for this model. Then draw them on a phase plane. (To plot them, find formulas for where the nullclines intercept the axes. You will also need to use that $\frac{r}{r+s} < 1$ and $\frac{u}{u+v} < 1$, regardless of the positive values of r, s, u, and v.)
- Give formulas for all equilibria of the model (still with $K = 1$).
- On the regions created in the phase plane by the nullclines, draw arrows suggesting the directions in which population orbits move. Which equilibria are likely to be stable?
- How does all this work change if $s = v = 0$?
- Using $r = s = u = v = 1$ for parameter values, linearize the model at each of the equilibria. Are each of the equilibria stable or unstable?
- Explain why your work does or does not support the ecological concept of *competitive exclusion*.

2. Investigate mutualism models, beginning with those treated in the text,

$$\Delta P = rP + sPQ,$$
$$\Delta Q = uQ + vPQ,$$

and

$$\Delta P = rP(1 - p/K) + sPQ,$$
$$\Delta Q = -uQ + vPQ,$$

and continuing on with your own model in which both populations face resource limitations.

Suggestions

- For the two models above, find all equilibria.
- Use a computer program such as `twopop` to experimentally investigate the behavior of the two models. You will have to choose reasonable

parameter values and may want to consider the resulting location of equilibria in doing so. Describe the behaviors you see and discuss whether they might be biologically relevant.

- Draw nullclines and analyze the phase plane for the two models presented.
- Determine the stability of equilibria.
- Create a model of mutualism in which both populations face resource limitations in the absence of the other. Explain why you have chosen your particular equations.
- For your model, draw nullclines and analyze the phase plane.
- Are there parameter values that lead to interesting equilibria in your model? That do not lead to interesting equilibria?
- If you find equilibria, what type are they?
- Create a new model in which the benefits each species receives from the other is limited in size, as in Problem 3.4.1 of this section. Investigate it fully.
- Discuss whether these models seem adequate to capture the basic dynamics of mutualism.

4

Modeling Molecular Evolution

Natural selection is the fundamental mechanism through which evolution occurs, but for selection to be possible there must be some underlying variability in genetic makeup within a species. Since selection usually acts to reduce variability, there must also be a source of new genetic variation. This is introduced at the molecular level, in the DNA of individuals, through what are viewed as random changes as the molecules are copied into new generations.

Depending on the nature of these changes in the DNA, offspring may be more, less, or equally viable than the parents. Many of the molecular changes are believed to be selectively neutral, and so are passed on to further descendents and preserved. The DNA within a particular gene may continue to mutate from generation to generation, gradually accumulating more differences from its ancestral form. Thus, several species arising from a common ancestor will have similar, but often not identical, DNA forming a particular gene. The similarities hint at the common ancestor, while the differences point to the evolutionary divergence of the descendents.

Since we can now "read" the structure of DNA with relative ease, a natural and compelling question arises: Can we reconstruct evolutionary relationships between several modern species by comparing the DNA sequences of their versions of a certain gene?

We, of course, expect that species that have more similar genetic sequences are probably more closely related. However, this observation really is not enough to make clear how to deduce an evolutionary tree relating a large number of different species, all with varying degrees of similarity in the chosen gene. In fact, we need to first decide what we might mean by a phrase like "degree of similarity."

In this chapter, we develop mathematical models of DNA mutation processes, that is, of *molecular evolution*. Because the language of probability is needed to describe random mutations, we will present the basics of that subject along the way. We will then see that probability naturally leads us to linear models to describe molecular evolution. The concept of a *phylogenetic*

113

distance as a measure of sequence similarity will emerge from these models. Then, in the next chapter, the material developed here will help address the issue of deducing evolutionary relationships.

4.1. Background on DNA

Genetic information is encoded by DNA molecules, which are passed from parent to offspring. For this transfer, the DNA must be copied. Despite rather elaborate mechanisms to ensure the correctness of the copying process, sections of the molecule may be altered in various ways. Before modeling the most important of these mutations, we need to review briefly the basic structure of DNA.

The DNA molecule forms a double helix, a twisted ladder-like structure. At each of the points where the ladder's upright poles are joined by a rung, one of four possible molecular subunits appears. These subunits, called *nucleotides* or *bases* – adenine, guanine, cytosine, and thymine – are denoted by the letters A, G, C, and T. Because of chemical similarity, adenine and guanine are called *purines*, while cytosine and thymine are called *pyrimidines*.

Each base has a complementary base with which it can form the rung of the ladder through a hydrogen bond. We always find either A paired with T or G paired with C. Thus, knowing one side of the ladder structure is enough to deduce the other. For example, if along one pole of the ladder we have a sequence of bases

$$AGCGCGTATTAG,$$

then the other would have the complementary sequence

$$TCGCGCATAATC.$$

Finally, the DNA molecule has a directional sense so that we can make a distinction between a sequence like $ATCGAT$ and the inverted sequence $TAGCTA$. The upshot of all this structure is that we will be able to think of DNA sequences mathematically simply as sequences composed of the four letters A, T, C, and G.

Some sections of DNA form *genes* that encode instructions for the manufacturing of proteins (though the production of the protein is accomplished through the intermediate production of messenger RNA). In these genes, triplets of consecutive bases form *codons*, with each codon specifying a particular amino acid to be placed in the protein chain according to the *genetic code*. For example, the codon TGC always means that the amino acid cysteine will occur at that location in the protein. Certain codons also signal the end of the protein sequence. Since there are $4^3 = 64$ different codons, and only 20 amino acids and one "stop" command, there is some redundancy in the genetic code. For instance, in many codons, the third base has no affect on the particular amino acid the codon specifies.

Although originally it was thought that genes always encoded for proteins via messenger RNA, we now know that some genes encode for the production of other types of RNA that are the "final products" of the gene, with no protein being produced. Finally, not all DNA is organized into the coding sections referred to as genes. About 97% of human DNA, for example, is believed to be noncoding. Some of this is likely to be meaningless raw material (sometimes called *junk DNA*), which may, of course, become meaningful in future generations through evolution. Other parts of the DNA molecules may serve regulatory purposes. The picture is quite complicated and still not fully understood.

When DNA is copied, the hydrogen bonds forming the rungs of the ladder are broken, leaving two single strands. Then new double strands are formed on these by assembling the appropriate complementary strands. The biochemical processes are elaborate, with various safeguards to ensure that few mistakes are made. Nonetheless, changes of an apparently random nature sometimes occur.

The most common mutation that is introduced in the copying of sequences of DNA is a *base substitution*. This is simply the replacement of one base for another at a certain site in the sequence. For instance, if the sequence $AATCGC$ in an ancestor becomes $AATGGC$ in a descendent, then a base substitution $C \to G$ has occurred at the fourth site. A base substitution that replaces a purine with a purine, or a pyrimidine with a pyrimidine, is called a *transition*, whereas an interchange of these classes is called a *transversion*. Transitions are often observed to occur more frequently than transversions, perhaps because the chemical structure of the molecule changes less under a transition than a transversion.

Other DNA mutations sometimes observed include the deletion of a base or consecutive bases, the insertion of a base or consecutive bases, and the inversion (reversal) of a section of the sequence. All these mutations tend to be seen more rarely in natural populations. Since these types of mutations usually have a dramatic effect on the protein for which a gene encodes, this is not too surprising. We will ignore such possibilities to make our modeling task both clearer and mathematically tractable.

Focusing solely on base substitutions, a basic problem to be addressed is how to deduce the amount of mutation that must have occurred during the evolutionary descent of DNA sequences. For instance, suppose we know that a descendent species S_2 descended from an intermediate species S_1, which in turn descended from an ancestral species S_0. Imagine that, for each of these, a certain gene included the sequences:

$$S_0 : \; AC\mathbf{C}TGCGCTA\ldots$$
$$S_1 : \; AC\mathbf{G}TGC\mathbf{A}CTA\ldots$$
$$S_2 : \; AC\mathbf{G}TGCGCTA\ldots.$$

Here, boldface marks the two sites among the first 10 sites where changes have occurred. (We will always assume the sequences have been *aligned* so that we can match ancestral and descendent sites. The mathematical methods by which this can be done could be the subject of another chapter or book.)

Now, if we only saw the sequences for S_0 and S_2, we would notice only one base substitution among the first 10 sites, the one appearing in the third site. It might seem reasonable that the ratio $\frac{1}{10}$ of mutations per site would be a good measure of how much mutation has occurred from S_0 to S_2.

However, because we have the sequence for S_1 as well, we know things are more complicated. At the seventh site, we notice that we have had the substitutions $G \rightarrow A \rightarrow G$. The original mutation has been *hidden* since a *back mutation* has occurred, leaving the final base the same as it initially was. Comparing S_0 with S_1 and then S_1 with S_2 has shown three mutations among the first 10 sites, leading to the much larger measure of $\frac{3}{10}$ mutations per site.

It could also happen that, at another site, substitutions such as $A \rightarrow T \rightarrow G$ occur. Here, even though there were two consecutive substitutions, we would notice only one if we only saw the initial and final sequences. Once again, a mutation has been hidden by a subsequent one.

Thus, a simple ratio of mutations per site obtained from comparing the first and last sequences may well give too low an estimate of the amount of mutation that actually occurred. Unless we believe that mutations have been quite rare, so that no hidden mutations occurred, we will need a mathematical model to be able to reconstruct the number of mutations that are likely to have occurred from those we see in comparing only the initial and final DNA sequences.

4.2. An Introduction to Probability

Describing the random mutation of DNA mathematically requires a facility with basic probability. Although we'll keep our discussion as informal as possible, we will need to be careful on a few points and that requires some terminology. Looking at some familiar nonbiological examples, such as coin flips and die tosses, will help make the ideas clearer.

Suppose we flip a coin or toss a die. When we refer to the *probability* of a certain outcome, such as getting a heads in the coin flip, or a 4 in the die toss, we mean a number $\mathcal{P} = \mathcal{P}(\text{outcome})$, with $0 \leq \mathcal{P} \leq 1$, that indicates the likelihood of that outcome occurring. For instance, if we flip a fair coin, we would say the probability of the outcome "heads" is

$$\mathcal{P} = \frac{1}{2}, \quad \text{or} \quad \mathcal{P}(\text{heads}) = \frac{1}{2},$$

because we expect to see heads in roughly 1 of every 2 tosses. This does not mean that if we flip the coin twice we will get one head and one tail, but rather that if we flipped it a very large number of times, we should find that in about $\frac{1}{2}$ of the tosses each outcome occurred. For the die toss, to express the chance of a 4 turning up, we would say that $\mathcal{P}(4) = \frac{1}{6}$, since we expect roughly 1 of every 6 in a large number of tosses to produce a 4.

We might say that a probability measures the chance of a "random" outcome occurring. Alternately, we may believe the outcome of a die toss is not random (it is, after all, governed by the deterministic laws of physics), but predicting it is too complicated to be practical. With this viewpoint, we are willing to give up trying to say exactly what will happen with any particular toss and instead accept a description of how often outcomes are likely to occur in the long run. More precisely, the probability $\mathcal{P}$ of an outcome gives our expectation of the percentage of trials in which that outcome will occur, assuming a very large number of trials are performed. The smaller $\mathcal{P}$ is, the less likely we believe an outcome is to occur in any given trial.

Usually, a probability will not indicate exactly what will happen in any trial. However, there are two exceptions. A probability of $\mathcal{P} = 1$ means an outcome is sure to happen – it will occur 100% of the time. Likewise, a probability of $\mathcal{P} = 0$ means the event is sure not to happen.

Do not assume that the probability of a heads in a coin flip is $\frac{1}{2}$ just because there are only two possible outcomes: heads and tails. For a weighted coin, there are still only two possible outcomes, but it might be that, with such a coin, we expect to get heads in 80% of the flips and so we have $\mathcal{P}(\text{heads}) = .8$. Such a coin is not "fair," but it is still capable of being described through probability. Similarly, for a fair die, the probability of any particular outcome is $\frac{1}{6}$, but for a weighted die, the probabilities of some of the outcomes might be more than $\frac{1}{6}$, while for others they are less than $\frac{1}{6}$.

Given a weighted coin, how can we determine the probability of it producing an outcome of heads? We simply perform many trials by flipping it repeatedly. After recording how often heads comes up in these trials, we can compute the estimate

$$\mathcal{P}(\text{heads}) \approx \frac{\text{no. of heads produced}}{\text{no. of trials}}.$$

For instance, if in 10 trials, we got 4 heads, we would estimate $\mathcal{P}(\text{heads}) \approx \frac{4}{10} = .4$. Performing 100 trials might turn up 56 heads, leading us to the improved estimate $\mathcal{P}(\text{heads}) \approx \frac{56}{100} = .56$. The more trials we perform, the more confidence we have in our estimate of the probability. Although we cannot prove a typical coin gives us heads and tails with probability $\frac{1}{2}$, we can gather evidence to back up that belief.

Example. To apply this language to a DNA sequence, suppose a 40-base sequence reads as follows:

$$AGCTTCCGATCCGCTATAATCGTTAGTTGTTACACCTCTG$$

What is the probability that the next base, in site 41, should be an A?

If we really know nothing about the function of this DNA, then we might proceed by imagining that the bases have been chosen at random. If each site is treated as a trial of some random selection process, we have the outcomes of 40 trials before us. A quick tally shows that there are 8 As, 7 Gs, 11 Cs, and 14 Ts. Thus, we estimate

$$\mathcal{P}(A) \approx \frac{8}{40} = .200, \quad \mathcal{P}(G) \approx \frac{7}{40} = .175,$$

$$\mathcal{P}(C) \approx \frac{11}{40} = .275, \quad \mathcal{P}(T) \approx \frac{14}{40} = .350.$$

We've used the frequency of the occurrence of the various bases to estimate the probabilities. Just as for the flip of a weighted coin, with a longer sequence of trials, we would have more confidence in our estimates. Nonetheless, with the limited number of trials at our disposal, we have done the best we can. Thus, we estimate the probability of an A in site 41 as .2.

Often, we'll need to group several outcomes into a set, which we call an *event*. For instance, for the coin flip, there are four possible events corresponding to the four ways we can make sets of the outcomes:

$$E_{\text{heads}} = \{\text{heads}\} \qquad E_{\text{either}} = \{\text{heads, tails}\}$$
$$E_{\text{tails}} = \{\text{tails}\} \qquad E_{\text{neither}} = \{\ \}.$$

We say an event occurs if any of the outcomes in the event is observed.

Example. In our DNA example, viewing each site as a trial, the possible basic outcomes are the appearance of the four bases. Events that might be of interest are "the base is a purine" and "the base is a pyrimidine," or even "the base is not *A*." In more formal notation,

$$E_{\text{purine}} = \{A, G\}, \quad E_{\text{pyrimidine}} = \{C, T\}, \quad E_{\text{not } A} = \{G, C, T\}.$$

When we know the probability of the basic outcomes, we can then assign probabilities to all events. For an event containing only a single outcome, the probability is simply the probability of that outcome. Thus, for the fair coin,

$$P(E_{\text{heads}}) = P(\text{heads}) = \frac{1}{2} \quad \text{and} \quad P(E_{\text{tails}}) = P(\text{tails}) = \frac{1}{2}.$$

Now, the event E_{either} means "either heads or tails" happens. Because this is a sure thing, its probability is 1 and so $P(E_{\text{either}}) = 1$. Similarly, the event E_{neither} means we get neither a head nor a tail, and this is sure *not* to occur, so its probability is 0.

Example. For the DNA sequence example, what should $P(E_{\text{purine}})$ be?

One way to estimate it is to go back to our data and simply tally the frequency with which purines occur. For instance, because in our 40-base sequence there were 8 *As* and 7 *Gs*, there were a total of 15 purines of the 40 bases, thus we estimate $P(E_{\text{purine}}) \approx \frac{15}{40} = .375$.

▶ Explain why $E_{\text{pyrimidine}} = .625$ and $E_{\text{not } A} = .800$.

There is another way we could estimate $P(E_{\text{purine}})$. Notice that

$$P(E_{\text{purine}}) = P(A) + P(G)$$
$$\frac{8 + 7}{40} = \frac{8}{40} + \frac{7}{40}.$$

The way fractions are added ensures that the probability of a purine appearing is the same as the sum of the probabilities of the bases *A* and *G* in the class of purines. In fact, we can generalize this example to the rule:

Addition Rule (Special case): The probability of any event is the sum of the probabilities of the individual outcomes making up that event.

Consider the toss of a fair die to make this clearer. Our basic one-outcome events are $E_1, E_2, \ldots, E_6$, where $E_i = \{\text{"the die shows an } i\text{"}\} = \{i\}$. The probabilities of getting any of the outcomes 1, 2, 3, 4, 5, or 6 are all $\frac{1}{6}$, because experience shows us that each outcome is equally likely and occurs in roughly 1 of 6 trials. Because the event $E = \{1, 2, 3, 4, 5, 6\}$ is a sure thing, its probability is 1. But now events such as "the die shows an odd number" can be given probabilities by

$$E_{\text{odd}} = \{1, 3, 5\}$$

so

$$P(E_{\text{odd}}) = P(1) + P(3) + P(5) = \frac{1}{6} + \frac{1}{6} + \frac{1}{6} = \frac{1}{2}.$$

▶ Explain why, for a toss of a fair die, the probability of the event "the die shows an even number" is $\frac{1}{2}$. What outcomes make up this event?

▶ What outcomes make up the event "the die shows a number ≤ 2"? What is the probability of this event for a fair die?

Mutually exclusive events and sums of probabilities. The rule we just used for assigning probabilities to events is actually an important special case of a more general rule that lets us use known probabilities of events to calculate probabilities of more complicated events.

Suppose we have two events, E and F, whose probabilities we know, and we are interested in knowing the probability that either E *or* F occurs. This new event, which is denoted by $E \cup F$, is the set of outcomes that appear in either E or F, or both. This new set is called the *union* of E and F. For example, the events "the die shows a number ≤ 4" and "the die shows an even number" have as their union the event "the die does not show a 5," as we see by

$$E_{\leq 4} \cup E_{\text{even}} = \{1, 2, 3, 4\} \cup \{2, 4, 6\} = \{1, 2, 3, 4, 6\} = E_{\text{not } 5}.$$

We'd like to understand how we can combine probabilities of several events to get the probability of the union.

This is most easily done when the events to be combined are *mutually exclusive*. Informally, two events are mutually exclusive if it is impossible for them to occur simultaneously; if one occurs, the other does not. If we have listed the outcomes in the events in sets, then we see they are mutually exclusive when the sets have no outcomes in common. That is, events are mutually exclusive when the sets are *disjoint*.

For instance, for a die toss, consider the three events: "the die shows an odd number," "the die shows a number ≤ 3," and "the die shows a number > 4." Writing out the outcomes in each of these events as

$$E_{\text{odd}} = \{1, 3, 5\}, \quad E_{\leq 3} = \{1, 2, 3\}, \quad \text{and} \quad E_{>4} = \{5, 6\},$$

we see the first two are not disjoint (both events will occur if the die shows a 1 or a 3), whereas the last two are disjoint (they cannot both occur at once).

For a coin toss, the events E_{heads} and E_{tails} are mutually exclusive, because one precludes the other. However, the composite event E_{either} and the event E_{heads} are not mutually exclusive: Knowing "heads or tails" was produced does not tell us that "heads" did not occur.

▶ Explain why in our DNA example, the events E_{purine} and $E_{\text{pyrimidine}}$ are mutually exclusive, whereas $E_{\text{pyrimidine}}$ and $E_{\text{not } A}$ are not.

Now suppose we consider any two events E and F that are mutually exclusive. Then, their probabilities can be combined according to

Addition Rule: *If events E and F are mutually exclusive, then the probability of the event "E or F," will be the sum of the probabilities of the two events:*

$$P(E \cup F) = P(E) + P(F), \qquad \text{if } E \text{ and } F \text{ are disjoint.}$$

Example. Consider a die toss, and the events $E_{\leq 2} =$ "the die shows a number ≤ 2" and $E_{\text{mult } 3} =$ "the die shows a multiple of 3."

▶ Explain why $P(E_{\leq 2}) = \frac{1}{3}$ by listing the outcomes that make up this event.

▶ Explain why $P(E_{\text{mult } 3}) = \frac{1}{3}$ by listing the outcomes that make up this event.

▶ Are these two events mutually exclusive?

Now, the probability of the event $E_{\leq 2} \cup E_{\text{mult } 3} =$ "the die shows either a number ≤ 2 or a multiple of 3" can be calculated with ease. Since $E_{\leq 2}$ and $E_{\text{mult } 3}$ are disjoint,

$$P(E_{\leq 2} \cup E_{\text{mult } 3}) = P(E_{\leq 2}) + P(E_{\text{mult } 3}) = \frac{1}{3} + \frac{1}{3} = \frac{2}{3}.$$

Of course, we could also have found this by listing all the outcomes in this event

$$E_{\leq 2} \cup E_{\text{mult } 3} = \{1, 2, 3, 6\},$$

and so

$$P(E_{\leq 2} \cup E_{\text{mult } 3}) = P(E_1) + P(E_2) + P(E_3) + P(E_6)$$
$$= \frac{1}{6} + \frac{1}{6} + \frac{1}{6} + \frac{1}{6} = \frac{2}{3}.$$

Example. Note that the events $E_{\text{mult 3}}$ and $E_{<4}$ are *not* mutually exclusive; it is possible for both to occur simultaneously if the outcome of the toss is a 3. Thus, we expect

$$\mathcal{P}(E_{\text{mult 3}} \cup E_{<4}) \neq \mathcal{P}(E_{\text{mult 3}}) + \mathcal{P}(E_{<4}).$$

In fact, since

$$E_{\text{mult 3}} \cup E_{<4} = \{1, 2, 3, 6\} = E_1 \cup E_2 \cup E_3 \cup E_6,$$

we find

$$\mathcal{P}(E_{\text{mult 3}} \cup E_{<4}) = \frac{2}{3} \neq \frac{1}{3} + \frac{1}{2} = \mathcal{P}(E_{\text{mult 3}}) + \mathcal{P}(E_{<4}).$$

There is a more general version of the addition rule that can be used on events such as these that are not mutually exclusive. You'll find it in the exercises.

As a final consequence of the addition rule of probabilities of disjoint events, we can understand the probability of an event *not* happening. If E is any event, let E' be the *complementary* event composed of all those outcomes not in E. For example, with a die toss

$$(E_{\leq 4})' = E_{>4}.$$

For any event E, note that E and E' are certainly exclusive (they cannot both happen at once). Then, by the addition rule

$$\mathcal{P}(E \cup E') = \mathcal{P}(E) + \mathcal{P}(E').$$

However, the event $E \cup E'$ is the event that anything at all happens, and because this is a sure thing, $\mathcal{P}(E \cup E') = 1$. Thus, $\mathcal{P}(E) + \mathcal{P}(E') = 1$, or

$$\mathcal{P}(E') = 1 - \mathcal{P}(E).$$

We now have a rule for calculating probabilities of complementary events.

Example. As an application to DNA, the event $E_{\text{pyrimidine}}$ is the same as E'_{purine}. Thus, $\mathcal{P}(E_{\text{pyrimidine}}) = 1 - \mathcal{P}(E_{\text{purine}})$. Of course, this is consistent with the example above where $\mathcal{P}(E_{\text{purine}}) = .375$ and $\mathcal{P}(E_{\text{pyrimidine}}) = .625$.

Independent events and products of probabilities. There is another important way we can combine events to get more complicated ones. If E and F are events, then $E \cap F$ denotes the event that both E *and* F occur. The set of outcomes $E \cap F$ is simply all outcomes appearing in both E and F. This is called the *intersection* of the sets. For instance,

$$E_{\leq 4} \cap E_{\text{mult } 2} = \{1, 2, 3, 4\} \cap \{2, 4, 6\} = \{2, 4\}.$$

Imagine flipping a coin and tossing a die together. Then, there are 12 possible outcomes: (heads, 1), (tails, 1), (heads, 2), (tails, 2), ..., (tails, 6). Assuming both the coin and die are fair, each of these outcomes should be equally likely. Since their probabilities must add to 1 (because they are disjoint, and it is certain that one of them occurs), each must have probability $\frac{1}{12}$.

▶ Explain why there are 12 possible outcomes.

Consider the event "the die shows a 5" and the event "the coin shows heads":

$E_5 = \{(\text{heads}, 5), (\text{tails}, 5)\},$

$E_{\text{heads}} = \{(\text{heads}, 1), (\text{heads}, 2), (\text{heads}, 3), (\text{heads}, 4), (\text{heads}, 5), (\text{heads}, 6)\}.$

The intersection of these two events is "the die shows a 5 and the coin shows heads,"

$$E_5 \cap E_{\text{heads}} = \{(\text{heads}, 5)\} = E_{\text{heads}, 5}.$$

How are the probabilities of these three events related?

▶ Explain why $\mathcal{P}(E_{\text{heads}}) = \frac{1}{2}$ and $\mathcal{P}(E_5) = \frac{1}{6}$ by thinking of each of them as a union of disjoint events and using the addition rule.

Because $\mathcal{P}(E_{\text{heads}, 5}) = \frac{1}{12}$, noting that

$$\frac{1}{2} \cdot \frac{1}{6} = \frac{1}{12}$$

shows

$$\mathcal{P}(E_{\text{heads}}) \cdot \mathcal{P}(E_5) = \mathcal{P}(E_{\text{heads}} \cap E_5).$$

At least in this example, the probability of an intersection of two events was simply the product of the probabilities of the two events. The reason that these probabilities behaved this way actually depended on a special feature of the events: the events E_5 and E_{heads} are independent.

Informally, we say two events are *independent* if knowledge that one of the events has occurred tells us absolutely nothing about whether the other has occurred. In other words, if we were told whether or not the first event occurred, that would have no effect on our belief about the likelihood of the second having occurred.

In this example, knowing whether the die shows a 5 or not, tells us nothing about the chance of seeing either of the coin outcomes, a head or a tail.

> **Multiplication Rule:** *If events E and F are independent, then the probability of the event "E and F" will be the product of the probabilities of the two events:*
>
> $$P(E \cap F) = P(E) \cdot P(F), \qquad \text{if } E \text{ and } F \text{ are independent.}$$

Example. Suppose we toss two fair dice in order. There are 36 equally likely outcomes such as (1, 1), (1, 2), etc., each with a probability of $\frac{1}{36}$. (Because we toss the dice and record what they show in order, the outcome (1, 2) is not the same as the outcome (2, 1).)

Consider the events

$$E_{d2=3} = \text{"the second die shows a 3,"}$$

$$E_{d1=even} = \text{"the first die is even."}$$

▶ Explain why $P(E_{d2=3}) = \frac{6}{36} = \frac{1}{6}$ by listing the 6 outcomes that make up the event.

▶ Explain why $P(E_{d1=even}) = \frac{18}{36} = \frac{1}{2}$ by listing the 18 outcomes that make up the event.

Now, intuitively, the events $E_{d1=even}$ and $E_{d2=3}$ are independent, since one tells us something about die 1 and the other about die 2. Knowledge about one die should communicate nothing about the other. Thus, the multiplication rule tells us

$$P(E_{d1=even} \cap E_{d2=3}) = \frac{1}{2} \cdot \frac{1}{6} = \frac{1}{12}.$$

We can confirm this by reasoning a different way. The compound event $E_{d1=even} \cap E_{d2=3}$ is the event that the first die is even and the second shows a 3. This means it is composed of the outcomes (2, 3), (4, 3), and (6, 3). Because each of these outcomes has probability $\frac{1}{36}$, we have

$$P(E_{d1=even} \cap E_{d2=3}) = \frac{1}{36} + \frac{1}{36} + \frac{1}{36} = \frac{1}{12}.$$

Example. Continuing with the toss of two dice in order, consider another event

$$E_{sum=9} = \text{"the sum of the results is 9."}$$

▶ Explain why $P(E_{\text{sum}=9}) = \frac{4}{36} = \frac{1}{9}$ by listing the 4 outcomes that make up the event.

Now, the events $E_{\text{sum}=9}$ and $E_{d2=3}$ are *not* independent. If we know the sum is a 9, then we know the outcome must have been one of $(6, 3)$, $(5, 4)$, $(4, 5)$, or $(3, 6)$. Since these are all equally likely, we see that knowledge that $E_{\text{sum}=9}$ occurred lets us say there is a 1 in 4 chance that $E_{d2=3}$ occurred. This is different than the 1 in 6 chance we would have without the knowledge that $E_{\text{sum}=9}$ occurred. Thus, knowledge of one event gave us some information about the other, so they are dependent.

To verify that the multiplication rule does not hold for this example, we check

$$P(E_{\text{sum}=9} \cap E_{d2=3}) = P((6, 3)) = \frac{1}{36},$$

whereas

$$P(E_{\text{sum}=9}) \cdot P(E_{d2=3}) = \frac{1}{9} \cdot \frac{1}{6} = \frac{1}{54}.$$

Although the definition of independent events given here has been an informal one, in the next section, we will be a bit more precise. Still, this informal way of thinking is often necessary, especially when probability is being used to model complicated processes.

The multiplication and addition rules are very useful in determining the probabilities of events. They allow us to calculate probabilities of complicated events by seeing how they are built from events we already understand by using the words "or," "and," and "not." An "or" means we add the probabilities, provided the events being combined are disjoint. An "and" means we multiply the probabilities, provided the events being combined are independent. A "not" means we compute the probability of the complementary event and subtract it from 1.

The key properties of probabilities we have discussed so far can be summarized as:

• The probability of any event E is a number $P = P(E)$ with $0 \leq P \leq 1$.
• If several events $E_1, E_2, \ldots, E_n$ are mutually exclusive, then the probability that any of them occur, i.e., the probability of $E = E_1 \cup E_2 \cup \cdots \cup E_n$, is $P(E) = P(E_1) + P(E_2) + \cdots + P(E_n)$, the sum of the individual probabilities.
• If several events $E_1, E_2, \ldots, E_n$ are independent, then the probability that they all occur, i.e., the probability of $E = E_1 \cap E_2 \cap \cdots \cap E_n$, is

$P(E) = P(E_1) \cdot P(E_2) \cdots P(E_n)$, the product of the individual probabilities.

- If the probability of an event E occurring is P, then the probability that E does not occur, i.e., the probability of the complementary event E', is $P(E') = 1 - P$.

Now let's apply these rules to a very simple model of DNA mutation. Suppose we focus on a particular site in a gene sequence, and on whether at that site a purine or a pyrimidine appears. We only care about these classes, not on the precise bases.

Suppose we also know that with each generation there is a 1.5% chance the base at this site undergoes a transversion, which we will call simply a "change." Thus, there is a 98.5% chance that there is no change (or a transition, which is treated as no change in this model). Then, for one generation

$$P(E_{\text{change}}) = .015, \quad P(E_{\text{no change}}) = .985.$$

While this probability of a change is much higher than is typically observed, we are not yet concerned with realism.

Now imagine what happens over two generations. There are four possibilities of interest:

$$\left\{ \begin{array}{l} \text{change} \\ \text{no change} \end{array} \right., \quad \text{followed by} \quad \left\{ \begin{array}{l} \text{change} \\ \text{no change} \end{array} \right..$$

What are their probabilities?

First, we make the important assumption that what happens in passing to the first generation is independent of what happens in passing to the second. This is reasonable if we think mutations are caused by errors and accidents, because the DNA should have no memory of what had happened before. With this assumption, we can use the multiplication rule for combining probabilities of independent events to get

$$P(E_{\text{change,change}}) = (.015)(.015) = .000225$$
$$P(E_{\text{change,no change}}) = (.015)(.985) = .014775$$
$$P(E_{\text{no change,change}}) = (.985)(.015) = .014775$$
$$P(E_{\text{no change,no change}}) = (.985)(.985) = .970225.$$

▶   What is the sum of these four probabilities? Why did it have to be that?

What is the probability of seeing no change from the original base in generation 0 to the descendent in generation 2? This event is actually composed of two events: either there was no change in each generation or there was a change in each generation producing no net change (i.e., the changes are hidden). Because these two events are mutually exclusive, we find the desired probability is

$$P(E_{\text{no change,no change}}) + P(E_{\text{change,change}}) = .970225 + .000225 = .97045.$$

Thus, the probability of observing no change when comparing a base across two generations is slightly greater than the chance of no change having actually occurred. Mutations followed by other mutations may result in no net observable change, yet they do affect the likelihood of what we observe.

Note that to deduce this result, we used both the multiplication rule for probabilities of independent events, and the addition rule for probabilities of disjoint events. This sort of analysis will form the basis of all of our modeling of molecular evolution. We just need to deal with very large numbers of generations and with all four of the bases.

Problems

4.2.1. Use a coin to conduct an experiment to determine the probability of it producing heads or tails when flipped.

    a. Flip the coin 10 times, recording your results. Use your data to estimate the probability of heads.

    b. Flip the coin 10 more times, for a total of 20 flips. Use your data to estimate the probability of heads.

    c. Flip the coin 20 more times, for a total of 40 flips. Use your data to estimate the probability of heads.

    d. If you believe your coin is fair, then you believe $P(\text{heads}) = .5$. Do your experiments support this? If your experiments did not exactly produce .5, should you be doubtful that the coin is fair? Which experiment produced the result closest to .5? Is that what you would have expected?

4.2.2. Suppose a fair coin is flipped 10 times (H = heads, T = tails).

    a. $HTTHTHHHTH$ is produced in 10 independent trials. What is the probability of this particular sequence of outcomes?

    b. $TTTTTTTTTT$ is produced in 10 independent trials. What is the probability of this particular sequence of outcomes?

    c. Your answers to parts (a) and (b) should be the same. Why might this be surprising to some people? Are you convinced they are equally likely?

4.2.3. Consider the 20-base sequence

$$AGGGATACATGACCCATACA.$$

a. Use the first five bases to estimate the four probabilities p_A, p_G, p_C, and p_T.
b. Repeat part (a) using the first 10 bases.
c. Repeat part (a) using all the bases.
d. Is there a pattern to the way the probabilities you computed in parts (a–c) changed? If so, what features of the original sequence does this pattern reflect?

4.2.4. Consider the 20-base sequence

$$CGGTTCGCCTGCGTAGTGCG$$

a. Give the best estimates you can for the probability that each base would appear at site 21.
b. Give the best estimates you can for the probabilities of a purine and of a pyrimidine at site 21.
c. Which base is most likely to appear at site 21? Is it a purine or a pyrimidine? Does this make sense in light of your answer to part (b)? Explain.

4.2.5. A simple model for human offspring is that each child is equally likely to be male or female. With this model, a three-child family can be thought of as three random determinations of sex, in order.
a. What are the 8 possible outcomes? What is the probability of each?
b. What outcomes make up the event "the oldest child is a daughter"? What is the event's probability?
c. What outcomes make up the event "the family has one daughter and two sons"? What is its probability?
d. What is the complement of the event in part (c)? List the outcomes in it and describe it in words. What is its probability?
e. What outcomes make up the event "the family has at least one daughter"? What is its probability?

4.2.6. For a coin toss, there are 2 possible outcomes, but 4 events listed in the text. More generally, if a trial has n possible outcomes, there will be 2^n events.
a. If one of the bases A, G, C, and T is chosen at random, so there are 4 possible outcomes, then there are $16 = 2^4$ different events. List them all.
b. Explain why, if there are n possible outcomes, then there are 2^n possible events.

4.2.7. Many genetic traits can be modeled using probability. Imagine picking a person at random from the world population. Then we can consider events such as "the person has brown eyes" or "the person is male." For each of the following pairs of events, decide whether the two events are mutually exclusive, and if it is reasonable to think of them as independent:

a. "the person is male" and "the person has brown eyes"

b. "the person has black hair" and "the person is an albino"

c. "the person has blue eyes" and "the person has blond hair"

4.2.8. If two events are mutually exclusive, can they also be independent? Explain.

4.2.9. The definition of "mutually exclusive" events given in the text was in words. Explain why it could be expressed more concisely as

E and F are mutually exclusive means $E \cap F = \{\ \ \}$.

4.2.10. There is a more general version of the addition rule for probabilities that does not require that events be mutually exclusive: For any events E and F,

$$P(E \cup F) = P(E) + P(F) - P(E \cap F).$$

a. Explain why, if E and F are disjoint, then this agrees with the addition rule in the text.

b. Show the general version holds in an example for a die toss using the events $E_{\text{mult 3}}$ and $E_{<4}$.

4.2.11. Explain informally why, if events E and F are independent, then the complementary events E' and F' must also be independent.

4.2.12. The text presents a model of DNA sequence mutation considering only the classes of purines and pyrimidines, and computes the probability of observing "no change" at a site when comparing an ancestral sequence and a sequence two generations later. Continue that discussion by answering:

a. What is the probability of observing a "change" when comparing an ancestral sequence and a sequence two generations later?

b. What 4 outcomes (ordered triples of "change"/"no change") make up the event "no change" is observed at a site when comparing an ancestral sequence and a sequence three generations later?

c. What is the probability of the event in part (b)?

4.3. Conditional Probabilities

When base substitutions occur in the evolution of DNA, the probability of a particular base appearing at a site in the descendent sequence might depend on the ancestral base. For example, if the ancestral base is a T, we would expect the probability of a T in the descendent to be high. If the ancestral base is a C, we would expect a lower probability of the descendent having a T, since a transition is less likely than no change. If the ancestral base is an A or G, we might expect an even lower probability that the descendent has a T, because transversions might be rarer than transitions.

To formalize this, we need the concept of *conditional probability*. This is the probability of one event given that we know another event has occurred. Letting S_0 refer to the ancestor and S_1 the descendent, we'll use notation like "$S_0 = C$" to mean that the ancestral site has base C, and "$S_1 = T$" to mean the descendent site has base T. Then,

$$\mathcal{P}(S_1 = T \mid S_0 = C) = .02$$

will mean that there is a 2% chance that the descendent base is a T given that the ancestral base is a C. Note that the vertical bar "|" in this conditional probability notation is read as "given that." We now have a good way to refer to the fact the probability of a "final" base appearing depends on the "initial" base that appeared.

▶   Taking into account the previous comments on the likelihood of transitions and transversions, which of $\mathcal{P}(S_1 = A \mid S_0 = C)$, $\mathcal{P}(S_1 = G \mid S_0 = C)$, $\mathcal{P}(S_1 = C \mid S_0 = C)$, and $\mathcal{P}(S_1 = T \mid S_0 = C)$ are likely to be smallest? Which is likely to be biggest?

The properties of probabilities discussed earlier carry over to the setting of conditional probabilities, as long as we keep in mind we are always assuming something particular happened – the given condition. For instance,

$$\mathcal{P}(S_1 = A \mid S_0 = C) + \mathcal{P}(S_1 = G \mid S_0 = C)$$
$$+ \mathcal{P}(S_1 = C \mid S_0 = C) + \mathcal{P}(S_1 = T \mid S_0 = C) = 1.$$

After all, given that $S_0 = C$, the four events $S_1 = A, G, C$, and T are mutually exclusive, yet certainly one of them will occur, and so the probabilities must add to 1.

Example. The conditional probability $\mathcal{P}(S_1 = T \mid S_0 = C)$ is not the same as the probability $\mathcal{P}(S_1 = T \text{ and } S_0 = C)$. To see this clearly, suppose we have aligned sequences

$S_0: AGCTTCCGATCCGCTATAATCGTTAGTTGTTACACCTCTG$
$S_1: AGCTTCTGATACGCTATAATCGTGAGTTGTTACATCTCCG.$

Then, of the 40 sites shown (which we think of as 40 trials), we find two sites with a T in S_1 and a C in S_0. Thus, we would estimate

$$P(S_1 = T \text{ and } S_0 = C) \approx \frac{2}{40} = .05.$$

However, of the 11 sites that have a C in S_0, we find only two of these have a T in S_1; so, we estimate

$$P(S_1 = T \mid S_0 = C) \approx \frac{2}{11} \approx .182.$$

Pay particular attention to this last calculation. We divided not by the total number of trials, but only by the number of trials that satisfied the given criterion $S_0 = C$. The trials in which $S_0 \neq C$ are irrelevant to the calculation of this conditional probability.

There is another way to find conditional probabilities, which is convenient if we have already computed some other probabilities. From this last example, we know the probability that both $S_0 = C$ and $S_1 = T$ is

$$P(S_1 = T \text{ and } S_0 = C) \approx \frac{2}{40} = .05.$$

Moreover, the probability that $S_0 = C$ can be found to be

$$P(S_0 = C) \approx \frac{11}{40} = .275.$$

Then

$$\frac{P(S_1 = T \text{ and } S_0 = C)}{P(S_0 = C)} \approx \frac{\frac{2}{40}}{\frac{11}{40}} = \frac{2}{11} \approx P(S_1 = T \mid S_0 = C).$$

The denominators of 40 canceled one another out, leaving us with the ratio we found above.

More formally, we can capture what has happened in this approach by the following general definition.

Definition of Conditional Probability: *If E and F are two events, then the conditional probability of F given E is defined by*

$$P(F \mid E) = \frac{P(F \cap E)}{P(E)}. \tag{4.1}$$

The concept of conditional probability also clarifies the notion of independence of events. Earlier, we informally said that events E and F were independent if knowledge that one had occurred gave us no information as to whether the other occurred. This could be expressed as

$$P(F \mid E) = P(F) \quad \text{and} \quad P(E \mid F) = P(E). \tag{4.2}$$

Using the definition of conditional probability, the first of these becomes

$$\frac{P(F \cap E)}{P(E)} = P(F),$$

or

$$P(F \cap E) = P(E)P(F).$$

▶ Explain why the second equation in (4.2) gives the same result.

This leads us to the formal mathematical definition of independence as

Definition of Independence: *Events E and F are said to be independent if*

$$P(E \cap F) = P(E)P(F).$$

Of course, this is essentially the same as the multiplication rule for independent events stated earlier. All the new definition really says is that the word "independent" is simply a concise way of saying the multiplication rule applies. In practice, to recognize whether events are independent or not, it is usually better to stick with the more informal definition given in the last section, which has been formalized in equations (4.2).

Example. Suppose a 40-base ancestral DNA sequence is

$S_0 : ACTTGTCGGATGATCAGCGGTCCATGCACCTGACAACGGT,$

and its descendent aligned sequence is

$S_1 : ACATGTTGCTTGACGACAGGTCCATGCGCCTGAGAACGGC.$

Table 4.1. *Frequencies of*
$S_1 = i$ *and* $S_0 = j$ *in 40-Site*
Sequence Comparison

$S_1 \backslash S_0$	A	G	C	T
A	7	0	1	1
G	1	9	2	0
C	0	2	7	2
T	1	0	1	6

Thinking of each site as a trial of the same probabilistic process, we can estimate 16 conditional probabilities describing the likelihood of observing different types of base substitutions when comparing the sequences of ancestor to descendent:

$$\mathcal{P}(S_1 = i \mid S_0 = j),$$

where $i, j = A, G, C, T$.

To do this, we begin by tallying the number of sites with an occurrence of each pair $S_0 = j$, $S_1 = i$ in the aligned sequences, recording the information in a *frequency array* such as Table 4.1.

▶ What is the sum of the 16 numbers in the table? Why?

If we add the numbers in a *column* of this table, we obtain the total number of sites with a particular base in S_0. For instance, the number of sites with $S_0 = A$ is $7 + 1 + 0 + 1 = 9$. In general, the number of sites with $S_0 = j$ is the sum of the entries in column j.

▶ What is the meaning of a row sum in the table?

Now, for any bases i, j, we estimate the conditional probabilities $\mathcal{P}(S_1 = i \mid S_0 = j)$ by dividing the number of sites with $S_1 = i$ and $S_0 = j$ by the number of sites with $S_0 = j$. That means we must divide the entry in row i, column j of the table by the sum of the entries in column j. We find all the conditional probabilities by dividing all table entries by their corresponding column sums. Rounding the results to 3 digits yields Table 4.2.

▶ What is the sum of the entries in any column of this new table? Why?
▶ If instead of dividing by column sums, you divided by row sums, would you get the same results? What conditional probabilities would you be calculating?

Table 4.2. *Estimates of Conditional*
Probabilities $\mathcal{P}(S_1 = i \mid S_0 = j)$

$S_1 \backslash S_0$	A	G	C	T
A	.778	0	.091	.111
G	.111	.818	.182	0
C	0	.182	.636	.222
T	.111	0	.091	.667

Problems

4.3.1. Assuming births of each sex are equally likely, a two-child family may have 4 outcomes in the sexes of the children.
   a. List the outcomes and give the probability of each.
   b. What is the probability that at least one child is a female?
   c. What is the probability that the youngest child is a female?
   d. What is the conditional probability that the youngest child is a female, given that at least one child is a female?
   e. What is the conditional probability that at least one child is a female, given that the youngest child is a female?
   f. Are the events in parts (b) and (c) independent? Explain.

4.3.2. Consider the toss of a single die.
   a. Show the events E_{odd} and $E_{\leq 2}$ are independent by using the formal definition.
   b. Show the events E_{odd} and $E_{\leq 3}$ are not independent by using the formal definition.
   c. Explain as intuitively as possible why the events of part (a) were independent, but those of part (b) were not.

4.3.3. Medical tests, such as those for diseases, are sometime characterized by their *sensitivity* and *specificity*. The sensitivity of a test is the probability that a diseased person will show a positive test result (a correct positive). The specificity of a test is the probability that a healthy person will show a negative test result (a correct negative).
   a. Both sensitivity and specificity are conditional probabilities. Which of the following are they:

$$\mathcal{P}(- \text{ result} \mid \text{disease}), \qquad \mathcal{P}(- \text{ result} \mid \text{no disease}),$$
$$\mathcal{P}(+ \text{ result} \mid \text{disease}), \qquad \mathcal{P}(+ \text{ result} \mid \text{no disease}).$$

   b. The other conditional probabilities listed in (a) can be interpreted as probabilities of false positives and false negatives. Which is which?

Table 4.3. *Data from Tuberculosis (TB) Diagnosis Study*

	Persons without TB	Persons with TB
Negative X-ray	1,739	8
Positive X-ray	51	22

   c. A study (Yerushalmy *et al.*, 1950) investigated the use of X-ray readings to diagnose tuberculosis. Diagnosis of 1,820 individuals produced the data in Table 4.3. Compute both the sensitivity and specificity for this method of diagnosis.

4.3.4. Ideally, the specificity and sensitivity of medical tests should be high (close to 1). However, even with a highly specific and sensitive test, screening a large population for a disease that is rare can produce surprising results.

   a. Suppose the sensitivity and specificity of a test for disease are both .99. The test is applied to everyone in a population of 100,000 individuals, only 100 of whom have the disease. Compute how many individuals with/without the disease you would expect to test positive/negative. Organize your results in a table like that in the preceeding problem.

   b. Use the table you produced in part (a) to compute the conditional probability that a person who tests positive actually has the disease.

4.3.5. In the text, data in Table 4.1 are used to compute the conditional probabilities $P(S_1 = i \mid S_0 = j)$.

   a. Use the same data to compute $P(S_0 = j \mid S_1 = i)$. Do you get the same results as in Table 4.2?

   b. Explain intuitively why you would usually not expect $P(S_1 = i \mid S_0 = j)$ and $P(S_0 = i \mid S_1 = j)$ to be the same.

4.3.6. In tables, such as Table 4.2, of conditional probabilities describing realistic DNA base substitutions between an ancestor and descendent, there is often a pattern to the sizes of the numbers.

   a. Which entries refer to no substitution occurring? Why are these likely to be the largest entries?

   b. Which entries refer to transitions? To transversions? Does Table 4.2 support the claim that transitions tend to be more common than transversions?

4.3.7. Using the data in Table 4.1:

   a. Compute each column sum and divide it by 40. These results can be interpreted as estimates of probabilities. What probabilities are being estimated?

    b. Compute the row sums and divide each by 40. What probabilities are being estimated?

4.3.8. For the two sequences S_0 and S_1 that are used in producing Table 4.1:

    a. Estimate the eight probabilities $P(S_0 = i)$ and $P(S_1 = j)$ for $i, j = A, G, C, T$.

    b. For each pair i, j, are the events $S_0 = i$ and $S_1 = j$ independent?

    c. Why does the fact that one sequence is descended from another help explain your answer to part (b)?

4.3.9. Two DNA sequences of the same length are chosen and labeled S_0 and S_1, but there is no ancestral relationship between the two.

    a. Why would you expect that for each pair i, j the events $S_0 = i$ and $S_1 = j$ would be independent?

    b. If the events $S_0 = i$ and $S_1 = j$ are independent, what would be the pattern in the entries in a table like Table 4.2?

4.3.10. Recall from the last section the two-class model of purine and pyrimidine sequence mutation. Modify the model so that, at each generation, the probabilities of mutation depend on the current class of the site according to Table 4.4:

    a. Explain intuitively why the formula

$$P(S_2 = pur \mid S_0 = pur)$$
$$= P(S_2 = pur \mid S_1 = pur) \cdot P(S_1 = pur \mid S_0 = pur)$$
$$+ P(S_2 = pur \mid S_1 = pyr) \cdot P(S_1 = pyr \mid S_0 = pur)$$

is reasonable. Write similar formulas for $P(S_2 = pyr \mid S_0 = pur)$, $P(S_2 = pur \mid S_0 = pyr)$, and $P(S_2 = pyr \mid S_0 = pyr)$.

    b. Using these formulas, compute numerical values for $P(S_2 = j \mid S_0 = i)$ for the four possible choices with $i, j = pur, pyr$.

Table 4.4. *Conditional Probabilities*
$P(S_{t+1} = i \mid S_t = j)$

$S_{t+1} \backslash S_t$	pur	pyr
pur	.98	.01
pyr	.02	.99

c. Using the definition of conditional probability, show that the formula in part (a) is valid. You will have to use the assumptions

$$\mathcal{P}(S_2 = pur \mid S_1 = pur \quad \text{and} \quad S_0 = pur)$$
$$= \mathcal{P}(S_2 = pur \mid S_1 = pur),$$
$$\mathcal{P}(S_2 = pur \mid S_1 = pyr \quad \text{and} \quad S_0 = pur)$$
$$= \mathcal{P}(S_2 = pur \mid S_1 = pyr).$$

These assumptions state that probabilities of substitutions between time 1 and time 2 are independent of the base at time 0.

4.3.11. Suppose E_1 and E_2 are two events, with E_2' being the event complementary to E_2. Recall that $\mathcal{P}(E_2) + \mathcal{P}(E_2') = 1$.

   a. Explain using your intuitive understanding of conditional probabilities why $\mathcal{P}(E_2 \mid E_1) + \mathcal{P}(E_2' \mid E_1) = 1$ should also hold.

   b. Show the formula in part (a) holds more formally by using the definition of conditional probability as a quotient of probabilities. You will need use that $(E_2 \cap E_1) \cup (E_2' \cap E_1) = E_1$.

4.3.12. MATLAB can be used to compare two sequences and produce a frequency array such as Table 4.1. Although the program `compseq` automates this, the individual steps are useful to know.

   a. Try the following command sequence and explain what each line does.

```
S0='AACTGCAGT'
S1='AGCCGCAGA'
S0=='A'
S1=='G'
(S0=='A') & (S1=='G')
sum( (S0=='A') & (S1=='G') )
```

   b. What one-line command would find the number of sites with a C in S_0 and a G in S_1?

   c. What one-line command would count the number of purines in S_0?

   d. What one-line command would give the number of sites with a purine in S_0 and a pyrimidine in S_1?

4.3.13. Suppose two sequences S_0 and S_1 have been compared, and a frequency table such as that in Table 4.1 has been produced and entered into MATLAB as a matrix F.

a. Explain why the sequence of commands

```
colsum=[1,1,1,1]*F, N=colsum*[1; 1; 1; 1], p0=colsum/N
```

will produce the fraction of sites with each base in S_0.

b. Give a sequence of commands to produce the fraction of sites with each base in S_1.

c. Try the MATLAB command D=diag(colsum) to see what it does. Then explain why if M denotes the matrix of estimated conditional probabilities such as in Table 4.2, that $F = M \times D$. Thus, M is easily computed by the command

$$M=F*inv(diag(colsum)).$$

4.4. Matrix Models of Base Substitution

We now can create a basic model of molecular evolution by making use of probability and matrix algebra.

We begin by modeling the ancestral sequence probabilistically. Each site in the sequence is one of the four bases A, G, C, or T, chosen randomly according to some probabilities $\mathcal{P}_A$, $\mathcal{P}_G$, $\mathcal{P}_C$, and $\mathcal{P}_T$. These four probabilities must satisfy

$$\mathcal{P}_A + \mathcal{P}_G + \mathcal{P}_C + \mathcal{P}_T = 1,$$

since one of the bases is certain to appear. For convenience, we will always use the order A, G, C, T for the bases (so the purines come first and then the pyrimidines) and put these four probabilities into a vector as

$$\mathbf{p}_0 = (\mathcal{P}_A, \mathcal{P}_G, \mathcal{P}_C, \mathcal{P}_T).$$

This vector describes the ancestral base distribution, with its entries giving the fraction of sites we would expect to be occupied by each of the four bases.

▶ To what extent is the assumption that all bases in the sequence are chosen "at random" reasonable? Would it matter whether the DNA sequence was coding or noncoding?

We model the mutation process over one time step, assuming that only base substitutions can occur – no deletions, insertions, or inversions are considered. We specify the 16 conditional probabilities of observing a base substitution, $\mathcal{P}(S_1 = i \mid S_0 = j)$, for $i, j = A, G, C$, and T. It will be convenient to put these numbers into a 4×4 matrix, using the ordering A, G, C, and T. In each column of the matrix are entries referring to the same ancestral base, and in each row are entries referring to the same descendent base. Using abbreviated notation, such as $\mathcal{P}_{i|j} = P(S_1 = i \mid S_0 = j)$, we let

$$
M = \begin{pmatrix}
\mathcal{P}_{A|A} & \mathcal{P}_{A|G} & \mathcal{P}_{A|C} & \mathcal{P}_{A|T} \\
\mathcal{P}_{G|A} & \mathcal{P}_{G|G} & \mathcal{P}_{G|C} & \mathcal{P}_{G|T} \\
\mathcal{P}_{C|A} & \mathcal{P}_{C|G} & \mathcal{P}_{C|C} & \mathcal{P}_{C|T} \\
\mathcal{P}_{T|A} & \mathcal{P}_{T|G} & \mathcal{P}_{T|C} & \mathcal{P}_{T|T}
\end{pmatrix}.
$$

▶ Why must the sum of the entries in any column of this matrix add to 1?

▶ How reasonable is it to assume only base substitutions occur? Why would you imagine that these might be the most common mutations, especially in coding regions of DNA?

Example. If we have two specific DNA sequences, such as those at the end of the last section, one the ancestor and the other the descendent after one time step, then all these probabilities can be estimated from the data. Data in the frequency array in Table 4.1 lead to

$$
\mathbf{p}_0 \approx (.225, .275, .275, .225) \quad \text{and} \quad M \approx \begin{pmatrix}
.778 & 0 & .091 & .111 \\
.111 & .818 & .182 & 0 \\
0 & .182 & .636 & .222 \\
.111 & 0 & .091 & .667
\end{pmatrix}. \quad (4.3)
$$

In fact, this estimate of M is just Table 4.2 treated as a matrix, and the estimate of $\mathbf{p}_0$ is just the column sums of Table 4.1 divided by the number of sites in the sequences.

▶ Explain why the calculation of $\mathbf{p}_0$ described here is the correct one to perform.

Expressing our model using a vector and matrix is more than just a concise notation; let's see what happens when we multiply them as

$$
M\mathbf{p}_0 = \begin{pmatrix}
\mathcal{P}_{A|A} & \mathcal{P}_{A|G} & \mathcal{P}_{A|C} & \mathcal{P}_{A|T} \\
\mathcal{P}_{G|A} & \mathcal{P}_{G|G} & \mathcal{P}_{G|C} & \mathcal{P}_{G|T} \\
\mathcal{P}_{C|A} & \mathcal{P}_{C|G} & \mathcal{P}_{C|C} & \mathcal{P}_{C|T} \\
\mathcal{P}_{T|A} & \mathcal{P}_{T|G} & \mathcal{P}_{T|C} & \mathcal{P}_{T|T}
\end{pmatrix} \begin{pmatrix}
\mathcal{P}_A \\
\mathcal{P}_G \\
\mathcal{P}_C \\
\mathcal{P}_T
\end{pmatrix}
$$

$$
= \begin{pmatrix}
\mathcal{P}_{A|A}\mathcal{P}_A + \mathcal{P}_{A|G}\mathcal{P}_G + \mathcal{P}_{A|C}\mathcal{P}_C + \mathcal{P}_{A|T}\mathcal{P}_T \\
\mathcal{P}_{C|A}\mathcal{P}_A + \mathcal{P}_{C|G}\mathcal{P}_G + \mathcal{P}_{C|C}\mathcal{P}_C + \mathcal{P}_{C|T}\mathcal{P}_T \\
\mathcal{P}_{G|A}\mathcal{P}_A + \mathcal{P}_{G|G}\mathcal{P}_G + \mathcal{P}_{G|C}\mathcal{P}_C + \mathcal{P}_{G|T}\mathcal{P}_T \\
\mathcal{P}_{T|A}\mathcal{P}_A + \mathcal{P}_{T|G}\mathcal{P}_G + \mathcal{P}_{T|C}\mathcal{P}_C + \mathcal{P}_{T|T}\mathcal{P}_T
\end{pmatrix}. \quad (4.4)
$$

To interpret this result, focus on the bottom entry

$$\mathcal{P}_{T|A}\mathcal{P}_A + \mathcal{P}_{T|G}\mathcal{P}_G + \mathcal{P}_{T|C}\mathcal{P}_C + \mathcal{P}_{T|T}\mathcal{P}_T.$$

Informally, we expect this to give the probability that a site in S_1 has base T, because we have multiplied the probability of each initial base occurring by the chance that base mutates to a T and summed over all possible initial bases. Checking this more formally, the first product appearing on the left is

$$\mathcal{P}_{T|A}\mathcal{P}_A = P(S_1 = T \mid S_0 = A)P(S_0 = A).$$

Using Eq. (4.1), this is the same as $P(S_1 = T \text{ and } S_0 = A)$. Applying similar reasoning to the other three products shows

$$\begin{aligned}
\mathcal{P}_{T|A}\,\mathcal{P}_A &+ \mathcal{P}_{T|G}\mathcal{P}_G + \mathcal{P}_{T|C}\mathcal{P}_C + \mathcal{P}_{T|T}\mathcal{P}_T \\
&= P(S_1 = T \text{ and } S_0 = A) + P(S_1 = T \text{ and } S_0 = G) \\
&\quad + P(S_1 = T \text{ and } S_0 = C) + P(S_1 = T \text{ and } S_0 = T).
\end{aligned}$$

Notice this is the sum of four probabilities of mutually exclusive events. By the addition rule, it gives the probability of the union of the four events, that is, of the event that $S_1 = T$:

$$\mathcal{P}_{T|A}\mathcal{P}_A + \mathcal{P}_{T|G}\mathcal{P}_G + \mathcal{P}_{T|C}\mathcal{P}_C + \mathcal{P}_{T|T}\mathcal{P}_T = P(S_1 = T).$$

If similar reasoning is applied to the other entries in the right-hand side of Eq. (4.4), we find $M\mathbf{p}_0 = \mathbf{p}_1$, where $\mathbf{p}_1$ is the vector of probabilities for various bases occurring in the sequence S_1. We can think of M as a *transition matrix* that tells us how the probabilities of each base in the ancestral sequence S_0 are transformed into the probabilities of each base in the descendent sequence S_1 one time step later.

What would be the meaning of $M\mathbf{p}_1$? For this to make sense biologically, we must assume the probabilistic mutation process over the first time step is identical to that over the next time step. Using the same transition matrix M of conditional probabilities means each type of base substitution has the same likelihood of occurring as it did before. Furthermore, what happens during the second time step depends only on what the base was at time $t = 1$ (the information in $\mathbf{p}_1$), and the conditional probabilities (the information in M). Whether that site experienced a substitution during the first time step is irrelevant.

To return to our numerical example with $\mathbf{p}_0$ and M coming from the data in Table 4.1, we can compute

$$\mathbf{p}_1 = M\mathbf{p}_0 = \begin{pmatrix} .225 \\ .275 \\ .300 \\ .200 \end{pmatrix}, \quad \mathbf{p}_2 = M\mathbf{p}_1 = \begin{pmatrix} .222 \\ .274 \\ .320 \\ .183 \end{pmatrix}.$$

▶ What is the sum of the entries in $\mathbf{p}_1$? In $\mathbf{p}_2$? (You may need to neglect an error due to rounding.) Why must this be the case?

Markov models. The model developed above is an example of a *Markov model*. In such a model, we describe a system that must be in one of n different *states*, but may switch from one state to another with time.

In the DNA substitution model, the system we describe is a site in a DNA sequence. That site is initially in one of 4 states (A, G, C, or T), according to the base that occupies it.

We specify initial probabilities that the system is in each of the states by giving a vector of these probabilities, $\mathbf{p}_0$. The entries of $\mathbf{p}_0$ must all be ≥ 0 (because they are probabilities) and must add to 1 (because we are certain the system is in one of the states).

We also specify conditional probabilities of the switch from every state to every state over one time step by giving a $n \times n$ *transition matrix M*. The entries of M must all be ≥ 0 (because they are probabilities), and each column must add to one (because the conditional probabilities in column j represent the probabilities of switching from state j to all states, and we are certain one of these will occur).

An important assumption is made in any Markov model: What happens to the system over a given time step depends only on the state the system is in at the start of that step and the transition probabilities. In particular, there is no "memory" of what state changes might have occurred during earlier time steps that has any effect. We say the conditional probabilities are *independent* of the past history.

▶ For a DNA substitution model, is it reasonable to assume this independence?

In our DNA model, we are also assuming that each site in the sequence behaves identically and independently of every other site. We used these assumptions to find the various probabilities we needed from our sequence data, by thinking of each site as an independent trial of the same probabilistic process.

This assumption is probably not very reasonable for DNA in some genes. For instance, because the genetic code allows for many changes in the third site of each codon to have no affect on the product of the gene, one could argue that substitutions in the third sites might be more likely than in the first two sites, violating the assumption that each site behaves identically. Moreover, since genes may lead to the production of proteins that are part of life's processes, the likelihood of change at one site may well be tied to changes at another, violating the assumption of independence.

Nonetheless, we must make simplifying assumptions to get anywhere with our model. Further work may find ways around these assumptions, allowing for different conditional probabilities for various sites. Or, we can be careful to take the assumptions into account when using the tools we develop on real data. For instance, we might ignore the third base of each codon in estimating information from our data, so that it is more reasonable to treat sites as independent and following identical processes.

A matrix whose entries are all ≥ 0 and whose columns sum to 1 is called a *Markov matrix*. Actually, you have seen an example of one before in the forest succession model of Chapter 2. That model can be reinterpreted as a Markov model, by imagining it describing one plot in the forest and tracking the likelihood of the plot being occupied by one type of tree or another.

There are quite a number of theorems concerning certain Markov models that are useful to know about, though we will not go into the proofs. Two that are relevant are:

Theorem. *A Markov matrix always has $\lambda_1 = 1$ as its largest eigenvalue and has all eigenvalues satisfying $|\lambda| \leq 1$. The eigenvector corresponding to λ_1 has all nonnegative entries.*

Unfortunately, this does not rule out -1 as an eigenvalue or having several different eigenvectors with eigenvalue 1. However, there is also:

Theorem. *A Markov matrix, all of whose entries are positive (i.e., nonzero), always has 1 as a strictly dominant eigenvalue. There will be only one eigenvector (up to scalar multiplication) associated with $\lambda = 1$.*

Note that we saw an example of this theorem for the tree model of Chapter 2, where we found the dominant eigenvector was $(5, 3)$, with eigenvalue 1. This explains why our numerical experiments with the model led to a stable distribution of $(A_t, B_t) \approx (625, 375)$, because $\dfrac{625}{375} = \dfrac{5}{3}$.

There are a few special Markov models of base substitutions used for DNA sequences that we can analyze very thoroughly.

The Jukes-Cantor model. The simplest Markov model of base substitution, the Jukes-Cantor model, adds several additional assumptions to the basic Markov model. First, it assumes all bases occur with equal probability in the ancestral sequence. Thus,

$$\mathbf{p}_0 = \left(\frac{1}{4}, \frac{1}{4}, \frac{1}{4}, \frac{1}{4} \right).$$

Second, in the Jukes-Cantor model, the conditional probabilities describing an observable base substitution from any base to any other base are all the same. Thus, all possible substitutions are equally likely; $A \leftrightarrow T, A \leftrightarrow C, A \leftrightarrow G$, $C \leftrightarrow T, C \leftrightarrow G$, and $T \leftrightarrow G$ have exactly the same chance of occurring. If we let $\frac{\alpha}{3}$ denote the conditional probability of a base substitution of any type occurring, so $\mathcal{P}(S_1 = i \mid S_0 = j) = \frac{\alpha}{3}$ for all $i \neq j$, then the 12 off-diagonal entries of the matrix M will all be $\frac{\alpha}{3}$.

▶ Since the entries in any column of M add to 1, what should the entries on the main diagonal be?

Therefore, for the Jukes-Cantor model, we use the transition matrix

$$M = \begin{pmatrix} 1 - \alpha & \frac{\alpha}{3} & \frac{\alpha}{3} & \frac{\alpha}{3} \\ \frac{\alpha}{3} & 1 - \alpha & \frac{\alpha}{3} & \frac{\alpha}{3} \\ \frac{\alpha}{3} & \frac{\alpha}{3} & 1 - \alpha & \frac{\alpha}{3} \\ \frac{\alpha}{3} & \frac{\alpha}{3} & \frac{\alpha}{3} & 1 - \alpha \end{pmatrix}.$$

The value of α will of course depend on the time step we use and features of the particular DNA sequence we are modeling.

▶ Why can you think of $1 - \alpha$ as the probability that no substitution is observed over a time step?

Although α is a probability, we can also interpret it as a rate: It is the rate at which observable base substitutions occur over one time step and is measured in units of (substitutions per site)/(time step). We emphasize that the *observable* mutations are those that we notice when comparing the ancestral and descendent sequences one time step later; several mutations may actually occur over the time step, but at most one is observable at any site. If back mutations occur during a time step, we may not observe a mutation, even though several occurred.

Mutation rates such as α for DNA in real organisms are not easily found. Ultimately, we will see how they can be deduced from data. Various researchers have given estimates of α around 1.1×10^{-9} mutations per site per year for certain sections of chloroplast DNA of maize and barley and around 10^{-8} mutations per site per year for mitochondrial DNA in mammals. The mutation rate for the influenza A virus has been estimated to be as high as .01 mutations per site per year. The rate of mutation is generally found to be a bit lower in coding regions of nuclear DNA than in noncoding DNA. At this point in the development of the model, however, we will treat α as an unknown constant.

In reality, the mutation rate may not be constant; it may change with time or with location within the DNA. Certainly, over the entire evolution of humans from primordial slime, it is unreasonable to think that mutation rates have always been the same. However, for shorter periods of time and for DNA serving a fixed purpose, the assumption of a constant mutation rate is sometimes reasonable. When mutation rates are constant, there is said to be a *molecular clock*.

To begin to understand the behavior of the Jukes-Cantor model, let's imagine we have a sequence evolving according to the model and ask ourselves some basic questions about what we will see happening. Remember, our initial sequence has equal proportions of each of the 4 bases, so

$$\mathbf{p}_0 = \left(\frac{1}{4}, \frac{1}{4}, \frac{1}{4}, \frac{1}{4} \right),$$

and for some small value of α, the base substitutions occur according to the transition matrix M given above.

Example. For the Jukes-Cantor model, in what proportion of the sites will each base appear after one time step?

To answer this, we merely compute

$$\mathbf{p}_1 = M\mathbf{p}_0 = \begin{pmatrix} 1-\alpha & \frac{\alpha}{3} & \frac{\alpha}{3} & \frac{\alpha}{3} \\ \frac{\alpha}{3} & 1-\alpha & \frac{\alpha}{3} & \frac{\alpha}{3} \\ \frac{\alpha}{3} & \frac{\alpha}{3} & 1-\alpha & \frac{\alpha}{3} \\ \frac{\alpha}{3} & \frac{\alpha}{3} & \frac{\alpha}{3} & 1-\alpha \end{pmatrix} \begin{pmatrix} \frac{1}{4} \\ \frac{1}{4} \\ \frac{1}{4} \\ \frac{1}{4} \end{pmatrix} = \begin{pmatrix} \frac{1}{4} \\ \frac{1}{4} \\ \frac{1}{4} \\ \frac{1}{4} \end{pmatrix}.$$

Thus we find the base composition of the sequence does not change under the Jukes-Cantor model. In the language of linear algebra, we would say that the vector $\left(\frac{1}{4}, \frac{1}{4}, \frac{1}{4}, \frac{1}{4} \right)$ is an eigenvector of M with eigenvalue 1. (In fact, it is the one promised by the two theorems on Markov matrices.) In this context, we might say that $\left(\frac{1}{4}, \frac{1}{4}, \frac{1}{4}, \frac{1}{4}\right)$ is an *equilibrium base distribution* for sequences under the Jukes-Cantor model. In earlier chapters, we might have called it a steady state for the model.

Example. What proportion of the sites will have a base A in the ancestral sequence and a T in the descendent one time step later? In other words, what is $p(S_0 = A$ and $S_1 = T)$?

To answer this, we note

$$\mathcal{P}(S_0 = A \text{ and } S_1 = T) = \mathcal{P}(S_1 = T \mid S_0 = A)\mathcal{P}(S_0 = A).$$

Now the conditional probability $\mathcal{P}(S_1 = T \mid S_0 = A) = \frac{\alpha}{3}$ can be found as the (4,1) entry in M, while $\mathcal{P}(S_0 = A) = \frac{1}{4}$ is an entry in $\mathbf{p}_0$. Thus, $\mathcal{P}(S_0 = A$ and $S_1 = T) = \frac{\alpha}{12}$.

Example. What is the probability that a base A in the ancestral sequence will have mutated to become a base T in the descendent sequence 100 time steps later? In other words, what is the conditional probability $\mathcal{P}(S_{100} = T \mid S_0 = A)$?

To answer this, we first observe that

$$\mathbf{p}_{100} = M^{100}\mathbf{p}_0. \tag{4.5}$$

Just as the formula $\mathbf{p}_1 = M\mathbf{p}_0$ holds because the entries of M are conditional probabilities of various substitutions occurring, the formula in Eq. (4.5) must mean that the entries of M^{100} are conditional probabilities of various net substitutions occuring in the passage from time 0 to time 100. We therefore need to find a certain entry of M^{100} – the entry in row 4, column 1 – and then we can answer the question.

Of course, finding all entries of M^t for all t is of more interest, since that will give us all the conditional probabilities of base substitutions over various numbers of time steps. We base our calculation of M^t on the insight of Chapter 2: Eigenvectors provide the best approach to understanding how powers of matrices behave.

Fortunately, the eigenvectors of the Jukes-Cantor matrix M are easily found. We have already seen one eigenvector (the equilibrium base distribution), but there are three more that can be found by trial and error or a long computation. The full set is

$$\mathbf{v}_1 = (1, 1, 1, 1) \qquad \lambda_1 = 1$$

$$\mathbf{v}_2 = (1, 1, -1, -1) \qquad \lambda_2 = 1 - \frac{4}{3}\alpha$$

$$\mathbf{v}_3 = (1, -1, 1, -1) \qquad \lambda_3 = 1 - \frac{4}{3}\alpha$$

$$\mathbf{v}_4 = (1, -1, -1, 1) \qquad \lambda_4 = 1 - \frac{4}{3}\alpha$$

▶ Check that these are correct by multiplying $M\mathbf{v}_i$ for each i.

Notice that the eigenvectors for the Jukes-Cantor model do not depend on the value of the mutation rate α, though the eigenvalues do.

To find the entries of M^t, we begin by focusing on the first column of M^t. The first column can be isolated by taking the product

$$M^t \begin{pmatrix} 1 \\ 0 \\ 0 \\ 0 \end{pmatrix} = \text{first column of } M^t.$$

Now we can express $(1, 0, 0, 0)$ in terms of the eigenvectors as

$$(1, 0, 0, 0) = \frac{1}{4}\mathbf{v}_1 + \frac{1}{4}\mathbf{v}_2 + \frac{1}{4}\mathbf{v}_3 + \frac{1}{4}\mathbf{v}_4.$$

Thus,

$$M^t \begin{pmatrix} 1 \\ 0 \\ 0 \\ 0 \end{pmatrix} = \frac{1}{4}M^t\mathbf{v}_1 + \frac{1}{4}M^t\mathbf{v}_2 + \frac{1}{4}M^t\mathbf{v}_3 + \frac{1}{4}M^t\mathbf{v}_4$$

$$= \frac{1}{4}1^t\mathbf{v}_1 + \frac{1}{4}\left(1 - \frac{4}{3}\alpha\right)^t\mathbf{v}_2 + \frac{1}{4}\left(1 - \frac{4}{3}\alpha\right)^t\mathbf{v}_3$$

$$+ \frac{1}{4}\left(1 - \frac{4}{3}\alpha\right)^t\mathbf{v}_4.$$

Substituting in the vectors $\mathbf{v}_i$, we find

$$M^t \begin{pmatrix} 1 \\ 0 \\ 0 \\ 0 \end{pmatrix} = \begin{pmatrix} \frac{1}{4} + \frac{3}{4}\left(1 - \frac{4}{3}\alpha\right)^t \\ \frac{1}{4} - \frac{1}{4}\left(1 - \frac{4}{3}\alpha\right)^t \\ \frac{1}{4} - \frac{1}{4}\left(1 - \frac{4}{3}\alpha\right)^t \\ \frac{1}{4} - \frac{1}{4}\left(1 - \frac{4}{3}\alpha\right)^t \end{pmatrix}.$$

The other columns of M^t are found similarly, giving

$$M^t =$$

$$\begin{pmatrix} \frac{1}{4} + \frac{3}{4}\left(1 - \frac{4}{3}\alpha\right)^t & \frac{1}{4} - \frac{1}{4}\left(1 - \frac{4}{3}\alpha\right)^t & \frac{1}{4} - \frac{1}{4}\left(1 - \frac{4}{3}\alpha\right)^t & \frac{1}{4} - \frac{1}{4}\left(1 - \frac{4}{3}\alpha\right)^t \\ \frac{1}{4} - \frac{1}{4}\left(1 - \frac{4}{3}\alpha\right)^t & \frac{1}{4} + \frac{3}{4}\left(1 - \frac{4}{3}\alpha\right)^t & \frac{1}{4} - \frac{1}{4}\left(1 - \frac{4}{3}\alpha\right)^t & \frac{1}{4} - \frac{1}{4}\left(1 - \frac{4}{3}\alpha\right)^t \\ \frac{1}{4} - \frac{1}{4}\left(1 - \frac{4}{3}\alpha\right)^t & \frac{1}{4} - \frac{1}{4}\left(1 - \frac{4}{3}\alpha\right)^t & \frac{1}{4} + \frac{3}{4}\left(1 - \frac{4}{3}\alpha\right)^t & \frac{1}{4} - \frac{1}{4}\left(1 - \frac{4}{3}\alpha\right)^t \\ \frac{1}{4} - \frac{1}{4}\left(1 - \frac{4}{3}\alpha\right)^t & \frac{1}{4} - \frac{1}{4}\left(1 - \frac{4}{3}\alpha\right)^t & \frac{1}{4} - \frac{1}{4}\left(1 - \frac{4}{3}\alpha\right)^t & \frac{1}{4} + \frac{3}{4}\left(1 - \frac{4}{3}\alpha\right)^t \end{pmatrix}.$$

$$(4.6)$$

This formula for M^t is actually quite simple, because it is of the Jukes-Cantor form itself. The value of the Jukes-Cantor parameter for it is just $\frac{3}{4} - \frac{3}{4}\left(1 - \frac{4}{3}\alpha\right)^t$.

Example. We can now easily answer questions such as: What is the probability that a site that initially has base A has base T after 100 time steps? This is the $(4,1)$ entry of M^{100}, which is

$$\frac{1}{4} - \frac{1}{4}\left(1 - \frac{4}{3}\alpha\right)^{100}.$$

The Kimura models. The Jukes-Cantor model is a *one-parameter* model of mutation, since it depends on the single parameter α to specify the mutation rate. Other models use several different parameters to specify mutation rates for several different types of mutations.

A good example of this is the Kimura 2-parameter model, which allows for different rates of transitions and transversions. Imagine that we have mutation rates β for transitions and γ for each of the possible transversions. If we assume these rates are independent of the initial base, then we are saying the off-diagonal entries of the transition matrix are given by:

$$M = \begin{pmatrix} * & \beta & \gamma & \gamma \\ \beta & * & \gamma & \gamma \\ \gamma & \gamma & * & \beta \\ \gamma & \gamma & \beta & * \end{pmatrix}.$$

▶ Why is it important to use the order A, G, C, T for the bases to get this matrix?

Because the columns must sum to 1, this means all the diagonal entries must be $1 - \beta - 2\gamma$. Notice that, if the probabilities of a transition and each transversion are equal so $\beta = \gamma$, then this model includes the Jukes-Cantor one as a special case with $\alpha = 3\beta = 3\gamma$.

An even more general model is the Kimura 3-parameter model, which assumes a transition matrix of the form

$$M = \begin{pmatrix} * & \beta & \gamma & \delta \\ \beta & * & \delta & \gamma \\ \gamma & \delta & * & \beta \\ \delta & \gamma & \beta & * \end{pmatrix}.$$

By appropriate choice of the parameters, this includes both the Jukes-Cantor and Kimura 2-parameter models as special cases.

Part of the Kimura models is the assumption that the initial base distribution vector is $\mathbf{p}_0 = \left(\frac{1}{4}, \frac{1}{4}, \frac{1}{4}, \frac{1}{4}\right)$. Because this vector is an eigenvector with eigenvalue 1 for both the Kimura 2- and 3-parameter matrices, sequences evolving according to these models have this uniform base distribution at all times. As you will see in the exercises, all the work done above for the Jukes-Cantor model can be performed for the Kimura 3-parameter model as well.

The general Markov model may well provide the most accurate description of the base substitutions that actually occur in evolution, because it assumes nothing special about the entries in the Markov matrix. It does not require any particular relationship between the various conditional probabilities. There are 12 parameters in picking a matrix for this model, since of the 16 entries we may freely pick 3 in each column, with the fourth determined by the condition that the columns sum to 1. If we also allow any initial base composition vector $\mathbf{p}_0$, then there are 3 additional parameters.

▶　　Why are there only 3 parameters for $\mathbf{p}_0$, even though it has 4 entries?

Unless we have specific parameter values in mind for the general Markov model, it is hard to derive detailed results for it of the sort we found for the Jukes-Cantor model. However, as long as all entries of the matrix are positive, the two theorems stated above do tell us that there must be an equilibrium base distribution. Furthermore, by applying the Strong Ergodic Theorem of Chapter 2, we know that, over time, the general Markov model will result in $\mathbf{p}_t$ approaching this equilibrium distribution, even if the initial base distribution is something else.

Problems

4.4.1. Review the forest succession model in the text of Chapter 2 to interpret it as a Markov model of a single plot in the forest.

　　a. What are the "states" for this model?

b. The matrix used in that model was $\begin{pmatrix} .9925 & .0125 \\ .0075 & .9875 \end{pmatrix}$. Explain why this is a Markov matrix.

c. Explain what conditional probabilities are given by each of the entries in this matrix.

d. In the text, we considered a forest that initially had 10 trees of species A and 990 trees of species B. What are the initial probabilities of a plot being in each of the states; that is, what is $\mathbf{p}_0$?

4.4.2. Recall the Leslie models of Chapter 2. The matrices used in these models are typically not Markov matrices. Why not?

4.4.3. Although the Jukes-Cantor model assumes $\mathbf{p}_0 = (.25, .25, .25, .25)$, a Jukes-Cantor matrix could describe mutations even with a different $\mathbf{p}_0$. Investigate the behavior of a model using a Jukes-Cantor matrix as you vary $\mathbf{p}_0$ by using a computer. For instance, with $\alpha = .03$, and $\mathbf{p}_0 = (.2, .3, .4, .1)$, you might use the MATLAB commands such as

```
a=.03, b=a/3
M=[1-a,b,b,b;b,1-a,b,b;b,b,1-a,b;b,b,b,1-a]
p=[.2; .3; .4; .1]
P=p
for i=1:10
p=M*p
P=[P p]
end
plot(P')
```

a. With the value of M and $\mathbf{p}_0$ suggested, do you see $\mathbf{p}_t$ approach its equilibrium value? Approximately how many time steps are necessary for all the $\mathbf{p}_t$ to be within .05 of the equilibrium? within .01?

b. Make several other choices of $\mathbf{p}_0$ and repeat step (a).

c. Using $\mathbf{p}_0 = (.25, .25, .25, .25)$, what do you observe? Why?

d. Using $\mathbf{p}_0 = (0, 1, 0, 0)$ what do you observe? What is the biological meaning of this $\mathbf{p}_0$?

4.4.4. Investigate the effect of varying α on the behavior produced by the Jukes-Cantor matrix. Let $\mathbf{p}_0 = (.2, .3, .4, .1)$ and use MATLAB commands such as those in the previous exercise to:

a. Compare the behavior of the model for $\alpha = .03$ and $\alpha = .06$. For which value of α does the model approach the equilibrium fastest?

b. Does your observation in part (a) hold for other initial choices of $\mathbf{p}_0$?

    c. Explain in intuitive terms why larger values of α should result in a quicker approach to the equilibrium.

4.4.5. The Markov matrices that describe real DNA mutation tend to have their largest entries along the main diagonal in the $(1,1)$, $(2,2)$, $(3,3)$, and $(4,4)$ positions. Why should this be the case?

4.4.6. Make up a 4×4 Markov matrix M with all positive entries and an initial $\mathbf{p}_0$. To be biologically realistic, make sure the diagonal entries of M are the largest.

    a. Use a computer to observe that, after many time steps, $\mathbf{p}_t = M^t \mathbf{p}_0$ appears to approach some equilibrium. Estimate the equilibrium vector as accurately as you can.

    b. Is your estimate in part (a) an eigenvector of M with eigenvalue 1? If not, does it appear to be close to having this property?

    c. Use a computer to compute the eigenvectors and eigenvalues of M, for instance with the MATLAB command [S D]=eig(M). Is 1 an eigenvalue? Is your estimate of the equilibrium close to its eigenvector?

    d. Are your computations in part (c) consistent with the two theorems about Markov matrices appearing in the text?

4.4.7. Express the Kimura 2-parameter model using a 4×4 matrix, but with the bases in the order A, C, G, T. Is this the same as the matrix in the text? Explain.

4.4.8. Consider the Markov matrix appearing in Eq. (4.3).

    a. Use a computer to find its eigenvectors and eigenvalues. Are they explained by the two theorems of this section?

    b. What is the equilibrium base distribution for this model? Be sure you give a vector whose entries sum to 1.

4.4.9. An ancestral DNA sequence of 40 bases was

$$CTAGGCTTACGATTACGAGGATCCAAATGGCACCAATGCT,$$

but in a descendent, it had mutated to

$$CTACGCTTACGACAACGAGGATCCGAATGGCACCATTGCT.$$

    a. Give an initial base distribution vector and a Markov matrix to describe the mutation process.

    b. These sequences were actually produced by a Jukes-Cantor simulation. Is that surprising? Explain. What value would you choose

Table 4.5. *Frequencies from 400 Site Comparisons for Two Pairs of Sequences*

$S_1 \backslash S_0$	A	G	C	T
A	92	15	2	2
G	13	84	4	4
C	0	1	77	16
T	4	2	14	70

$S_1' \backslash S_0'$	A	G	C	T
A	90	3	3	2
G	3	79	8	2
C	2	4	96	5
T	5	1	3	94

for the Jukes-Cantor parameter α to approximate your matrix by a Jukes-Cantor one?

4.4.10. Data from two comparisons of 400-base ancestral and descendent sequences are shown in Table 4.5.

a. For one of these pairs of sequences a Jukes-Cantor model is appropriate. Which one and why?

b. What model would be appropriate for the other pair of sequences? Explain.

4.4.11. In MATLAB, type load seqdata to read in some simulated sequence data. The three pairs of sequences, s0 and s1, t0 and t1, u0 and u1, are simulated ancestor and descendent sequences produced according to three different models. Which one was made according to the Jukes-Cantor model? The Kimura 2-parameter model? A general Markov model? Explain how you can tell. To easily compare sequences by producing a frequency array, use a command like compseq(s0,s1).

4.4.12. Suppose we wish to model molecular evolution not at the level of DNA sequences, but rather at the level of the proteins that genes encode.

a. Create a simple one-parameter mathematical model (similar to the Jukes-Cantor model) describing the process. You will need to know that there are 20 different amino acids from which proteins are constructed in linear chains.

b. In this situation, how many parameters would the general Markov model have?

Table 4.6. *Frequencies of* $S_\beta = i$ *and*
$S_\alpha = j$ *in 1,000-Site Sequence*
Comparison

$S_\beta \backslash S_\alpha$	A	G	C	T
A	105	25	35	25
G	15	175	35	25
C	15	25	245	25
T	15	25	35	175

4.4.13. The MATLAB program `mutate` can be used to simulate the mutation of a DNA sequence according to a Markov model. It will allow you to specify a 4×4 Markov matrix M and initial base distribution vector $\mathbf{p}_0$, as well as the number of bases you would like in your sequences.

   a. Use the MATLAB program `mutate` to perform a 10-base simulation for the Jukes-Cantor model with $\alpha = .1$ and $\mathbf{p}_0 = (.25, .25, .25, .25)$. Now imagine that the results of your simulation were two data sequences. Use them to estimate probabilities for an initial base distribution vector and a Markov matrix. (The program `compseq` will be useful for this.) Are your estimates close to what you began with?

   b. Repeat part (a), but using sequences of length 100 and then of length 1,000.

   c. The difference between a probabilistic model's description and what actually happens under that model when only a finite number of trials are performed is sometimes called *stochastic error*. What conclusions can you draw from parts (a) and (b) about the stochastic error for short sequences as opposed to long ones?

4.4.14. Repeat the last problem, but using your own choice of a 4×4 Markov model and initial base distribution. Are the results similar?

4.4.15. Suppose you have compared two sequences S_α and S_β of length 1,000 sites and obtained the data in Table 4.6 for the number of sites with each pair of bases.

   a. Assuming S_α is the ancestral sequence, find an initial base distribution $\mathbf{p}_0$ and a Markov matrix M to describe the data. Is your matrix M Jukes-Cantor? Is $\mathbf{p}_0$ an equilibrium distribution for M?

   b. Assuming S_β is the ancestral sequence, find an initial base distribution $\mathbf{p}_0'$ and a Markov matrix M' to describe the data. Is your matrix M' Jukes-Cantor? Is $\mathbf{p}'_0$ an equilibrium distribution for M'? You should have found that one of your matrices was Jukes-Cantor and the other was not. This cannot happen if both S_α and S_β have base distribution (.25, .25, .25, .25).

4.4.16. The formula for M^t for the Jukes-Cantor model can be used to show that powers of M approach a certain matrix as $t \to \infty$.

a. For $0 < \alpha \leq 1$, explain why $-\frac{1}{3} \leq 1 - \frac{4}{3}\alpha < 1$.

b. Use this to explain how $\left(1 - \frac{4}{3}\alpha\right)^t$ behaves as $t \to \infty$, and thus why

$$M^t \to \begin{pmatrix} .25 & .25 & .25 & .25 \\ .25 & .25 & .25 & .25 \\ .25 & .25 & .25 & .25 \\ .25 & .25 & .25 & .25 \end{pmatrix}.$$

Note that each of the columns of this matrix is the equilibrium distribution.

c. Why did we exclude $\alpha = 0$ from our analysis?

4.4.17. Based on the last problem, one might conjecture that powers of a Markov matrix all of whose entries are positive approach a matrix whose columns are the equilibrium distribution. On a computer, investigate this experimentally by creating a Markov matrix, computing very high powers of it to see if the columns become approximately the same, and then checking whether this column is an eigenvector with eigenvalue 1 of the original matrix.

4.4.18. Show the product of two Jukes-Cantor matrices is again a Jukes-Cantor matrix as follows: Let $M(\alpha_1)$ be the Jukes-Cantor matrix with parameter α_1, and $M(\alpha_2)$ the Jukes-Cantor matrix with parameter α_2. Compute $M(\alpha_1)M(\alpha_2)$ to show it has the form $M(\alpha_3)$. Give a formula for α_3 in terms of α_1 and α_2.

4.4.19. Show the product of two Kimura 3-parameter matrices is again a Kimura 3-parameter matrix.

4.4.20. Show the Kimura 3-parameter matrix has the same eigenvectors as those given in the text for the Jukes-Cantor matrix. What are the eigenvalues?

4.4.21. Use the results of the last problem to give formulas for the entries of the first column of M^t, where $M = M(\beta, \gamma, \delta)$ is the Kimura 3-parameter matrix. (The other columns could be handled similarly, leading to the result that $M(\beta, \gamma, \delta)^t = M(\beta', \gamma', \delta')$ where

$$\beta' = \frac{1}{4} + \frac{1}{4}(1 - 2\gamma - 2\delta)^t - \frac{1}{4}(1 - 2\beta - 2\delta)^t - \frac{1}{4}(1 - 2\beta - 2\gamma)^t$$

$$\gamma' = \frac{1}{4} - \frac{1}{4}(1 - 2\gamma - 2\delta)^t + \frac{1}{4}(1 - 2\beta - 2\delta)^t - \frac{1}{4}(1 - 2\beta - 2\gamma)^t$$

$$\beta' = \frac{1}{4} - \frac{1}{4}(1 - 2\gamma - 2\delta)^t - \frac{1}{4}(1 - 2\beta - 2\delta)^t + \frac{1}{4}(1 - 2\beta - 2\gamma)^t. \,)$$

4.4.22. The Jukes-Cantor model can be presented in a different form as a 2×2 Markov model. Let q_t represent the fraction of sites that agree between the ancestral sequence and the descendent sequence at time t, and p_t the fraction that differ, so $q_0 = 1$ and $p_0 = 0$. Assume that over each time step, the probability that a base substitution occurs is α, and that each of the three possible base substitutions is equally likely. Then

$$\begin{pmatrix} q_{t+1} \\ p_{t+1} \end{pmatrix} = \begin{pmatrix} 1 - \alpha & \frac{\alpha}{3} \\ \alpha & 1 - \frac{\alpha}{3} \end{pmatrix} \begin{pmatrix} q_t \\ p_t \end{pmatrix}, \quad \begin{pmatrix} q_0 \\ p_0 \end{pmatrix} = \begin{pmatrix} 1 \\ 0 \end{pmatrix}.$$

a. Explain why each entry in the matrix has the value it does. (Observe that $1 - \frac{\alpha}{3} = (1 - \alpha) + \frac{2\alpha}{3}$.)

b. Compute the steady state of the model by finding the eigenvector with eigenvalue 1.

c. Find the other eigenvalue and eigenvector for the matrix.

d. Use parts (b) and (c), together with the initial conditions $(q_0, p_0) = (1, 0)$, to give a formula for q_t and p_t as functions of time.

4.4.23. This exercise will derive one of the entries in Eq. (4.6) another way, in the style of Chapter 1. Let q_t denote the probability that the base at a fixed site at time t is the same as it was at time 0, and let α denote the probability of a substitution in a single time step for the Jukes-Cantor model.

a. Explain why

$$q_{t+1} = (1 - \alpha)q_t + \frac{\alpha}{3}(1 - q_t).$$

(You will need to think about two ways the base at time $t + 1$ might agree with that at time 0: Either it agreed at time t and did not change, or did not agree at time t and changed back to the original base.) What value should q_0 have? Investigate the behavior of this model in MATLAB using onepop.

The equation in part (a) simplifies to

$$q_{t+1} = \frac{\alpha}{3} + \left(1 - \frac{4\alpha}{3}\right) q_t.$$

Note that this model is a little different from those we dealt with in Chapter 1. If we graphed q_{t+1} as a function of q_t, we would get a straight line, but because the form of the equation is $q_{t+1} = s + rq_t$ rather than just $q_{t+1} = rq_t$, we cannot call it linear. (The term "linear" in this context requires that there be no constant term.) Instead, a model of the form $q_{t+1} = s + rq_t$ is called an *affine* model. Affine models can be converted to linear models and analyzed as outlined in the next few steps:

b. Find the equilibrium q^* of the model by solving $q^* = \frac{\alpha}{3} + \left(1 - \frac{4\alpha}{3}\right) q^*$.

c. Let $q_t = q^* + \epsilon_t$ to focus on the perturbation ϵ_t from equilibrium. Substitute this and a similar expression for q_{t+1} into the model equation, and simplify to get an equation expressing ϵ_{t+1} in terms of ϵ_t. Your result should be linear.

d. What is q_0? Use this value to give the value of the initial perturbation ϵ_0.

e. Based on your work in parts (c) and (d), give a formula for ϵ_t in terms of t.

f. From parts (c) and (e), show that

$$q_t = \frac{1}{4} + \frac{3}{4}\left(1 - \frac{4}{3}\alpha\right)^t.$$

4.5. Phylogenetic Distances

With a model of DNA mutation in hand, we can better understand how to relate the amount of mutation that we observe in comparing an ancestral and descendent sequence to the amount of mutation that must have actually occurred. We will be able to uncover the amount of hidden mutation that was obscured by subsequent mutations at the same site.

To frame the issue we want to address more clearly, let's consider the Jukes-Cantor example of the last section. There, we imagined modeling sequence mutation by the Jukes-Cantor matrix

$$M = M(\alpha) = \begin{pmatrix} 1-\alpha & \frac{\alpha}{3} & \frac{\alpha}{3} & \frac{\alpha}{3} \\ \frac{\alpha}{3} & 1-\alpha & \frac{\alpha}{3} & \frac{\alpha}{3} \\ \frac{\alpha}{3} & \frac{\alpha}{3} & 1-\alpha & \frac{\alpha}{3} \\ \frac{\alpha}{3} & \frac{\alpha}{3} & \frac{\alpha}{3} & 1-\alpha \end{pmatrix},$$

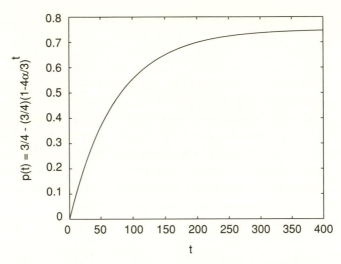

Figure 4.1. The Jukes-Cantor model, $\alpha = .01$: Fraction of differing sites at time t.

and computed the entries of M^t for $t = 0, 1, 2, 3, \ldots$. The diagonal entries of M^t turned out to all be

$$\frac{1}{4} + \frac{3}{4}\left(1 - \frac{4}{3}\alpha\right)^t.$$

Now, the diagonal entries of M^t give conditional probabilities that the base at time t is the same as the base at time 0. In other words, they indicate the probability of observing no change when the site at time 0 is compared with the site at time t. Because all these diagonal entries are equal, this means that at time step t we would expect to observe that the fraction of sites that agreed with their initial base was given by the formula

$$q(t) = \frac{1}{4} + \frac{3}{4}\left(1 - \frac{4\alpha}{3}\right)^t.$$

The fraction of sites that are different, then, will be

$$p(t) = 1 - q(t) = \frac{3}{4} - \frac{3}{4}\left(1 - \frac{4\alpha}{3}\right)^t.$$

▶   Why could you also get the formula for $p(t)$ by adding the three off-diagonal entries in any column of M^t?

In the graph of $p(t)$ in Figure 4.1, we of course see that $p(0) = 0$, because at time $t = 0$, no substitutions have yet occurred. More interestingly, we see that the fraction of sites that differ from their original base gradually increases with t, approaching the value $\frac{3}{4}$. This fraction never exceeds $\frac{3}{4}$, however.

▶   Even if so much mutation has occurred that the two sequences appear to be completely unrelated, you would expect to find agreement at $\frac{1}{4}$ of the sites. Why?

The graph also illustrates that, for each time t, $p(t)$ has a different value. This means that given any value $0 \le p \le 3/4$, we should be able to find a t with $p(t) = p$. That is, from the proportion of sites that differ between two sequences, we should be able to recover the number of elapsed time steps (assuming we know α). For real sequence data, p is easily estimated, although the elapsed time t and the mutation rate α usually are not known. Recovering them from data is now our goal.

The Jukes-Cantor distance. Suppose we have records of an original DNA sequence and a mutated version of it from some later time. Suppose we also believe the Jukes-Cantor model describes the mutation process that occurred, but we do not know either the mutation rate α or the number of elapsed time steps t.

From the DNA sequence data, we can estimate $p = p(t)$ by comparing many sites before and after mutation and using the proportion of sites that disagree in the two sequences as an estimate. For instance, if the original sequence were $ATTGAC$ and the final one $ATGGCC$, we would estimate $p(t) = 2/6 \approx .333$. Of course with real data, it is best to have much longer DNA sequences so that we have more confidence in our estimate.

With $p = p(t)$ estimated, how do we recover information on the mutation rate α and the amount of elapsed time t? Since

$$p = \frac{3}{4} - \frac{3}{4}\left(1 - \frac{4}{3}\alpha\right)^t,$$

solving for t yields

$$t = \frac{\ln\left(1 - \frac{4}{3}p\right)}{\ln\left(1 - \frac{4}{3}\alpha\right)}. \tag{4.7}$$

To go further, we need to realize that our choice of a step size for time in formulating our model affects both the value of the mutation rate α, and the number of elapsed time steps between ancestor and descendent. We cannot really expect to recover both of these. However, the product of the two does have a meaning that is more intrinsic to what we are modeling: Let

$$d = t\alpha$$
$$= \text{(no. of time steps)(mutation rate)}$$
$$= \text{(no. of time steps)(no. of substitutions per site/time step)}$$
$$= \text{(expected no. of substitutions per site during the elapsed time).}$$

We emphasize that this expected number of substitutions includes even those we do not observe because they are hidden by subsequent substitutions.

To extract $d = t\alpha$ from Eq. (4.7), we must use an approximation. Now $\ln(1 + x) \approx x$ when x is near 0 (see the Problems section). Furthermore, we can be sure $-\frac{4}{3}\alpha$ is near 0 if we assume that we have chosen a time step that is very small, so that the mutation rate per time step, α, is also very small. Thus,

$$\ln\left(1 - \frac{4}{3}\alpha\right) \approx -\frac{4}{3}\alpha.$$

Substituting this into Equation (4.7) gives

$$t \approx \frac{\ln\left(1 - \frac{4}{3}p\right)}{-\frac{4}{3}\alpha}$$
$$\approx -\frac{3}{4\alpha}\ln\left(1 - \frac{4}{3}p\right),$$

or

$$d = t\alpha \approx -\frac{3}{4}\ln\left(1 - \frac{4}{3}p\right).$$

If our time steps are made smaller, so the mutation rate α is also smaller, the approximation used for the logarithm is increasingly accurate. We therefore define the *Jukes-Cantor distance* between DNA sequences S_0 and S_1 as

$$d_{JC}(S_0, S_1) = -\frac{3}{4}\ln\left(1 - \frac{4}{3}p\right),$$

where p is the fraction of sites that disagree in comparing S_0 with S_1. Provided the Jukes-Cantor model accurately describes the evolution of one sequence into another, it is an estimate of the total number of substitutions per site that occurred during the evolution.

"Distance" here is an abstract notion of how different the sequences are because of mutations. Recall that if the mutation rate is constant over an evolutionary history, we say there is a molecular clock. Provided a molecular clock hypothesis is valid, the distance computed here is proportional to the amount of elapsed time, with the constant of proportionality being the mutation rate. Thus, the distance can be thought of as a measure of how much time was required for one sequence to mutate into the other. If the molecular clock hypothesis does not hold, it is still a reconstruction of the average number of substitutions that occurred at any one site. The larger it is, the greater the evolutionary change.

Although we were unable to recover either the mutation rate α or the number of elapsed time periods t by themselves, we could at least recover the product of the two from comparing sequences. If there is some other data (such as a geological record) suggesting the time involved, then the mutation rate can be found from d_{JC}. This is one way that real DNA mutation rates are estimated.

Example. Consider the two 40-base sequences at the end of Section 4.3. From Table 4.1, we find that 11 of the sites have undergone a substitution, so $p = 11/40 = .2750$. Thus,

$$d_{JC}(S_0, S_1) = -\frac{3}{4} \ln \left(1 - \frac{4}{3} \frac{11}{40} \right) \approx .3426.$$

Therefore, while we observed .2750 substitutions per site on average, we estimate that in the course of evolution .3426 substitutions per site occurred. Hidden mutations account for the difference.

The Kimura distances. Given any Markov model of base substitution, we could hope to imitate the steps above to derive an appropriate formula reconstructing the amount of mutation that has occurred. For the Kimura models, you will find an exercise that steps you through the procedure. The final formula for the Kimura 3-parameter model is

$$d_{K3} = -\frac{1}{4} \left(\ln(1 - 2\beta - 2\gamma) + \ln(1 - 2\beta - 2\delta) + \ln(1 - 2\gamma - 2\delta) \right),$$

where β, γ, and δ are estimates of parameters for a Kimura 3-parameter matrix describing the mutation of the initial sequence to the final.

Of course, if $\gamma = \delta$, this also gives a distance for the Kimura 2-parameter model. In that case, β is the probability of a transition, while $\gamma + \delta = 2\gamma$ is the probability of a transversion. Thus, if from sequence data we estimate the probability of a transition p_1 by counting all transitions and dividing by the length of the sequence, and the probability of a transversion p_2 similarly, we have

$$d_{K2} = -\frac{1}{2}\ln(1 - 2p_1 - p_2) - \frac{1}{4}\ln(1 - 2p_2).$$

If sequence data seems to indicate that transitions and each transversion type did not proceed at equal rates, then the Jukes-Cantor model is a poor one, and so the Kimura distance formulas are better choices for estimating the total amount of mutation.

Additive and symmetric distances: Log-det. The distance formulas given so far assume the data are consistent with either the Jukes-Cantor model or a Kimura 2- or 3-parameter model. Because these models do not necessarily describe all sequence data well, it is natural to ask for a distance formula for the general Markov model.

To motivate such a formula, we will not focus on reconstructing the total number of base substitutions that occurred, but rather on a property shared by both the Jukes-Cantor and Kimura distances.

This property concerns the behavior of the distance formula when we consider two successive mutation processes. Imagine an ancestral sequence S_0 from which has evolved S_1, from which in turn has evolved S_2, as shown schematically in Figure 4.2.

Let $M_{0\to1} = M(\alpha_1)$ and $M_{1\to2} = M(\alpha_2)$ be two Jukes-Cantor matrices describing the two mutation processes as shown. Then, we can calculate a mutation matrix $M_{0\to2}$ for the full passage from S_0 to S_2 as the product

$$M_{0\to2} = M_{1\to2}M_{0\to1}.$$

▶   Why are the matrices multiplied in this order?

A short calculation shows that $M_{0\to2}$ is also a Jukes-Cantor matrix, $M_{0\to2} = M(\alpha_3)$, with

$$\alpha_3 = \alpha_1 + \alpha_2 - \frac{4}{3}\alpha_1\alpha_2.$$

As part of the Jukes-Cantor model, suppose the base distribution for each of S_0, S_1, and S_2 is the equilibrium $(1/4, 1/4, 1/4, 1/4)$. Then, in passing from one sequence to the next by the Jukes-Cantor matrices $M(\alpha_i)$, we find

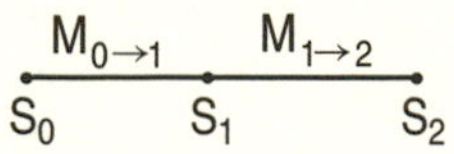

Figure 4.2. Three sequences in evolutionary order.

the fraction of sites that change is $p = \alpha_i$, and so the Jukes-Cantor distances are

$$M(\alpha_1): \quad -\frac{3}{4}\ln\left(1 - \frac{4}{3}\alpha_1\right)$$

$$M(\alpha_2): \quad -\frac{3}{4}\ln\left(1 - \frac{4}{3}\alpha_2\right)$$

$$M(\alpha_1)M(\alpha_2) = M(\alpha_3): \quad -\frac{3}{4}\ln\left(1 - \frac{4}{3}(\alpha_1 + \alpha_2 - \frac{4}{3}\alpha_1\alpha_2)\right).$$

But a little algebra shows

$$-\frac{3}{4}\ln\left(1 - \frac{4(\alpha_1 + \alpha_2 - \frac{4}{3}\alpha_1\alpha_2)}{3}\right) = \left(-\frac{3}{4}\ln\left(1 - \frac{4\alpha_1}{3}\right)\right)$$
$$+ \left(-\frac{3}{4}\ln\left(1 - \frac{4\alpha_2}{3}\right)\right).$$

This means that *multiplying* two Jukes-Cantor matrices corresponds to *adding* the associated distances.

We can see why this had to be the case if we recall that the Jukes-Cantor distance is recovering the total number of substitutions per site that must have occurred, including hidden ones. If we imagine a sequence mutating first according to one matrix and then the other (i.e., according to the product of the matrices), then the total number of substitutions per site would be the sum of those described by each individual matrix.

Returning to the general Markov model, we would like a definition of distance between sequences that has the *additive property* that

$$d(S_0, S_2) = d(S_0, S_1) + d(S_1, S_2)$$

in situations described by Figure 4.2.

To define such a distance, suppose F is the 4×4 frequency array obtained by comparing sites in sequences S_0 and S_1. Let $\mathbf{f}_0$ and $\mathbf{f}_1$ be the frequency vectors for the bases in S_0 and S_1, respectively. For instance, F might be the entries in Table 4.1 with $\mathbf{f}_0$ and $\mathbf{f}_1$ its column and row sums. Then, one version of the *log-det distance* (also called the *paralinear distance* in this form) between S_0 and S_1 is defined by

$$d_{LD}(S_0, S_1) = -\frac{1}{4}\left(\ln\left(\det(F)\right) - \frac{1}{2}\ln(g_0 g_1)\right),$$

where g_i is the product of the 4 entries in $\mathbf{f}_i$. Recall from Chapter 2 that "det" denotes the determinant of a matrix. Because the argument for why this distance is additive depends on some knowledge of linear algebra beyond this text, we leave it to the exercises.

The meaning of the log-det distance is harder to interpret than the other distances we have discussed. Unlike those, it usually is not just the total number of mutations per site that must have occurred over the evolutionary history. Still, you should think of it as some measure of the amount of mutation that has occurred. In special circumstances, such as when the Jukes-Cantor or Kimura models apply exactly, it gives the same result as they do, as you will also see in the exercises.

The key fact that all the phylogenetic distances we have discussed are additive will be extremely useful in the next chapter, when we turn to constructing phylogenetic trees relating many species.

Another useful property of all of these distances is *symmetry*. Although we thought of having ancestral and descendent sequences in discussing the various distances, in fact none of the final formulas depend on knowing which one of the sequences was the ancestral one. For instance, the Jukes-Cantor distance is calculated from first finding the fraction of sites that differ in the two sequences. If we had the same sequences, but switched which one we imagined was ancestral, we would calculate exactly the same distance. This means

$$d(S_0, S_1) = d(S_1, S_0).$$

This property will also be very valuable to us, because usually we do not have an ancestral sequence and a descendent one, but rather two descendents. In the exercises, you will see how symmetry helps us use a distance formula in this circumstance.

Problems

4.5.1. Calculate $d_{JC}(S_0, S_1)$ for the two 40-base sequences

S_0 : $CTAGGCTTACGATTACGAGGATCCAAATGGCACCAATGCT$
S_1 : $CTACGCTTACGACAACGAGGATCCGAATGGCACCATTGCT.$

4.5.2. Ancestral and descendent sequences of 400 bases were simulated according to the Jukes-Cantor model. A comparison of aligned sites gave the frequency data in Table 4.7.
   a. Compute the Jukes-Cantor distance to 10 decimal digits, showing all steps.

Table 4.7. *Frequencies of $S_1 = i$ and $S_0 = j$ in 400-Site Sequence Comparison*

$S_1 \backslash S_0$	A	G	C	T
A	90	3	3	2
G	3	79	8	2
C	2	4	96	5
T	5	1	3	94

b. Compute the Kimura 2-parameter distance to 10 decimal digits, showing all steps.

c. Are the answers to parts (a) and (b) identical? Explain.

4.5.3. Ancestral and descendent sequences of 400 bases were simulated according to the Kimura 2-parameter model with $\gamma = \beta/5$. A comparison of aligned sites gave the frequency data in Table 4.8.

a. Compute the Jukes-Cantor distance to 10 decimal digits, showing all steps.

b. Compute the Kimura 2-parameter distance to 10 decimal digits, showing all steps.

c. Which of these is likely to be a better estimate of the number of substitutions per site that actually occurred? Explain.

4.5.4. Compute the Kimura 3-parameter and log-det (paralinear) distance for the sequences of the last two problems.

4.5.5. Graph d_{JC} as a function of p.

a. Why does $d_{JC} = 0$ if two sequences are identical?

b. Why does d_{JC} not make sense if two sequences differ in 3/4 or more of the sites? Should this cause problems when trying to use the formula on real data?

Table 4.8. *Frequencies of $S_1 = i$ and $S_0 = j$ in 400-Site Sequence Comparison*

$S_1 \backslash S_0$	A	G	C	T
A	92	15	2	2
G	13	84	4	4
C	0	1	77	16
T	4	2	14	70

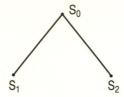

Figure 4.3. An evolutionary tree.

    c. Explain in biological terms why, if two sequences differ in just under 3/4 of the sites, the value of d_{JC} should be very large.

4.5.6. Complete the gaps in the derivation of the formulas for d_{JC} in the text by doing all the necessary algebra on the equation

$$p = \frac{3}{4} - \frac{3}{4}\left(1 - \frac{4}{3}\alpha\right)^t$$

to find the formula in Eq. (4.7) for t in terms of p and α.

4.5.7. The Jukes-Cantor distance formula is sometimes stated as

$$d_{JC} = -\frac{3}{4}\ln\left(\frac{4q - 1}{3}\right),$$

where q is the proportion of bases that are the same in the "before" and "after" sequences. Derive this formula from the one in the text.

4.5.8. Give numerical evidence that the approximation $\ln(1 + x) \approx x$ is valid for small x by making a table of values of x and $\ln(1 + x)$ for x close to 0. Give graphical evidence by plotting $y = \ln(1 + x)$ and $y = x$.

4.5.9. (Calculus) Show the approximation $\ln(1 + x) \approx x$ is valid for x near 0 by using calculus to find the tangent line approximation to $y = \ln(1 + x)$ at the point where $x_0 = 0$.

4.5.10. In practice, when applying a distance formula to real DNA sequence data, it is uncommon to have sequences for both an ancestor and a descendent. Instead, we usually have two descendent DNA sequences S_1 and S_2 that mutated from a common, yet unknown, ancestral sequence S_0, as in Figure 4.3. From the data, we can only compute $d(S_1, S_2)$. Show that, for an additive, symmetric distance, this is the same as $d(S_0, S_1) + d(S_0, S_2)$.

4.5.11. When transitions are more frequent than transversions, the Kimura 2-parameter distance often gives a larger value than the Jukes-Cantor distance. Explain this informally by explaining why hidden mutations are more likely under this circumstance.

4.5.12. The Jukes-Cantor distance is an estimate of the number of mutations that occurred per site over the course of one sequence evolving from another. A simpler estimate for this number is just p, the proportion of sites that have changed from the initial to final sequence.

    a. Explain why multiple mutations at the same site would cause p to be less reliable. Does it give an overestimate or underestimate of the true amount of mutation?

    b. Give an intuitive explanation of why, if p is relatively small, so that the sequences have few differences, this simpler estimate might be reasonable anyway.

    c. Explain why part (b) is consistent with the Jukes-Cantor model. That is, explain why for small p

$$-\frac{3}{4}\ln\left(1-\frac{4}{3}p\right)\approx p$$

by using the approximation for $\ln(1+x)$ valid for small x.

    d. It has been claimed that, if p is less than .1, it can be used as a reasonable approximation of the Jukes-Cantor distance. Do you agree? Illustrate by graphing both $d = -\frac{3}{4}\ln\left(1-\frac{4}{3}p\right)$ and $d = p$ for $0 \le p < 3/4$.

4.5.13. Show that the formula for the Jukes-Cantor distance can be recovered from the formula for the Kimura 3-parameter distance by letting β, γ, and δ all be $\alpha/3$.

4.5.14. Use the MATLAB program `mutate` to simulate a 100-base sequence evolving acording to the Jukes-Cantor model for $t = 400$ time steps, using a matrix with parameter $\alpha = .001$ for each time step. Compute a frequency array of base combinations with `F=compseq(Sinit,Sfinal)` and then compute the Jukes-Cantor distance with `distJC(F)`. Is the computed distance $\alpha t = .4$? If not, explain why not.

4.5.15. In MATLAB, type `load seqdata` to read in some simulated sequence data. Type `who` to see the names of the things you just loaded.

    a. Compute all six Jukes-Cantor distances between the sequences $a1$, $a2$, $a3$, and $a4$. You can compute a frequency array for base combinations with `F=compseq(a1,a2)` and then compute the distance with `distJC(F)`.

b. Suppose these sequences came from currently living species whose evolutionary relationships we would like to deduce. Draw the evolutionary tree that you believe best describes the relationship. Explain how you have used the distance data in your reasoning. *Note*: This problem is the subject of the next chapter.

4.5.16. a. Show that the formula $d_{K2} = -\frac{1}{2}\ln(1 - 2\beta - 2\gamma) - \frac{1}{4}\ln(1 - 4\gamma)$ for the Kimura 2-parameter model can be derived from the formula for d_{K3} by setting $\delta = \gamma$.

b. Suppose we have two aligned sequences S_0 and S_1. Explain why the proportion p_1 of sites that undergo transitions is a good estimate for β and why the proportion of sites p_2 that undergo transversions is a good estimate for 2γ. Use this to derive the Kimura 2-parameter distance formula for sequence data that is given in the text.

4.5.17. Derive the formula for the Kimura 3-parameter distance as follows. Refer to exercise 4.4.21 of the last section, which gives formulas for the parameters β', γ', and δ' in $M(\beta', \gamma', \delta') = M(\alpha, \beta, \gamma)^t$.

a. Show that

$$1 - 2\beta' - 2\delta' = (1 - 2\beta - 2\delta)^t,$$
$$1 - 2\beta' - 2\gamma' = (1 - 2\beta - 2\gamma)^t,$$

and

$$1 - 2\gamma' - 2\delta' = (1 - 2\gamma - 2\delta)^t.$$

b. Use part (a) to show

$$\ln(1 - 2\beta' - 2\delta') + \ln(1 - 2\beta' - 2\gamma') + \ln(1 - 2\gamma' - 2\delta')$$
$$= t\left(\ln(1 - 2\beta - 2\delta) + \ln(1 - 2\beta - 2\gamma) + \ln(1 - 2\gamma - 2\delta)\right).$$

c. Assuming β, γ, and δ are all small, use the approximation $\ln(1 + x) \approx x$ for $x \approx 0$ in the last equation to get

$$\ln(1 - 2\beta' - 2\delta') + \ln(1 - 2\beta' - 2\gamma') + \ln(1 - 2\gamma' - 2\delta')$$
$$\approx -4t\left(\beta + \gamma + \delta\right).$$

d. Explain why $\beta + \gamma + \delta$ should be interpreted as the total rate of base substitution, and thus why it is reasonable to define

$$d_{K3} = -\frac{1}{4}\left(\ln(1 - 2\beta' - 2\delta') + \ln(1 - 2\beta' - 2\gamma')\right.$$
$$\left. + \ln(1 - 2\gamma' - 2\delta')\right).$$

4.5.18. (Linear Algebra) The goal of this problem is to show that the Jukes-Cantor distance is a special case of the log-det distance. You will need to know the following two facts about determinants of $k \times k$ matrices that are proved in a Linear Algebra course:

i) $\det(cA) = c^k \det(A)$.

ii) $\det(A) = $ the product of A's k eigenvalues.

   a. Suppose two sequences, S_0 and S_1, of length N, were compared and the frequency table F was found to be *exactly* described by a Jukes-Cantor matrix $M(\alpha)$ with base distribution $(1/4, 1/4, 1/4, 1/4)$ for S_0. Show that $F = \frac{N}{4} M(\alpha)$.

   b. Explain why $\mathbf{f}_0 = \mathbf{f}_1 = (N/4, N/4, N/4, N/4)$.

   c. Use the facts above to show that, in this case, $d_{LD}(S_0, S_1) = d_{JC}(S_0, S_1)$.

4.5.19. (Linear Algebra) Proceeding as in the last problem, show that the Kimura 3-parameter distance is a special case of the log-det distance.

4.5.20. (Linear Algebra) Show the log-det distance formula is symmetric and additive through the following steps. You will need to know the following three facts about determinants of $k \times k$ matrices that are proved in a Linear Algebra course:

i) $\det(A^T) = \det(A)$, where A^T, the *transpose* of A, is a matrix whose (i, j) entry is the (j, i) entry of A.

ii) $\det(AB) = \det(A) \det(B)$.

iii) If the (i, j) entries of D are all zero for $i \neq j$, then

$$\det(D) = D(1, 1) \cdot D(2, 2) \cdots D(k, k).$$

   a. Use fact (i) to show the log-det distance is symmetric.

   b. For the situation in Figure 4.2, with initial base distribution $\mathbf{p}_0$ in S_0, explain why $\mathbf{p}_1 = M_{0 \to 1} \mathbf{p}_0$ and $\mathbf{p}_2 = M_{1 \to 2} \mathbf{p}_1$ are the base distributions in S_1 and S_2, respectively.

   c. For the vector $\mathbf{p}_i = (a, b, c, d)$, let

$$D_i = \begin{pmatrix} \sqrt{a} & 0 & 0 & 0 \\ 0 & \sqrt{b} & 0 & 0 \\ 0 & 0 & \sqrt{c} & 0 \\ 0 & 0 & 0 & \sqrt{d} \end{pmatrix}.$$

Then, for each pair i, j with $0 \leq i < j \leq 2$, define the matrix

$$N_{i \to j} = D_j^{-1} M_{i \to j} D_i.$$

Show $N_{1\to2}N_{0\to1} = N_{0\to2}$, and use fact (ii) to conclude

$$\ln(\det(N_{1\to2})) + \ln(\det(N_{0\to1})) = \ln(\det(N_{0\to2})).$$

d. Show the relative frequency array for comparing S_i to S_j is $G_{i\to j} = D_j N_{i\to j} D_i$, and then use fact (ii) to show

$$\ln(\det(G_{i\to j})) = \ln(\det(N_{i\to j})) + \ln(\det(D_i)) + \ln(\det(D_j)).$$

e. Combine parts (c) and (d), and fact (iii) to show the log-det distance is additive.

Projects

1. Investigate how the Jukes-Cantor distance formula performs on simulated sequence data produced according to the Jukes-Cantor model.

   The MATLAB program `mutate` can be used to simulate DNA mutations according to any specified Markov model of base substitution. Then, the Jukes-Cantor distance can be computed for the sequences so produced. However, seldom does the Jukes-Cantor distance exactly recover the value of αt used in the simulation.

   Explore the performance of the Jukes-Cantor distance formula for recovering αt on data that is produced by the Jukes-Cantor model under varying circumstances.

Suggestions
- So that the derivation of the Jukes-Cantor distance formula is valid, pick a small value of the mutation rate, such as $\alpha = .001$, to use in all your simulations.
- Use `compseq` to compare sequences and `distJC` to compute distances.
- For some fixed value of t, say around 300, perform a number of simulations for various values of N, and compute the Jukes-Cantor distances. Compare these to αt. Do you see a pattern in how accuracy varies with N?
- For some fixed value of N, say around 400, perform a number of simulations for various values of t, and compute the Jukes-Cantor distances. Compare these to αt. Do you see a pattern in how accuracy varies with t?
- In comparing the computed values of d_{JC} to αt, you could consider either $d_{JC} - \alpha t$ or $d_{JC}/(\alpha t)$. What do each of these mean?

- For a fixed choice of N and t, do many simulations. Then present your results by plotting a histogram of the values you find for d_{JC}. Do they appear to cluster around αt? How spread out is the histogram? Compute means and standard deviations.
- Repeat the last step, changing N or t to a new value.
- What conclusions can you draw about using the Jukes-Cantor formula for real data? Does it appear to be most accurate when sequences are long or short? When the length of elapsed time is large or small? Can you give an intuitive explanation of why this should be the case?

2. Investigate how the various distance formulas perform on simulated sequence data produced according to models *different* from the one underlying the distance formula.

    This is an important issue, because real sequence evolution is at best only approximately described by any of these models.

    If, for instance, one sequence evolved from another according to the Kimura 2-parameter model with $\beta \neq \gamma$, then we should not expect the Jukes-Cantor distance to be a valid reconstruction of the amount of mutation. However, it should be close if β and γ are close to one another.

    As in the project above, the MATLAB program mutate can simulate DNA mutations according to a specified Markov model of base substitution. Then the various distances can be computed for the sequences so produced, using compseq, distJC, distK2 and distLD.

Suggestions
- Explore the performance of the Jukes-Cantor and Kimura 2-parameter distance formulas for recovering the total amount of mutation on data that is produced by various Kimura 2-parameter matrices. You should keep the parameters β and γ very small, and use a large number of time steps for mutate. If $\beta/\gamma = \kappa$, how different can κ be from 1 for the Jukes-Cantor formula to be close to correct?
- Explore the performance of the Jukes-Cantor distance formula for recovering αt on data that is produced by a model using Jukes-Cantor matrices, but with an initial base distribution other than the Jukes-Cantor one. (Keep α small and use a large value of t for mutate.) How different can the initial base distribution be before the distance seems unreliable?
- Repeat the last item for the Kimura 2-parameter model.
- If data are produced according to some general Markov model, how close to a Kimura 2-parameter matrix must the model matrix be for the log-det distance and Kimura 2-parameter distance to be close?

- Even if data are produced according to the Jukes-Cantor model, do the Kimura distances and log-det distances give the same results as the Jukes-Cantor distance? Shouldn't they?
- Explain how you would decide which distance formula to use if you were given two sequences.

5

Constructing Phylogenetic Trees

Having modeled the evolution of DNA in the last chapter, we are now ready to use these models to make important deductions from real DNA data. We will see how a model of molecular evolution, together with some new mathematical techniques, can be used to deduce evolutionary history.

Let's consider a well-studied, yet still compelling question: What is the relationship of humans to the modern apes? More specifically, which of the gorilla, chimpanzee, orangutan, and gibbon are our closest evolutionary kin, or are all these apes more closely related to each other than they are to us?

Early evolutionists viewed the chimpanzee and gorilla as our closest relatives. Humans and these African apes were believed to form one evolutionary grouping, which had split from other ape lineages in the more distant past. A bit later, the dominant view became that all the modern apes were more closely related to one another than to humans. Two possible diagrams that represent more detailed versions of these competing views of hominoid evolution are shown in Figure 5.1.

▶ Since the chimpanzee and gorilla are African, whereas the orangutan and gibbon are Asian, what, if anything, would each of these trees indicate about the likely location of the appearance of the first humans?

How can we choose which of these or the many other possible evolutionary trees is the best description of hominoid descent? One approach involves

Figure 5.1. Two possible hominoid phylogenies.

171

first choosing a particular gene that all the apes and humans share, but whose DNA sequence shows variation from species to species. Assuming this gene is shared and similar in all the hominoids because it arose from a common ancestor (i.e., the sequences are *orthologous*), then the variations in the sequences among the species should contain information about their evolutionary history.

For instance, the 898 base-pair *Hind*III sequences of mitochondrial DNA from these hominoids and seven other primates have been reported in (Hayasaka *et al.*, 1988), which draws on work in (Anderson *et al.*, 1981) and (Brown *et al.*, 1982). These sequences are in agreement at between 67% and 97% of the sites, depending on which two are compared. (To see the sequences for yourself, type `primatedata` in MATLAB, and then `who` to see the names of the variables in which they are stored.)

We would like to infer a *phylogenetic tree*, such as one of those in Figure 5.1, showing how all the apes evolved from a common ancestor. But how does the data indicate a "best" tree, or even a good tree, to describe the evolutionary descent?

Of course scientists have drawn trees showing suspected evolutionary relationships between species since well before the advent of DNA sequencing. Morphological similarities between species are one source of clues as to which trees accurately describe the descent. The identification of common ancestors from fossils is another. Now sequence data provides a new source of information about evolutionary history, but using it to infer phylogenetic trees requires the development of new mathematical tools.

5.1. Phylogenetic Trees

Before we begin developing methods to deduce phylogenetic trees, though, we will need some terminology. Because the sequences we might want to relate could come from different species, as in the hominoid example, or instead from different subspecies, populations, or even individuals, we will refer to each source of the DNA sequence as a *taxon* (pl. *taxa*). An equivalent term in common use is *operational taxonomic unit*, usually abbreviated as OTU.

We hope to draw a diagram consisting of line segments that represents the evolutionary history of the taxa. Each of the line segments in the diagram is referred to as an *edge*. A diagram such as those above, in which there are no loops formed by the edges, is called a *tree*.

> ► Why is it reasonable to assume evolutionary relationships can be modeled by drawing trees? What would it mean if there were a loop of edges?

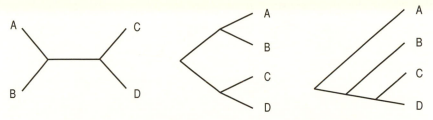

Figure 5.2. An unrooted tree (L) and two rooted versions (C,R).

Because *lateral gene transfer* can occur, for instance when viral DNA is permanently incorporated into that of a host, trees cannot describe all evolutionary relationships. They provide the simplest model, which is nonetheless fully adequate for most uses.

The meeting point of several edges is called an *interior vertex* (pl. *vertices*), while the end of an edge at a taxon is called a *terminal vertex* or a *leaf.* The vertex where the common ancestor of all the taxa would be located is referred to as the *root.*

A tree is said to be *bifurcating* if at each interior vertex three edges meet and at the root two edges meet, as in the trees in Figure 5.1. Although it is conceivable biologically that a tree other than a bifurcating one might describe an evolutionary lineage, it is usual to ignore that possibility.

▶   What would the evolutionary meaning be of a vertex in a tree where four edges meet (i.e., where one edge splits into three)? Can you think of plausible circumstances under which several species might diverge in this way?

Although ideally every phylogenetic tree would have a root showing the common ancestor of the taxa, sometimes we have to do without one. Some methods of phylogenetic tree construction yield *unrooted* trees. For example, Figure 5.2 shows an unrooted tree and several of the rooted trees that agree with it. The two trees on the right could each be bent and stretched to look like the tree on the left; only the location of the root distinguishes them.

Topological trees. A tree relating a number of taxa can actually specify several different types of information about their relationships. First, if we do not specify the lengths of edges, and hence only look at the branching structure, we are considering only the *topology* of the tree. We consider two trees to be topologically the same if we can bend and stretch the edges of either one to get the other. We are not, however, allowed to cut off an edge and reattach it elsewhere; doing that may give us a tree that is topologically distinct from the original one.

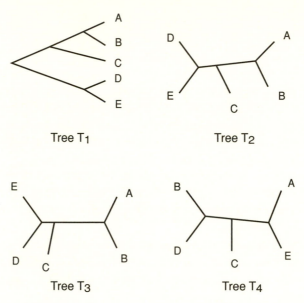

Figure 5.3. Four topological trees; as unrooted trees, all but lower right are identical.

In Figure 5.3, trees T_1, T_2, and T_3 are all topologically the same as unrooted trees, because if any of these figures were made of rubber it could be deformed into the other ones without either cutting or gluing pieces of it together. Tree T_4, on the other hand, is topologically distinct from T_1, T_2, and T_3.

For rooted trees, we use a similar concept. Two rooted trees are topologically the same if one can be deformed into the other without moving the root. Edge lengths can be changed, but not the branching structure.

▶   In Figure 5.3, where can you place a root on T_2 so that it is *not* topologically the same as T_1 as a *rooted* tree? So that it is topologically the same as T_1 as a rooted tree?

A topological tree, even an unrooted one, tells us quite a lot about the evolutionary history of the taxa it relates. For instance, all the trees in Figure 5.2 indicate that taxa A and B are related by a single split in lineage, as are C and D. However, several bifurcations of lineage occurred during the course of A and D evolving from a common ancestor, as two other taxa arose in the process.

Knowing the location of the root conveys more information and may give a better sense of the ordering of events in time. For instance, the tree on the far right in Figure 5.2 clearly indicates the order in which bifurcations occurred:

The common ancestor gave rise to two taxa, one of which may have evolved further to become A. The other subsequently gave rise to B and a third taxon. This third taxon then gave rise to both C and D.

The tree in the middle of Figure 5.2 can be interpreted similarly. The common ancestor gave rise to two taxa, one of which gave rise to both A and B, while the other gave rise to C and D. Note, however, that with only a topological tree, we cannot say which of these last two bifurcations occurred first: Did the most recent common ancestor of A and B exist more recently than that of C and D? We have no way to tell from this tree.

The number of different topological trees that might relate several terminal taxa increases rapidly with the number of taxa. For instance, there is only 1 unrooted topological tree relating 3 taxa, but there are 3 unrooted topologically distinct trees relating 4 taxa.

▶ Draw the one unrooted topological tree that might relate terminal taxa A, B, and C. Draw the three unrooted topological trees that might relate terminal taxa A, B, C, and D.

For 5 terminal taxa, there are 15 such trees. Thus, ignoring the root location, there are 13 more trees that might relate the 5 hominoids than were presented in the chapter introduction. For 6 terminal taxa, there are more than 100 possible unrooted trees. As the number of taxa increases, the number of trees quickly grows to astronomical size. In the exercises, you will find precise formulas giving the number of unrooted and rooted trees relating n taxa. You will also see just how large these numbers are, even for a relatively small number of taxa. The large number of trees is unfortunate, because it means some approaches to finding a good tree to relate taxa will be slow. If a method finds the "best" tree by looking individually at each possible tree, then using it will be extremely time-intensive if there are more than a handful of taxa involved.

Metric trees. In addition to a topological structure, a tree may have a *metric* structure; each edge may be assigned a certain length. This metric structure might be specified by writing numbers for the lengths next to the edges (see Figure 5.4 (L)), or it may be merely suggested by drawing the tree with edges of those lengths, yet not explicitly numbering them. Thus, a topological tree and an unlabeled metric tree can be hard to tell apart. (For clarity, in this book, we will always label edges with their lengths when the tree is intended to be a metric one.)

Generally, the lengths of edges in a phylogenetic tree constructed from DNA sequence data somehow represent the amount of mutation that occurred

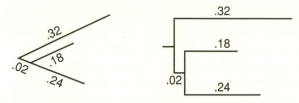

Figure 5.4. Alternate depictions of the same metric tree.

between splittings of the lineage. The longer an edge is, the more the DNA sequence mutated in the course of the evolution that edge represents.

If, for instance, the Jukes-Cantor model of base substitution adequately described the evolution of several taxa, then the edge length in a tree relating them might be the Jukes-Cantor distance between the sequences at the two ends of the edge. As we saw in Chapter 4, this distance is the average number of base substitutions per site that occurred in the descent. Included in this are the mutations obscured by other mutations that the distance formula was designed to estimate. Because the Jukes-Cantor distance is additive and symmetric, the total distance between two taxa along a tree should be the Jukes-Cantor distance between them.

If the molecular clock assumption holds for the evolution of the sequences being related, then the distances in a tree have a more direct meaning. Recall that a molecular clock just means that the mutation rate is constant for all lineages under consideration. If μ denotes the mutation rate, measured in (base substitutions per site)/year, for instance, and t denotes a time in years, then the amount of mutation that will occur during this time is

$$d = \mu t \text{ base substitutions per site.}$$

Thus, a molecular clock means that the amount of mutation along any edge is proportional to the elapsed time, with the constant of proportionality being the constant rate of mutation. Under the assumption of a molecular clock, then, whether we draw edge lengths representing amount of mutation or elapsed time, we draw exactly the same figure, up to scaling by this constant.

If the molecular clock hypothesis holds for a rooted metric tree, then every leaf will be located the same total distance from the root. This is because distances from the root are proportional to the elapsed time since the taxa began to diverge from the common ancestor. Every taxon has had the same amount of time to evolve from the root ancestor, so each will have accumulated the same amount of mutation.

Without a molecular clock, the relationship between the amount of mutation along an edge and the amount of time may be complicated. Suppose that, along one edge of a phylogenetic tree, the mutation rate was quite small, and along another, the mutation rate was large. Then, even though both edges might correspond to the same amount of time, considerably more mutation would occur along one. Without somehow getting additional information about the rate of mutation – perhaps by comparisons to the fossil record – we usually do not have ways of determining elapsed times associated to tree edges.

Metric trees are sometimes drawn in a "square" manner so that it is easier to compare distance along various evolutionary paths. As an example, the two trees in Figure 5.4 both represent the same information. In the tree on the left, the edges have specified lengths, and in the tree on the right, the horizontal edges have those same lengths. Thus, the vertical edges on the right-hand tree are read as contributing nothing to the amount of mutation; they serve solely to separate the various lineages for increased readability.

Problems

5.1.1. Consider the trees in Figure 5.5.
    a. Which of them are the same, as rooted metric trees?
    b. Which of them are the same, as unrooted metric trees?

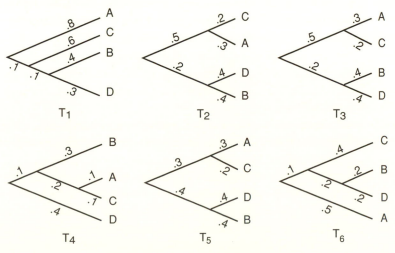

Figure 5.5. Trees for Problem 5.1.1.

    c. Which of them are the same, as rooted topological trees?

    d. Which of them are the same, as unrooted topological trees?

    e. For which trees does a molecular clock appear to be operating?

5.1.2. a. Draw the single topologically distinct unrooted bifurcating tree that could describe the relationship between 3 taxa.

    b. Draw the three topologically distinct rooted bifurcating trees that could describe the relationship between 3 taxa.

5.1.3. a. Draw all 3 topologically distinct unrooted bifurcating trees that could describe the relationship between 4 taxa.

    b. Draw all 15 topologically distinct rooted bifurcating trees that could describe the relationship between 4 taxa.

5.1.4. For n terminal taxa, the number of unrooted bifurcating trees is

$$1 \cdot 3 \cdot 5 \cdots (2n - 5) = \frac{(2n - 5)!}{2^{n-3}(n - 3)!}.$$

Make a table of values and graph this function for $n \leq 10$.

5.1.5. For n terminal taxa, the number of rooted bifurcating trees is

$$1 \cdot 3 \cdot 5 \cdots (2n - 3) = \frac{(2n - 3)!}{2^{n-2}(n - 2)!}.$$

Make a table of values and graph this function for $n \leq 10$.

5.1.6. In this problem, we will step through the reasoning behind the formulas for the number of topologically distinct trees, rooted and unrooted.

    a. Suppose we already know that an unrooted tree with n terminal vertices is made up of e edges. Explain why an unrooted tree with $n + 1$ terminal vertices will have $e + 2$ edges. (*Hint*: Think about how adding one more terminal vertex to an existing tree affects the number of edges.)

    b. Because an unrooted tree with 2 terminal vertices has 1 edge, explain from part (a) why an unrooted tree with n terminal vertices will have $1 + 2(n - 2) = 2n - 3$ edges.

    c. Suppose we know there are m unrooted trees with n terminal vertices. Explain why there will be $(2n - 3)m$ unrooted trees with $n + 1$ terminal vertices. (*Hint*: Think about how many different ways you can add one more terminal vertex to an existing tree.)

    d. Because there is only 1 unrooted tree with 2 terminal vertices, explain from part (c) why there are $1 \cdot 3 \cdot 5 \cdots (2n - 5)$ unrooted trees with n terminal vertices when $n > 2$.

e. Explain why

$$1 \cdot 3 \cdot 5 \cdots (2n - 5) = \frac{(2n - 5)!}{2^{n-3}(n - 3)!}.$$

f. Why is the number of rooted trees with n terminal vertices the same as the number of unrooted trees with $n + 1$ terminal vertices?

g. Conclude that the formulas of Problems 5.1.4 and 5.1.5 are correct.

5.1.7. Because mitochondrial DNA in humans is inherited solely from the mother, it can be used to construct a tree relating any number of humans from different ethnic groups, assuming we all descended from a single first human female. Depending on the clustering pattern of the ethnic groups, this might give insight into the physical location of this woman sometimes called Mitochondrial Eve.

In (Cann *et al.*, 1987), a work that first purported to locate Mitochondrial Eve in Africa, supporting the "out of Africa" theory of human origins, a rooted tree was constructed that was claimed to show the relationships between 147 individuals. How many topologically different trees would need to be looked at if every possibility was really examined? (You may need to use Stirling's formula: $n! \sim \sqrt{2\pi n} \, n^{n+\frac{1}{2}} e^{-n}$. Here, the symbol "$\sim$" can be interpreted as "is approximately.")

See (Gibbons, 1992) for the fall-out from the difficulty of considering so many trees.

5.1.8. The phylogeny of four terminal taxa A, B, C, and D are related according to a certain metric tree. The total distances between taxa along the tree have been found to be as in Table 5.1.

a. Using any approach you wish, determine the correct unrooted tree relating the taxa, as well as all edge lengths. Explain how you rule out other topological trees.

b. Can you determine the root from this data? Explain why or why not.

Note: Techniques for this sort of problem are the subject of the next few sections.

Table 5.1. *Distances Between Taxa for Problem 5.1.8*

	A	B	C	D
A		.6	.6	.2
B			.4	.6
C				.6

5.2. Tree Construction: Distance Methods – Basics

In constructing a phylogenetic tree, the taxa we wish to relate are usually ones currently living. We have information, such as DNA sequences, from the terminal taxa and no information from the ones represented by internal vertices. Indeed, we do not even know which internal vertices should exist, because we do not yet know the tree topology.

The first class of methods for constructing phylogenetic trees that we will discuss are *distance methods*. These attempt to build a tree using information that we believe describes the total distances between terminal taxa along the tree.

To see how we might obtain these distances, imagine trying to find the evolutionary relationship of four species: S1, S2, S3, and S4. Choosing a particular orthologous stretch of DNA from their genomes, we obtain and align sequences from each. If the Jukes-Cantor model of base substitution discussed in Chapter 4 seems appropriate for the data, we then compute Jukes-Cantor distances between each pair of sequences. These are our estimates of distances along the tree, which we organize in Table 5.2.

Table 5.2. *Distances Between Taxa*

	S1	S2	S3	S4
S1		.45	.27	.53
S2			.40	.50
S3				.62

Depending on the sequence data, we might instead adopt a different model of base substitution, leading us to use a different distance formula, such as the Kimura 2-parameter or the log-det distance. Regardless, the distance we calculate between sequences is believed to be a measure of the amount of mutation that has occurred. If these distances were an exact measure of the amount of mutation that occurred, they would match up with the total distances between terminal taxa in the metric tree we would to find.

We do not really expect to find a tree that this data fits exactly; after all, the distances are inferred from sequence data and are not expected to be exactly correct. Moreover, the method of inferring the distances depended on a model that involved assumptions that are certainly not met in real organisms. We hope that however we construct a tree will not be too sensitive to these sorts of errors in the distances.

UPGMA. The first method we consider is called the average distance method, or, more formally, the unweighted pair-group method with arithmetic means (UPGMA). This method produces a rooted tree and assumes a molecular clock. The easiest way to understand the algorithm is by following an example of its use.

With the data table above, we pick the two closest taxa, S1 and S3. Because they are .27 apart, we draw Figure 5.6 with each edge $.27/2 = .135$ long.

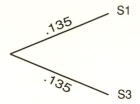

Figure 5.6. UPGMA; step 1.

We then combine S1 and S3 into a group, and average the distances of S1 and S3 to each different taxon to get the distance from the group to that taxon. For example, the distance between S1–S3 and S2 is $(.45 + .40)/2 = .425$, and the distance between S1–S3 and S4 is $(.53 + .62)/2 = .575$. Our table thus collapses to Table 5.3.

Table 5.3. *Distances Between Groups; UPGMA, Step 1*

	S1–S3	S2	S4
S1–S3		.425	.575
S2			.50

Now, we simply repeat the process, using the distances in the collapsed table. Because the closest taxa and/or groups in the new table are S1–S3 and S2, which are .425 apart, we draw Figure 5.7.

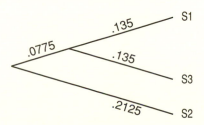

Figure 5.7. UPGMA; step 2.

The edge to S2 must have length $.425/2 = .2125$, while the other new edge must have length $(.425/2) - .135 = .0775$, because we already have the edges of length .135 to account for some of the distance between S2 and the other taxa.

Again combining taxa, we form a group S1–S2–S3, and compute its distance from S4 by averaging *the original distances from S4 to each of S1, S2, and S3*. This gives us $(.53 + .5 + .62)/3 = .55$. (Note that this is *not* the same as averaging the distance from S4 to S1–S3 and to S2.) Because a new collapsed distance table would have this as its only entry, there is no need to give it. We draw Figure 5.8, estimating that S4 is $.55/2 = .275$ from the root. The final edge has length .0625, since that places the other taxa .275 from the root as well.

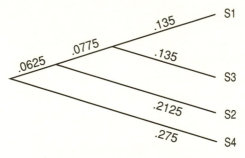

Figure 5.8. UPGMA; step 3.

As we suspected, the tree we have constructed for the data does not exactly fit the data. The distance on the tree from S3 to S4, for instance, is .55, although according to the original data, it should be .62. Nonetheless, the tree distances are at least reasonably close to the distances given by the data.

If we had more taxa to relate, we would have to do more steps to complete the UPGMA process, but there would be no new ideas involved. At each step, we join the two closest taxa or groups together, always placing them at equal distances from a common ancestor. We then collapse the joined taxa into a group, using averaging to compute a distance from that group to the taxa and groups still to be joined. The one point to be particularly careful about is that when the distances between two groups are computed, we must average *all* the distances from members of one group to another – if one group has n members and another has m members, we have to average nm distances. Each step of the algorithm reduces the size of the distance table by one, so that after enough steps, all of the taxa are joined into a single tree.

Notice that the molecular clock assumption is implicit in UPGMA. In this example, when we placed S1 and S3 at the ends of equal length branches, we assumed that the amount of mutation each underwent from their common ancestor was equal. UPGMA always places all the taxa at the same distance from the root, so that the amount of mutation from the root to any taxon is identical.

Fitch-Margoliash algorithm. This method is a bit more complicated than UPGMA, but builds on its basic approach. However, it attempts to drop the molecular clock assumption of UPGMA.

Before giving the algorithm, we need a few mathematical observations. First, if we attempt to put 3 taxa on an unrooted tree, then there is only one topology that needs to be considered. Furthermore, for 3 taxa, we can assign lengths to the edges to fit the data *exactly*. To see this, consider the tree in Figure 5.9. If we have some distance data d_{AB}, d_{AC}, and d_{BC}, then

$$
\begin{aligned}
x + y \quad &= d_{AB}, \\
x + \quad z &= d_{AC}, \\
y + z &= d_{BC}.
\end{aligned}
$$

These equations can be solved either by writing the system as a matrix equation and finding an inverse, or by substituting formulas for one variable obtained from one of the equations into the others. Either way leads to the solution

$$
\begin{aligned}
x &= (d_{AB} + d_{AC} - d_{BC})/2, \\
y &= (d_{AB} + d_{BC} - d_{AC})/2, \qquad (5.1) \\
z &= (d_{AC} + d_{BC} - d_{AB})/2.
\end{aligned}
$$

We will refer to these formulas as the *3-point formulas* for fitting taxa to a tree. Unfortunately, with more than 3 taxa, exactly fitting data to a tree is usually not possible. The Fitch-Margoliash (cited in tables as FM) algorithm uses the 3 taxa case, however, to handle more taxa.

Now we explain the operation of the algorithm with an example. We'll use the distance data in Table 5.4.

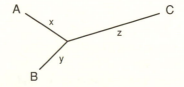

Figure 5.9. The unrooted 3-taxon tree.

Table 5.4. *Distances Between Taxa*

	S1	S2	S3	S4	S5
S1		.31	1.01	.75	1.03
S2			1.00	.69	.90
S3				.61	.42
S4					.37

We begin by choosing the closest pair of taxa to join, just as we did with UPGMA. Looking at our distance table, S1 and S2 are the first pair to join. In order to join them *without* placing them at an equal distance from a common ancestor, we temporarily reduce to the 3-taxa case by combining *all other* taxa into a group. For our data, we thus introduce the group S3–S4–S5. We find the distance from each of S1 and S2 to the group by averaging their distances to each group member. The distance from S1 to S3–S4–S5 is thus $d(S1, S3-S4-S5) = (1.01 + .75 + 1.03)/3 = .93$, whereas the distance from S2 to S3–S4–S5 is $d(S2, S3-S4-S5) = (1.00 + .69 + .90)/3 = .863$. This gives us Table 5.5.

Table 5.5. *Distances Between Groups; FM Algorithm, Step 1a*

	S1	S2	S3–S4–S5
S1		.31	.93
S2			.863

With only three taxa in this table, we can exactly fit the data to the tree using the 3-point formulas to get Figure 5.10. The key point here is that the 3-point formulas, unlike UPGMA, can produce unequal distances of taxa from a common ancestor.

Figure 5.10. FM algorithm; step 1.

We now keep only the edges ending at S1 and S2 in Figure 5.10 and return to our original data. Remember, the group S3–S4–S5 was only needed temporarily so we could use the 3-point formulas; we did not intend to join those taxa together yet. Because we have joined S1 and S2, however, we combine them into a group for the rest of the algorithm, just as we would have done with UPGMA. This gives us Table 5.6.

Table 5.6. *Distances Between Groups; FM Algorithm, Step 1b*

	S1–S2	S3	S4	S5
S1–S2		1.005	.72	.965
S3			.61	.42
S4				.37

We again look for the closest pair (now S4 and S5) and join them in a similar manner. We combine everything but S4 and S5 into a single temporary group S1–S2–S3 and compute d(S4, S1–S2–S3) = (.75 + .69 + .61)/3 = .683 and d(S5, S1–S2–S3) = (1.03 + .90 + .42)/3 = .783. This gives us Table 5.7. Applying the 3-point formulas to Table 5.7 produces Figure 5.11.

Table 5.7. *Distances Between Groups; FM Algorithm, Step 2a*

	S1–S2–S3	S4	S5
S1–S2–S3		.683	.783
S4			.37

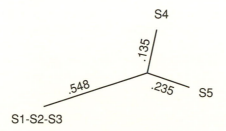

Figure 5.11. FM algorithm; step 2.

We keep the edges joining S4 and S5 in Figure 5.11, discarding the edge leading to the temporary group S1–S2–S3. Thus we now have two joined groups, S1–S2 and S4–S5. To compute a new table containing these two groups we have found, we average d(S1–S2, S4–S5) = (.75 + 1.03 + .69 + .90)/4 = .8425 and d(S3, S4–S5) = (.61 + .42)/2 = .515. We have already computed d(S1–S2, S3) so we produce Table 5.8. At this point, we can fit a

Table 5.8. *Distances Between Groups;*
FM Algorithm, Step 2b

	S1–S2	S3	S4–S5
S1–S2		1.005	.8425
S3			.515

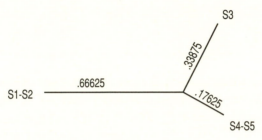

Figure 5.12. FM algorithm; step 3.

tree exactly to the table by a final application of the 3-point formulas, yielding Figure 5.12.

Now we replace the groups in this last diagram with the branching patterns we have already found for them. This gives Figure 5.13.

Our final step is to fill in the remaining lengths a and b, using the lengths in Figure 5.12. Because S1 and S2 are on average $(.1885 + .1215)/2 = .155$ from the vertex joining them and S4 and S5 are on average $(.135 + .235)/2 = .185$ from the vertex joining them, we compute $a = .66625 - .155 = .51125$ and $b = .17625 - .185 = -.00875$ to assign lengths to the remaining sides.

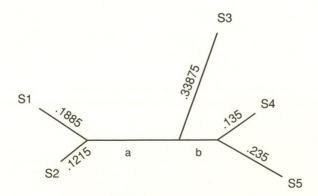

Figure 5.13. FM algorithm; completion.

Notice that one edge has turned out to have negative length. Because this cannot really be meaningful, many practitioners would choose to simply reassign the length as 0. If this happens, however, we should at least check that the negative length was close to 0 or we would worry about the quality of the data.

Although it may seem surprising at first, both the Fitch-Margoliash algorithm and UPGMA will produce exactly the same topological tree when applied to a data set. The reason for this is that, when deciding which taxa or groups to join at each step, both methods consider exactly the same collapsed data table and both choose the pair corresponding to the smallest entry in the table. It is only the metric features of the resulting trees that will differ. This undermines a bit the hope that the Fitch-Margoliash algorithm is much better than UPGMA. Although it may produce a better metric tree, topologically it never differs.

Fitch and Margoliash (Fitch and Margoliash, 1967) actually proposed their algorithm not as an end in itself, but rather as a heuristic method for producing a tree likely to have a certain optimality property (see the Problems section). We are viewing it here, like UPGMA, as a step toward the Neighbor Joining algorithm of the next section. Familiarity with UPGMA and the Fitch-Margoliash algorithm will aid us in understanding that more elaborate method.

Of course, both UPGMA and the Fitch-Margoliash algorithm are better done by computer programs than by hand. However, a few hand calculations are necessary to understand fully how the methods function and what assumptions go into them.

Rooting a tree. Although the Fitch-Margoliash algorithm has allowed us to obtain unequal branch lengths in our trees, we have paid a price – the trees it constructs are unrooted. However, since finding a root is often desirable, a clever idea can get around this deficiency.

When applying any phylogenetic tree method that produces an unrooted tree, an additional taxon can be included. This extra taxon is chosen so that it is known to be more distantly related to each of the taxa of interest than they are to each other, and is known as an *outgroup*. For instance, if we are trying to relate species of ducks to one another, we might include a different type of bird as the outgroup. Once an unrooted tree is constructed, we locate the root where the edge to the outgroup joins the rest of the tree. Biological knowledge that the outgroup must have diverged from the other taxa before they split from one another gives us the location in the tree of the common ancestor.

Problems

5.2.1. For the tree in Figure 5.8 constructed by UPGMA, compute a table of distances between taxa along the tree. How does this compare with the original data table of distances?

5.2.2. Suppose four sequences S1, S2, S3, and S4 of DNA are separated by phylogenetic distances as in Table 5.9. Construct a rooted tree showing the relationships between S1, S2, S3, and S4 by UPGMA.

5.2.3. Perform UPGMA on the distance data in Table 5.4 that was used in the text in the example of the Fitch-Margoliash (FM) algorithm. Does UPGMA produce the same tree as the FM algorithm topologically? Metrically?

5.2.4. The FM algorithm utilizes the fact that distance data relating three terminal taxa can be exactly fit by the single unrooted tree relating them.

    a. Derive the 3-point formulas of Eq. (5.1).

    b. If the distances are $d_{AB} = .634$, $d_{AC} = 1.327$, and $d_{BC} = .851$, what are the lengths x, y, and z?

5.2.5. Use the FM algorithm to construct an unrooted tree for the data in Table 5.9 that was also used in Problem 5.2.2. How different is the result?

5.2.6. Suppose three terminal taxa are related by an unrooted metric tree.

    a. If the three edge lengths are .1, .2, and .3, explain why a molecular clock hypothesis must be invalid, no matter where the root is located.

    b. If the three edge lengths are .1, .1, and .2, explain why the molecular clock hypothesis *might* be valid. If it is, where would the root be located?

    c. If the three edge lengths are .1, .2, and .2, explain why the molecular clock hypothesis must be invalid, no matter where the root is located.

Table 5.9. *Distance Data for*
Problems 5.2.2 and 5.2.5

	S1	S2	S3	S4
S1		1.2	.9	1.7
S2			1.1	1.9
S3				1.6

5.2.7. While distance data for 3 terminal taxa can be exactly fit to an unrooted tree, if there are 4 (or more) taxa, this is usually not possible.

   a. Draw an unrooted tree with terminal taxa A, B, C, and D. Denote the lengths of the five edges by r, s, t, u, and v.

   b. Denoting distances between terminal taxa with notation like d_{AB}, write down equations for each of the 6 such distances in terms of r, s, t, u, and v. Explain why, if you are given numerical distances between terminal taxa, these equations are not likely to have an exact solution.

   c. Give a concrete example of values of the 6 distances between terminal taxa so that the equations in part (b) cannot be solved exactly. Give another example of values where the equations can be solved.

5.2.8. A number of different measures of goodness of fit between distance data and metric trees have been proposed. Let d_{ij} denote the distance between taxa i and j obtained from experimental data, and let e_{ij} denote the distance from i to j along the tree. A few of the measures that have been proposed are:

$$s_{FM} = \left(\sum_{i,j} \left(\frac{d_{ij} - e_{ij}}{d_{ij}} \right)^2 \right)^{\frac{1}{2}} \qquad \text{(Fitch and Margoliash, 1967)}$$

$$s_F = \sum_{i,j} |d_{ij} - e_{ij}| \qquad \text{(Farris, 1972)}$$

$$s_{TNT} = \left(\sum_{i,j} (d_{ij} - e_{ij})^2 \right)^{\frac{1}{2}} \qquad \text{(Tateno } et~al., \text{ 1982)}$$

In all these measures, the sums include terms for each distinct pair of taxa, i and j.

   a. Compute these measures for the tree constructed in the text using the FM algorithm, as well as the tree constructed from the same data using UPGMA in Problem 5.2.3. According to each of these measures, which of the two trees is a better fit to the data?

   b. Explain why these formulas are reasonable ones to use to measure goodness of fit. Explain how the differences between the formulas make them more or less sensitive to different types of errors.

Note: Fitch and Margoliash proposed choosing the optimal metric tree to fit data as the one that minimized s_{FM}. The FM algorithm was introduced in an attempt to get an approximately optimal tree.

5.2.9. Read the data file seqdata.mat into MATLAB by typing load seqdata. Then investigate the performance of UPGMA with the Jukes-Cantor distance to construct a tree for the sequences a1, a2, a3, and a4. All the distances between the sequences can be computed most easily by putting the sequences into rows of an array with the command a=[a1;a2;a3;a4] and then using the command [DJC DK2 DLD]=distances(a). Although this command computes distances using each of the Jukes-Cantor, Kimura 2-parameter, and log-det formulas, for this problem, use only the Jukes-Cantor distances.

   a. Draw the UPGMA tree for the 4 taxa, labeling each edge with its length.
   b. From your edge lengths, compute the distances between taxa along the tree. Are these close to the original distances?

   Note: This data was simulated according to a Jukes-Cantor model with a molecular clock.

5.2.10. Repeat the last problem, but use the FM algorithm instead of UPGMA. Is the tree you produce "better" then the one produced before? Explain?

5.2.11. Investigate the performance of UPGMA with the Jukes-Cantor distance to construct a tree for the sequences b1, b2, b3, b4, and b5 in the data file seqdata.mat. See Problem 5.2.9 for useful MATLAB commands.

   a. Draw the UPGMA tree for the 5 taxa, labeling each edge with its length.
   b. From your edge lengths, compute the distances between taxa along the tree. Are these close to the original data?

   Note: This data was simulated according to a Jukes-Cantor model, but *without* a molecular clock.

5.2.12. Repeat the last problem, but use the FM algorithm instead of UPGMA. Is the tree you produce "better" than the one produced before? Explain?

5.2.13. Constructing a tree by UPGMA assumes a molecular clock. Suppose the unrooted metric tree in Figure 5.14 correctly describes the evolution of taxa A, B, C, and D.

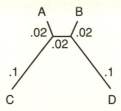

Figure 5.14. Tree for Problem 5.2.13.

a. Explain why, regardless of the location of the root, a molecular clock could not have operated.
b. Give the array of distances between each pair of the four taxa. Perform UPGMA on that data.
c. UPGMA did not reconstruct the correct tree. Where did it go wrong? What was it about this metric tree that led it astray?
d. Explain why the FM algorithm will also not reconstruct the correct tree.

5.3. Tree Construction: Distance Methods – Neighbor Joining

In practice, UPGMA and the Fitch-Margoliash algorithm are seldom used for tree construction, because there is a distance method that tends to perform better than either. Nonetheless, the ideas behind them help motivate the popular Neighbor Joining algorithm that we will focus on next.

To see why UPGMA, or the Fitch-Margoliash algorithm, might be flawed, consider the metric tree with 4 taxa in Figure 5.15. Here, x and y represent specific lengths, with x much smaller than y. We say the vertices S1 and S3 in this tree are *neighbors*, because the edges leading from them join. Similarly, S2 and S4 are neighbors, but S1 and S2 are not.

Suppose the metric tree of Figure 5.15 describes the true phylogeny of the taxa. Then, perfect data would give us the distances in Table 5.10.

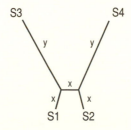

Figure 5.15. A 4-taxon metric tree with distant neighbors, $x \ll y$.

Table 5.10. *Distances Between Taxa in*
Figure 5.15

	S1	S2	S3	S4
S1		$3x$	$x + y$	$2x + y$
S2			$2x + y$	$x + y$
S3				$x + 2y$

But, if y is much bigger than x (in fact, $y > 2x$ is good enough), then the closest taxa by distance are S1 and S2, which are not neighbors. Thus, UPGMA or the Fitch-Margoliash algorithm, by choosing the closest taxa, chooses nonneighbors to join. The very first joining step will be incorrect, and once we join nonneighbors, we will not recover the true tree. The essence of the problem is that if no molecular clock is operating, as with the tree in Figure 5.15, then the closest taxa by distance are not necessarily neighbors on the tree.

▶   If x is much less than y, why do you know that no molecular clock operates in the evolution described by the tree in Figure 5.15?

Choosing the closest taxa to join has misled us; we need a more sophisticated criterion for choosing the taxa to join. To develop one, imagine a tree in which taxa S1 and S2 are neighbors joined at vertex V, with V somehow joined to the remaining taxa S3, S4, ..., SN, as in Figure 5.16.

If our data exactly fit this metric tree then for every $i, j = 3, 4, \ldots N$, our tree would include a subtree like the one in Figure 5.17. But, in that figure, we can see that

$$d(S1, S2) + d(Si, Sj) < d(S1, Si) + d(S2, Sj),$$

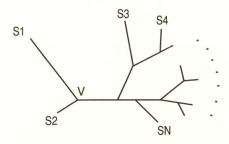

Figure 5.16. Tree with S1 and S2 neighbors.

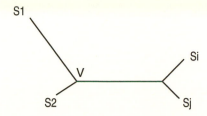

Figure 5.17. Subtree of the tree in Figure 5.16.

since the quantity on the left includes only the lengths of the four edges leading from the leaves of the tree, whereas the quantity on the right includes all of those and, in addition, twice the central edge length. This inequality is called the *4-point condition* for neighbors. If S1 and S2 are neighbors, it holds for any choice of i, j between 3 and N.

The 4-point condition is the basis for Neighbor Joining, but we have more work to do to get it into an easy-to-use form. For fixed i, there are $N - 3$ possible choices of j with $3 \leq j \leq N$ and $j \neq i$. If we add up the 4-point inequalities for these j, we get

$$(N - 3)d(\text{S}1, \text{S}2) + \sum_{\substack{j=3 \\ j \neq i}}^{N} d(\text{S}i, \text{S}j) < (N - 3)d(\text{S}1, \text{S}i) + \sum_{\substack{j=3 \\ j \neq i}}^{N} d(\text{S}2, \text{S}j).$$

$$(5.2)$$

To simplify this, define the total distance from taxon Si to all other taxa as

$$R_i = \sum_{j=1}^{N} d(\text{S}i, \text{S}j),$$

where the distance $d(\text{S}i, \text{S}i)$ in the sum is interpreted as 0, naturally. Then, adding $d(\text{S}i, \text{S}1) + d(\text{S}i, \text{S}2) + d(\text{S}1, \text{S}2)$ to each side of inequality (5.2) allows us to write it in the simpler form

$$(N - 2)d(\text{S}1, \text{S}2) + R_i < (N - 2)d(\text{S}1, \text{S}i) + R_2.$$

Subtracting $R_1 + R_2 + R_i$ from each side of this then gives it the more symmetric form

$$(N - 2)d(\text{S}1, \text{S}2) - R_1 - R_2 < (N - 2)d(\text{S}1, \text{S}i) - R_1 - R_i.$$

If we apply the same argument to Sn and Sm, rather than $S1$ and $S2$, we are led to define

$$M(Sn, Sm) = (N - 2)d(Sn, Sm) - R_n - R_m.$$

Then, if Sn and Sm are neighbors, we have that

$$M(Sn, Sm) < M(Sn, Sk)$$

for all $k \neq m$.

This gives us the criterion used for Neighbor Joining: From the distance data $d(Si, Sj)$, compute a new table of values for $M(Si, Sj)$. Then, choose to join the pair of taxa with the smallest value of $M(Si, Sj)$. The argument above shows that if Si and Sj are neighbors, their corresponding M value will be the smallest of the values in the ith row and jth column of the table. A more complicated argument (see (Studier and Keppler, 1988)) shows that if data perfectly fit a tree, then the smallest entry in the entire table of M values will indicate a pair of taxa that are neighbors.

Since the full Neighbor Joining algorithm is fairly complicated, here is an outline of the method:

Step 1: Given distance data for N taxa, compute a new table of values of M. Choose the smallest value to determine which taxa to join. (This value may be, and usually is, negative; so, "smallest" means the negative number with the greatest absolute value.)

Step 2: If Si and Sj are to be joined at a new vertex V, temporarily collapse all other taxa into a single group G, and determine the lengths of the edges from Si and Sj to V by using the 3-point formulas of the last section on Si, Sj, and G, as in the Fitch-Margoliash algorithm.

Step 3: Determine distances from each of the taxa Sk in G to V by applying the 3-point formulas to the distance data for the 3 taxa Si, Sj, and Sk. Now include V in the table of distance data, and drop Si and Sj.

Step 4: The distance table now includes $N - 1$ taxa. If there are only 3 taxa, use the 3-point formulas to finish. Otherwise, go back to step 1.

As you can see already, Neighbor Joining is tedious to do by hand. Even though the steps are relatively straightforward, it is easy to get lost in the process with so much arithmetic to do. In the exercises, you will find an example partially worked that you should complete to be sure you understand the steps. After that, we suggest you use a computer program to avoid mistakes.

The accuracy of various tree construction methods – the three outlined so far in this text and many others – has been tested primarily through simulating DNA mutation according to certain specified phylogenetic trees and then applying the methods to see how often they recover the correct tree. Some studies have also been done with real taxa related by a known phylogenetic tree; the trees constructed from DNA sequences using various methods could then be compared with the tree known to be correct. These tests have lead researchers to be more confident of the results given by Neighbor Joining than of the other methods we have discussed so far. Although UPGMA or the Fitch-Margoliash algorithm may be reliable in some circumstances, Neighbor Joining works well on a broader range of data. For instance, if no molecular clock is operating, Neighbor Joining is superior, because it makes no implicit assumptions about a molecular clock. Since there is now much data indicating the molecular clock hypothesis is often violated, Neighbor Joining has become the distance method of choice for tree construction.

Problems

5.3.1. Before working through an example of Neighbor Joining, it is helpful to derive formulas for steps 2 and 3 of the algorithm. Suppose we have chosen to join Si and Sj in step 1.

a. Show that for step 2, the distances of Si and Sj to the internal vertex V can be computed by

$$d(Si, V) = \frac{d(Si, Sj)}{2} + \frac{R_i - R_j}{2(N-2)}$$

$$d(Sj, V) = \frac{d(Si, Sj)}{2} + \frac{R_j - R_i}{2(N-2)}.$$

Then show the second of these formulas can be replaced by

$$d(Sj, V) = d(Si, Sj) - d(Si, V).$$

b. Show that for step 3, the distances of Sk to V, for $k \neq i, j$, can be computed by

$$d(Sk, V) = \frac{d(Si, Sk) + d(Sj, Sk) - d(Si, Sj)}{2}.$$

Table 5.11. *Taxon Distances for Problem 5.3.2*

	S1	S2	S3	S4
S1		.83	.28	.41
S2			.72	.97
S3				.48

5.3.2. Consider the distance data of Table 5.11. Use the Neighbor Joining algorithm to construct a tree as follows:

a. Compute R_1, R_2, R_3, and R_4, and then a table of values for M for the taxa S1, S2, S3, and S4. To get you started

$$R_1 = .83 + .28 + .41 = 1.52 \quad \text{and}$$
$$R_2 = .83 + .72 + .97 = 2.52,$$

so

$$M(S1, S2) = (4 - 2).83 - 1.52 - 2.52 = -2.38.$$

b. If you did part (a) correctly, you should have a tie for the smallest value of M. One of these smallest values is $M(S1, S4) = -2.56$, so let's join S1 and S4 first.

For the new vertex V where S1 and S4 join, compute $d(S1, V)$ and $d(S4, V)$ by the formulas in part (a) of the previous problem.

c. Compute $d(S2, V)$ and $d(S3, V)$ by the formulas in part (b) of the previous problem.

Put your answers into the new distance Table 5.12.

d. Because there are only 3 taxa left, use the 3-point formulas to fit V, S2, and S3 to a tree.

e. Draw your final tree by attaching S1 and S4 to V with the distances given in part (b).

Table 5.12. *Group Distances for Problem 5.3.2*

	V	S2	S3
V		?	?
S2			.72

Table 5.13. *Taxon Distances for Problem 5.3.3*

	S1	S2	S3	S4
S1		.3	.4	.5
S2			.5	.4
S3				.7

5.3.3. Consider the distance data in Table 5.13, which is exactly fit by the tree of Figure 5.15, with $x = .1$ and $y = .3$.

   a. Use UPGMA to reconstruct a tree from this data. Is it correct?

   b. Use Neighbor Joining to reconstruct a tree from this data. Is it correct?

5.3.4. Perform the Neighbor Joining algorithm on the distance data used in the examples in the text of Section 5.2. To use MATLAB to do this for the first example, enter the distance array as

```
D=[0 .45 .27 .53; 0 0 .40 .50; 0 0 0 .62; 0 0 0 0]
```

and taxa names as

```
Taxa={'S1','S2','S3','S4'}
```

Then type `nj(D,Taxa{:})`.

   a. Does Neighbor Joining on the 4-taxon example produce the same tree as UPGMA?

   b. Does Neighbor Joining on the 5-taxon example produce the same tree as the Fitch-Margoliash algorithm?

5.3.5. Use the Jukes-Cantor distance and the Neighbor Joining program `nj` to construct trees for the following simulated sequence data in `seqdata.mat`. Compare your results with that produced by other methods in Problems 5.2.9 through 5.2.12 of the last section. How has whether a molecular clock operated in the simulation affected the results?

   a. a1, a2, a3, and a4 (molecular clock)

   b. b1, b2, b3, b4, and b5 (no molecular clock)

5.3.6. The sequences c1, c2, c3, c4, and c5 in `seqdata.mat` were simulated using a Kimura 2-parameter model.

   a. Even without knowing what model was used, how might comparing some of these sequences suggest that the Kimura 2-parameter distance was a good choice for these sequences?

   b. Construct the Neighbor Joining tree using the Kimura 2-parameter
      distance.
   c. Does your tree roughly support a molecular clock hypothesis? Ex-
      plain.
5.3.7. The sequences d1, d2, d3, d4, d5, and d6 are in `seqdata.mat`.
   a. Choose a distance formula to use for these sequences and explain
      why your choice is reasonable.
   b. Construct a Neighbor Joining tree from the data.
   c. One of the 6 taxa is an outgroup that was included to provide a
      rooted tree for the others. Which one is the outgroup? Draw the
      rooted metric tree relating only the other taxa.

5.4. Tree Construction: Maximum Parsimony

One criticism of distance methods for tree construction is that because they
begin by reducing the full DNA sequence data to a collection of pairwise
distances between taxa, they may not use all the information in the original
sequences.

The method of Maximum Parsimony is a rather different approach to
tree construction that uses the entire sequences. Among all possible trees
that might relate the taxa, it looks for the one that would require the fewest
possible mutations to have occurred. To assess the number of mutations, we
never compute distances, but instead consider how mutations occur at each
separate site in the sequences.

The plan is this: For a given tree, somehow count the smallest number of
mutations that would have been required if the sequences had arisen from
a common ancestor according to that tree. We refer to this number as the
parsimony score of the tree. One by one, consider all the trees that might
relate our taxa and compute a parsimony score for each. Then choose the tree
that has the smallest parsimony score. This tree, the most parsimonious one,
is the one the method considers to be optimal for our sequence data.

As a first step to implementing this plan, we need a way of computing
the parsimony score for a specific tree and sequences. For a first example,
suppose we look at a single site in the DNA for each of our taxa and have, for
example,

$$S1: A, \quad S2: T, \quad S3: T, \quad S4: G, \quad S5: A.$$

If we imagined these were related by the tree in Figure 5.18, then we can
trace backward up the tree to determine what base might have been at this

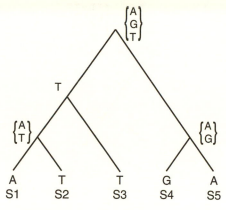

Figure 5.18. Computing the parsimony score for a tree at one site.

site at each interior vertex, assuming the fewest possible mutations occurred. For instance, above S1 and S2, we could have had either an *A* or a *T*, but not a *C* or a *G*, and at least 1 mutation had to have occurred. We label that vertex with the two possibilities {*A*, *T*} and have a mutation count of 1 so far. However, given what appears at S3, at the vertex joining S3 to S1 and S2, we should have a *T*; no additional mutation is necessary, beyond the one we already counted. We have now labeled two interior vertices and still have a mutation count of 1.

We continue to trace backward through the tree, placing a base or set of possible bases at each vertex. If below the vertex are two different bases (or sets of bases that do not overlap), we need to increase our mutation count by 1 and combine the two bases (or take the union of the sets) into a single larger set of possible bases at the higher vertex. If the two lower bases agree (or the sets have common elements), then we label the higher vertex with that base (or the intersection of the two sets). In this case, no additional mutation needs to be counted. When all the vertices of the tree are labeled, the final value of the mutation count gives the minimum number of mutations needed if that tree correctly described the evolution of the taxa. Thus, the tree in Figure 5.18 would have a minimum mutation count, or parsimony score, of 3.

Actually, there are several important facts that we have not proved here. First, it is not really obvious that this method gives the minimum possible mutations needed for the tree. While it should seem reasonable, and is in fact true, that you cannot assign bases to internal vertices in a way that requires fewer mutations, we will not go into the proof. As you will see in the exercises, there can be assignments of bases to the internal vertices that are not consistent with the assignment this method produces, yet that still achieve the same

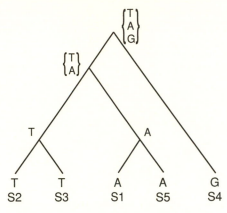

Figure 5.19. A more parsimonious tree.

minimum number of mutations. This means you cannot interpret our method of computing the parsimony score as unambiguously "reconstructing" the sequences of the taxa's ancestors.

Second, the parsimony score of a tree does not depend on the location of the root. If the same tree is used, but the root is moved, our counting method may lead us to put different bases or sets of bases at each of the vertices. However, it is possible to prove we still get the same parsimony score. Thus, while our counting procedure requires temporarily inserting a root, we are really judging the fitness of an unrooted tree. (We could, however, include an outgroup as was discussed with distance methods if the location of a root is desired.)

Finally, because the method does not reliably construct the sequences at internal vertices, we have no way of knowing along which edges mutations occurred. That means we cannot assign a precise length to an edge by using the number of mutations occurring along it. Therefore, the method of maximum parsimony is one that really focuses on using unrooted topological trees to relate taxa.

Now that we have evaluated the parsimony score of the tree in Figure 5.18, let us consider another tree, in Figure 5.19, that might relate the same 1-base sequences. Keep in mind, the tree is drawn with a root only for convenience. Applying the previous method to produce the labeling at the internal vertices, we find this tree has a parsimony score of 2; only two mutations are needed. Thus, the tree in Figure 5.19 is more parsimonious than that of Figure 5.18.

To find the most parsimonious tree for these taxa, we would need to consider all 15 possible topologies of unrooted trees with 5 taxa and compute the

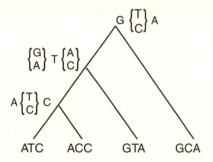

Figure 5.20. Computing a parsimony score for a tree at three sites.

minimum number of mutations for each. Rather than methodically go through the 13 remaining trees, let's try to think about what trees are likely to have low parsimony scores. If the score is low, S1 and S5 are likely to be near one another, as are S2 and S3, but S4 might be anywhere.

- ▶ For the 5 taxa here, draw a few (unrooted) trees that are topologically distinct from that of Figure 5.19, but also have a parsimony score of 2.
- ▶ Explain why no tree relating these 5 taxa could have a parsimony score of 1. (*Hint*: If only one mutation were required for the tree, what would the bases on the leaves have to look like?)

For this example, there are several trees (five, in fact, have a parsimony score of 2) that are tied for most parsimonious. When this happens, use of the parsimony method requires reporting *all* trees that achieve the minimum score, because they are all equally good by our selection criterion.

When dealing with real sequence data, we of course need to count the number of mutations required for a tree among *all* sites in the sequences. This can be done in the same manner as previously, just treating each site in parallel. An example is in Figure 5.20.

Proceeding up the tree beginning with the 2 taxa sequences, *ATC* and *ACC* on the far left, we see we do not need mutations in either the first or third site, but do in the second. Thus, the mutation count is now 1, and the ancestor vertex is labeled as shown. At the vertex where the edge from the third taxa joins, we find the first site needs a mutation, the second does not, and the third does. This increases the mutation count by 2 to give us 3 so far. Finally, at the root, we discover we need a mutation only in the second site, for a final parsimony score of 4.

Although this is not hard to do by hand with only a few sites, as more sites are considered it quickly becomes too big a job. Even worse, if we have more than a few taxa, the number of tree topologies that must be considered is huge. Thus, the parsimony method is really only practically done on a computer. In fact, with a large number of taxa, the number of possible trees is so large that often computer programs only check certain ones to choose the most parsimonious. Good software, operated by knowledgeable users, can often find what are likely to be the most parsimonious trees, but there is no guarantee. (This has caused some embarrassment to researchers publishing trees without understanding the operation of the software they used to produce those trees.)

We can save some effort in using the parsimony method if we make the observation that not all sites will affect the number of mutations needed for a tree. The obvious case is that if all sequences have the same base at a particular site, then all trees will need 0 mutations for that site. Thus, we can eliminate that site from our sequences before applying the algorithm. A less obvious case is when at a site all sequences have the same base (say A), except for at most one sequence each with the other bases (C, T, and G). In this case, *regardless of the tree topology*, if we put an A at every interior vertex, then we have the minimum possible number of mutations. That means such a site will not influence what tree we pick as most parsimonious. This leads to:

Definition. An *informative site* is one at which at least two different bases occur at least twice each among the sequences being considered.

Before applying the parsimony algorithm, we can eliminate all noninformative sites from our sequences, because they will not affect the choice of most parsimonious tree. In the previous examples, you will note only informative sites have been used.

The Maximum Parsimony method does not use the Jukes-Cantor model of molecular evolution, nor any other explicit model of DNA mutation. Instead, it carries an implicit assumption that mutation is rare, and the best explanation of evolutionary history is the one that requires the least mutation. There has been a vigorous, and at times acrimonious, debate between researchers advocating model-based methods of tree reconstruction and those advocating parsimony. Rather than join a philosophical argument, we simply point out that when there are few mutations obscuring previous mutations, both distance and parsimony methods seem to work well in practice. The assumptions of both can be justifiably criticized, and much work is still being done to find better methods.

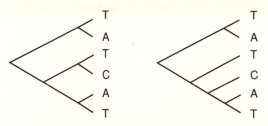

Figure 5.21. Trees for Problem 5.4.1.

Problems

5.4.1. a. Compute the minimum number of base changes needed for the trees in Figure 5.21.

b. Give at least three trees that tie for most parsimonious for the one-base sequences used in part (a). (*Remember*: You can list the taxa in a different order.)

c. For trees tracing evolution at only one site as in parts (a) and (b), why can we always find a tree requiring no more than three substitutions no matter how many taxa are present?

5.4.2. a. Find the parsimony score of the trees in Figure 5.22. (Only informative sites in the DNA sequences are shown.)

b. Draw the third possible (unrooted) topological tree relating these sequences and find its parsimony score. Which of the three trees is most parsimonious?

5.4.3. Consider the following sequences from four taxa.

$$
\begin{array}{ll}
\text{S1:} & AATCGCTGCTCGACC \\
\text{S2:} & AAATGCTACTGGACC \\
\text{S3:} & AAACGTTACTGGAGC \\
\text{S4:} & AATCGTGGCTCGATC
\end{array}
$$

a. Which sites are informative?

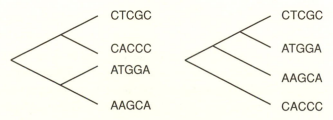

Figure 5.22. Trees for Problem 5.4.2.

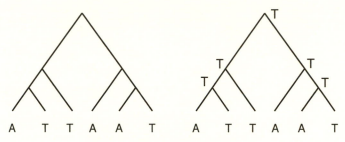

Figure 5.23. Trees for Problem 5.4.5.

b. Use the informative sites to determine the most parsimonious unrooted tree relating the sequences.

c. If S4 is known to be an outgroup, use your answer to part (b) to give a rooted tree relating S1, S2, and S3.

5.4.4. Although noninformative sites do not affect which tree is judged to be the most parsimonious, they do affect the parsimony score. Explain why, if P_{all} and P_{info} are the parsimony scores for a tree using all sites and just informative sites, then

$$P_{all} = P_{info} + n_1 + 2n_2 + 3n_3,$$

where, for $i = 1, 2, 3$, by n_i we denote the number of sites with all taxa in agreement, except for i taxa that are all different. (*Note*: Whereas P_{all} and P_{info} may be different for different topologies, $n_1 + 2n_2 + 3n_3$ does not depend on the topology.)

5.4.5. For the first tree in Figure 5.23, calculate the minimum number of base changes required, labeling the interior vertices according to the algorithm of the text. Then show that the second tree requires exactly the same number of base changes, even though it is not consistent with the way you labeled the interior vertices on the first tree. (The moral of this problem is that the algorithm we are using for counting the minimum number of base changes needed for a tree does not necessarily show all the ways that minimum might be achieved.)

5.4.6. If sequences are given for 3 terminal taxa, there can be no informative sites. Explain why this is the case, and why it does not matter.

5.4.7. The bases at a particular site in aligned sequences from different taxa form a *pattern*. For instance, in comparing $n = 5$ sequences at a site, the pattern $(ATTGA)$ means A appears at that site in the first taxon's sequence, T in the second's, T in the third's, G in the fourth's, and A in the fifth's.

    a. Explain why, in comparing sequences for n taxa, there are 4^n possible patterns that might appear.

    b. Some patterns are not informative. Easy examples are the four patterns showing the same base in all sequences. Explain why there are $(4)(3)n$ noninformative patterns that have all sequences, but one in agreement.

    c. How many patterns are noninformative because 2 bases each appear once, and all the others agree?

    d. How many patterns are noninformative because 3 bases each appear once, and all others agree?

    e. Combine your answers to calculate the number of informative patterns for n taxa. For large n, are most patterns informative?

5.4.8. A computer program that computes parsimony scores might operate as follows: First compare sequences and count the number of sites f_{pattern} for each informative pattern that appears. Then, for a given tree, calculate the parsimony score p_{pattern} for each of these patterns. Finally, use this information to compute the parsimony score for the tree using the entire sequences. What formula is needed to do this final step? In other words, give the parsimony score of the tree in terms of the f_{pattern} and p_{pattern}.

5.4.9. Parsimony scores can be calculated even more efficiently by using the fact that several different patterns always give the same score. For instance, in relating 4 taxa, the patterns $(ATTA)$ and $(CAAC)$ will have the same score.

    a. Using this observation, for 4 taxa, how many different informative patterns must be considered to know the parsimony score for all?

    b. Repeat part (a) for 5 taxa.

5.4.10. Use the maximum parsimony method to construct an unrooted tree for the simulated sequences `a1`, `a2`, `a3`, and `a4` in the data file `seqdata.mat`. First, put the sequences into the rows of an array with `a=[a1;a2;a3;a4]`. Then, find the informative sites with `infosites=informative(a)`. Finally, extract the informative sites with `ainfo=a(:,infosites)`.

    a. What percentage of the sites are informative?

    b. How many different trees must be considered to find the most parsimonious one relating the four taxa?

    c. You may find it too difficult to use all informative sites for a hand calculation. If so, use at least the first 10 informative sites to pick the most parsimonious tree.

d. Does your tree agree topologically with the one produced by UPGMA and/or Neighbor Joining using the Jukes-Cantor distance?

5.4.11. In this problem, you will attempt to use the maximum parsimony method to construct an unrooted tree for the simulated sequences d1, d2, d3, d4, d5, and d6 in the data file `seqdata.mat`. Begin by finding the informative sites as in the last problem.

a. What percentage of the sites are informative?

b. Compute the number of unrooted trees that must be examined if we really consider all possibilities.

c. Use Neighbor Joining, with the log-det distance computed from the full sequences, to get a tree that is a good starting point for a search for the most parsimonious. Compute its parsimony score using only the first 10 informative sites.

d. Again, using only the first 10 informative sites, compute parsimony scores of at least 4 other trees that are similar to the one in part (c). Can you find a more parsimonious one?

e. How confident are you that the most parsimonious tree you found is actually the most parsimonious? What percentage of the possible trees did you compute parsimony scores for? What percentage of the informative sites did you use?

5.5. Other Methods

There are actually many other approaches to phylogenetic tree construction. The list of proposed methods is quite long and grows longer each year, as researchers continue to investigate the problem.

In addition to distance methods and parsimony, there is a third major class of approaches called *Maximum Likelihood* methods. The basic approach of maximum likelihood is to first specify a particular model of molecular evolution (such as Jukes-Cantor, Kimura 2- or 3-parameter, or a more elaborate one). Then, we consider a specific tree that is a candidate for relating our taxa. Assuming our model of evolution and specific tree are correct, we can calculate the probability that the DNA sequence in our data could have been produced. This is called the *likelihood* of the tree, given our data. We repeat this process for all other trees, getting a likelihood value for each. Then, we choose the tree with the greatest likelihood as the tree we feel best fits our data.

To many researchers, maximum likelihood approaches, which follow a long-established tradition in statistics, offer the best hope for good tree construction. However, they face several problems. First, they depend on choosing a specific model of evolution, and if that model does not describe the real process well, one could question their validity. Second, as with parsimony, they require considering all possible trees, and so they require heavy computational effort. For each tree topology considered, a time-consuming calculation is needed to find optimal model parameters consistent with the data. If the number of taxa is large, it is impossible to search through all possible trees, optimizing model parameters for each, and so in practice heuristic shortcuts are taken. Although these seem to perform well in practice, maximum likelihood remains more computationally intensive than other approaches.

Another way of thinking of phylogenetic tree construction methods is to split them into two classes: those that pick a tree based on some optimality criterion, and those that are algorithms that produce a tree. Both Maximum Parsimony and Maximum Likelihood are based on optimality criteria, whereas the distance methods discussed here are algorithmic. Some researchers argue that optimality criteria methods are inherently superior, because they at least make explicit on what the tree choice is based. However, since searching through a large number of trees for an optimal one can be computationally infeasible, computer implementations of parsimony and likelihood methods sometimes begin by considering trees produced by an algorithm (such as Neighbor Joining), and variants of it obtained by moving a few branches around.

One of the difficulties of picking a method to use is that you can find good arguments for and against them all. Nonetheless, the need to construct trees to investigate biological problems is simply too great not to make use of them. The cautious approach is to always use a number of different methods on your data. Rather than trusting a single method to give an accurate tree, look to see if different methods give roughly the same results. They often do, and if they do not, it is worth investigating why they don't. It is simply not enough to just run a computer program on your data and accept the tree produced.

Even when a tree has been chosen by one method or another, it would be desirable to quantify how confident we should be of it. A partial answer to this can be given by the statistical technique of *bootstrapping*. In the bootstrap procedure, the true data sequences are used to create a set of new *pseudo-replicate* sequences of the same length. The bases at a particular site in the new sequences are chosen to be the bases appearing in a randomly chosen site in the original sequences. A tree is constructed for the phylogeny of the pseudo-replicates and recorded. This procedure is then repeated many times, giving a large collection of bootstrap trees. If a high percentage of the bootstrap trees are in agreement with the one produced using the original data, then we may be more confident of it.

An important caveat on using boostrapping, however, is that the technique only helps assess the effects on tree construction of variability *within* the sequences. Boostrapping says nothing about the fundamental soundness of the method by which we choose a tree – it only indicates how variability in the data affects the outcome of the method.

Computer software is essential for using any of the methods on more than a few taxa. Two widely used packages that implement a variety of methods are PAUP* (Swofford, 2002) and PHYLIP (Felsenstein, 1993). If you have access to either, exploring their capabilities is worthwhile.

5.6. Applications and Further Reading

Let's return to the question of hominoid phylogeny that introduced this chapter. What tree can be inferred from the mitochondrial DNA data? Although we could give you an answer, we would rather you found it for yourself. In the exercises, you will have a chance to apply some of the methods of this chapter to the data, starting either with the raw sequences or with some distances already computed from the sequences.

The analysis of the data in (Hayasaka *et al.*, 1988) rests primarily on use of the Neighbor Joining algorithm, as will the analysis you can easily do with MATLAB. If you have access to software designed for maximum parsimony, maximum likelihood, or other methods, we urge you to see if those methods give similar results.

Also, keep in mind the analysis you do is based on one particular stretch of DNA. Studies based on other orthologous sequences might give different results. Furthermore, there are many approaches to phylogenetic inference that are not sequence-based. The evidence of all should be weighed before making too strong a statement about the hominoid phylogeny.

As methods of phylogenetic tree construction from DNA sequence data have developed, they have been used to study a number of interesting questions. Even a quick survey of a general research journal like *Science* turns up a large number of papers in which genetic sequences are used to investigate the evolution of various species from a common ancestor. Here are just a few examples of some recent applications.

1. Investigating whether the evolution of several species parallel one another: For instance, the evolution of hosts and parasites can be studied by constructing separate phylogenetic trees for each. The similarity of the tree topologies can indicate whether the parasites evolved with

the host, or if parasites "jumped" from one host species to another (Hafner *et al.*, 1994). Likewise, trees for two symbiotic species, such as fungus-growing ants and the fungus they grow, help indicate how far back in evolutionary history the symbiotic partnership stretches (Chapela *et al.*, 1994; Hinkle *et al.*, 1994).

2. Determining likely infection sources of human immunodeficiency virus (HIV) by constructing trees from HIV sequences from a number of infected individuals: There have been several forensic applications of this, to the Florida Dentist AIDS cases (Altman, 1994; Ou *et al.*, 1992) and to the case of a doctor accused of intentionally injecting HIV into a former lover (Vogel, 1997, 1998).

3. Studying whether genes have entered the genome of a species through lateral transfer (Andersson *et al.*, 2001; Salzberg *et al.*, 2001): When a tree is constructed from DNA sequences for a gene, it is really a "gene tree" showing gene relationships that may or may not be the same as taxa relationships. Because some human genes are believed to have been obtained by lateral transfer from bacteria that infected us, for certain genes we may appear to be more closely related to some bacteria than other mammals. If a gene is suspected to have arisen in a eukaryote through lateral transfer from bacteria, then a tree can be constructed using gene sequences from both eukaryotes and bacteria. The clustering pattern should help indicate whether genes were transferred laterally or not.

4. Monitoring restrictions on whale hunting: DNA samples from whale meat sold as food and from whales in the wild were used to construct a tree, indicating not only the species of whales being sold, but even the ocean of origin (Baker and Palumbi, 1994).

5. Investigating the "Out of Africa" hypothesis of human origins: The clustering pattern on a tree constructed from human DNA sequences from ethnic groups around the world should help indicate how human populations are related and hence how and from where they spread (Cann *et al.*, 1987; Gibbons, 1992).

Because sequences used in most published research are readily accessible via the internet in databases such as GenBank, it is possible to investigate a dataset from these or other studies on your own.

Sequence-based phylogenetic methods are still being actively researched, by biologists, statisticians, computer scientists, and mathematicians. There are many problems, approaches, and techniques that we have not touched on here. How DNA sequences are identified as good data on which to base a phylogeny, how those sequences are aligned, and how we might measure the confidence we should have in a tree are only three of the topics we have ignored. For more comprehensive overviews, good references are (Hillis *et al.*, 1996) and (Li, 1997).

Problems

Before attempting these problems, type `primatedata` in MATLAB to gain access to the sequences and distance arrays mentioned, all of which come from (Hayasaka *et al.*, 1988). Type who to see the names of the variables this m-file creates.

5.6.1. The distance array `Dist_primates` is a 12 × 12 matrix, with distances computed from a 6-parameter model of base substitution. The names of the taxa in the order of the matrix entries are in `Names_primates`. Perform the neighbor-joining algorithm on this data with the command

$$nj \, (\texttt{Dist\_primates}, \texttt{Names\_primates}\{:\}) \, .$$

Draw the resulting metric tree.

5.6.2. Use biological knowledge and your answer to the last problem to draw a rooted topological tree that might describe the evolutionary history of the five hominoids mentioned in the introduction.

5.6.3. How many possible unrooted topological trees might describe the evolution of the 12 primates? How many possible rooted topological trees might describe the evolution of the five hominoids of the chapter introduction?

5.6.4. The commands

```
Names_hominoids=Names_primates(1:5),
Dist_hominoids=Dist_primates(1:5,1:5)
```

will extract the names and distances between the first five primates, the hominoids of the introduction of this chapter. Use the program `nj` on the distance data for these five only, drawing the resulting metric tree. Does the topology agree with that produced in Problem 5.6.1? Does the metric structure agree? Explain how any discrepancies you notice might have been produced.

5.6.5. Use the command `Seq_hominoids=Seq_primates([1:5],:)` to extract the sequences for the hominoids. Some of the sequences have gaps, indicated by the "–" character. Sites where any sequence has a gap should be removed before computing distances. The commands

```
gaps=(Seq_hominoids =='-')
gapsites=find(sum(gaps))
Seq_nogaps=Seq_hominoids
Seq_nogaps(:,gapsites)=[ ]
```

will find and delete those sites. Using the gapless sequences, compute the Jukes-Cantor, Kimura 2-parameter, and log-det distances. Recall that [DJC, DK2, DLD]=distances(Seq_nogaps) will do this easily.

a. How similar are these distances to those in the array Dist_primates?

b. Use each distance array you produce to construct a tree by Neighbor Joining. Are they all the same topologically? Metrically?

5.6.6. Investigate how reasonable the Jukes-Cantor and Kimura models of base substitution are for the descent of the hominoids from a common ancestor. Do this by considering two sequences at a time, using compseq to calculate a frequency array of bases in the two sequences. Then, compute base distributions for each sequence and Markov matrices that would describe the evolution of one into the other. Are these close to those of a Jukes-Cantor or Kimura model? Does the choice of a different model in (Hayasaka *et al.*, 1988) seem necessary? Explain.

5.6.7. Repeat Problem 5.6.5, but use all 12 primate sequences. Which of the distances do you think is most valid to use? Explain.

5.6.8. From the sequences of the hominoids, isolate the first 10 informative sites. Use these to compute the parsimony score (by hand) of each of the trees at the beginning of this chapter, as well as the one with neighbor pairs (chimpanzee, gorilla) and (orangutan, gibbon). Which of the three is most parsimonious?

5.6.9. Repeat the last problem, but using 10 informative sites chosen to be equally spaced among the informative sites. Do you think this choice of informative sites should be more or less sound than that of the last problem? Explain. (Obviously using all informative sites would be preferable, but that cannot be done easily by hand, because there are 90 of them for these 5 taxa.)

5.6.10. If you have access to software that will attempt to find the most parsimonious tree, use it on the full sequences for the five primates.

(*Note*: these sequences are in a sample data file distributed with (Swofford, 2002).)

5.6.11. The vectors `codingsites` and `noncodingsites` contain the indices of the coding and noncoding sites in the primate sequences. The coding sites can be extracted with the command `Seq_coding=Seq_primates(:,codingsites)`.

   a. Compute frequency arrays of bases in the coding sequences for the primates by comparing sequences two at a time. Does the Jukes-Cantor model or a Kimura model seem reasonable, or do you think a different model would be needed?

   b. Repeat part (a) for the noncoding regions of the sequences. Do you think the same model might apply to both the coding and noncoding regions? Explain, referring to the data.

5.6.12. Because the coding and noncoding sites might evolve differently, they might lead to inferring different trees.

   a. Using only the coding sites, and the log-det distance, find the Neighbor Joining tree for the 12 primates. Does it agree topologically with the tree made the same way using all sites?

   b. Using only the noncoding sites, and the log-det distance, find the Neighbor Joining tree for the 12 primates. Does it agree topologically with the tree made the same way using all sites?

Projects

1. *Dental transmission of HIV*

   In 1990, it was reported in the CDC's *Morbidity and Mortality Weekly Report* that a young woman in Florida had most likely been infected with HIV by her dentist. This conclusion was based primarily on a lack of alternative explanations for the infection. The dentist, who was HIV-positive, then publicly requested that his patients be tested. Altogether, seven patients were found to be HIV-positive.

   Of course, a patient who was HIV-positive was not necessarily infected by the dentist. A large dental practice might be expected to have some infected patients whose infection had nothing to do with their dental care. An epidemiological investigation tried to assess other risk factors for the patients. For some, nondental infection scenarios seemed likely, while for others, nondental infection seemed unlikely. Because of the difficulties of getting accurate answers from patients on high-risk behaviors, however, the results of such an investigation cannot be considered conclusive.

Because no other possible dental infection cases had ever been recorded, some doubt remained concerning the Florida cases.

In 1992, the paper (Ou *et al.*, 1992) appeared in *Science*. This work took a completely different approach using DNA evidence to try to establish the likelihood of a dental infection route for the patients. Because HIV mutates so quickly into quasi-species, one would expect people recently infected from a common source to have more similar quasi-species than those whose common source of infection was more removed. The researchers therefore decided to sequence the highly variable envelope gene of HIV from each patient, the dentist, and some other HIV-infected people living nearby who were not expected to have had any close contact with the cases being studied (i.e., local controls). They then used the sequences to construct a phylogenetic tree, and by the clustering pattern, identified which patients they believed had been infected by the dentist.

Some of the DNA sequences in the paper have been downloaded from GenBank for you to use. In MATLAB, run the m-file `flhiv` to read the sequences. This will create sequences with names:

`dnt, lc1, lc5, ptb, ptc, ptd.`

These refer to the dentist, local control 1, local control 5, patient b, patient c, and patient d in the *Science* paper. Although these sequences are already aligned, they are of differing lengths, so you will have to find the shortest and cut off the ends of the others to compare them.

Construct phylogenetic trees using these sequences and draw conclusions as to which patients were likely to have been infected by the dentist.

Suggestions
- It is best to try several different tree construction methods.
- In deciding to use UPGMA or Neighbor Joining (or perhaps both), consider the assumptions these methods make.
- In selecting a distance formula to use, make sure you look at the data to see which model seems most appropriate. If different distance formulas give different trees, which one are you most confident of? Why?
- If you use a method that produces an unrooted tree, where should you place the root?
- Before using the method of maximum parsimony, compute how many different trees would need to be considered if all were to be examined.
- Because parsimony is not really practical to do by hand for a large number of trees, you should use as many informative sights as you find

bearable and compute the parsimony of a handful of different trees. One of these trees should be the one produced by a distance method, and the others should be trees that you think might also be good candidates.

- How confident do you feel about the validity of your results and why? If you dismiss your work constructing trees as not being rigorous enough, do you feel more confident in just accepting the word of the patients involved about their HIV risk factors? Give an honest appraisal of how valuable you think phylogenetic methods are.

6

Genetics

We have all observed that offspring tend to have physical traits in common with their parents. In humans, similarity in hair color, eye color, height, and build often quite clearly run in families. That selective breeding might enhance traits must have been noticed long ago in our history, as domesticated animals and crops have strongly developed features that we find useful.

On the other hand, the traits of offspring are generally not completely predictable from observing those of the parents. A child might have a trait, such as hemophilia, that neither parent exhibits, though such a trait might occur more commonly within one family than another. Thus, despite patterns to inheritance, chance also appears to be involved. Creating a mathematical model of heredity requires capturing both of these aspects.

The first decisive step was taken by the Augustinian monk Gregor Mendel in the latter half of the nineteenth century. Experimenting with some carefully chosen traits in peas, he was led to propose what we now call a *gene* as the basic unit of inheritance. Though it is perhaps surprising to the modern student, at that time the gene was an entirely abstract concept, with no proposed physical basis, such as the DNA sequences we now immediately imagine.

Recognizing the value of quantitative analysis, Mendel created a mathematical model for the transmission of heritable traits, based on the concepts of probability. His genius was in both identifying simple enough traits to be able to formulate a good model and then modeling the inheritance of those traits successfully. Though subsequent work has added many new features to our models, and we now know much more about the chemical and biological mechanisms behind genetics, Mendel's simple model remains the basic core of our understanding of how many organisms pass on traits to their offspring.

6.1. Mendelian Genetics

In 1865, Mendel presented his findings from breeding experiments with garden peas (Mendel, 1866) to a small group of scientists in Brünn, in the

215

modern-day Czech Republic. Although the world scientific community largely failed to notice until the turn of the century, Mendel's genetic theory was a major advance. Let's consider some of his experiments carefully to understand how the model describes what he observed.

Mendel isolated seven characteristics of pea plants: stem length, seed shape, seed color, flower color, pod shape, pod color, and flower position for study. Each of these characteristics appeared in the peas in one of two forms we'll call *traits*. For instance, stem length might be tall or dwarf, while seed shape could be round or wrinkled. By selective breeding, he then developed *true-breeding* lines of peas for these traits – strains of pea plants that produced progeny, all of which were identical to the parents. Thus, all the descendents of a true-breeding line for tall plants would be tall, and all the descendents of a true-breeding dwarf line would be dwarf.

For each of the characteristics, Mendel cross-bred the two true-breeding lines. For example, true-breeding tall plants were crossed with true-breeding dwarf plants, and true-breeding plants with smooth seeds were crossed with true-breeding plants with wrinkled seeds. Thus, inheritance could be studied one characteristic at a time, and the influence of pure parental traits on the progeny observed. Mendel discovered that, in these crosses, the progeny displayed only one of the traits of the parental generation: The progeny of tall and dwarf plants were all tall; the progeny of plants with round seeds and those with wrinkled seeds all had round seeds. Since the same trait from the parental generation was exhibited by all the progeny, Mendel called such a trait *dominant* and the hidden trait *recessive*. The dominant traits discovered by Mendel's crosses are given in Table 6.1.

Mendel furthered experimented by allowing the offspring of these first generation crosses, or F_1, to self-pollinate and produce a second generation F_2. (The symbols F_1 and F_2 are the standard notations in genetics for the first and second filial generations.) Interestingly, the recessive traits, absent in F_1,

Table 6.1. *Mendel's F_1 Data*

Parental Traits	Dominant Trait
Tall, dwarf plants	Tall
Round, wrinkled seeds	Round
Yellow, green seeds	Yellow
Purple, white flowers	Purple
Inflated, constricted pods	Inflated
Green, yellow pods	Green
Axial, terminal flowers	Axial

Table 6.2. *Mendel's F_2 Data*

Cross Producing F_1	F_2	Ratio
Tall × dwarf plants	787 tall, 277 dwarf	2.84:1
Round × wrinkled seeds	5,474 round, 1,850 wrinkled	2.96:1
Yellow × green seeds	6,022 yellow, 2,001 green	3.01:1
Purple × white flowers	705 purple, 224 white	3.15:1
Inflated × constricted pods	882 inflated, 299 constricted	2.95:1
Green × yellow pods	428 green, 152 yellow	2.82:1
Axial × terminal flowers	651 axial, 207 terminal	3.14:1

reappeared in F_2. Mendel's data for the frequency of each observed trait is shown in Table 6.2.

The last column of Table 6.2 shows the ratio *(number of plants with dominant trait):(number of plants with recessive trait)* in the F_2 plants. These ratios are all remarkably close to 3:1 for each of the seven traits under study. (In fact, they are so close to 3:1 that some believe Mendel may have selectively reported his data at a time when scientific standards were less developed.)

▶ Is noticing this 3:1 ratio enough to help you create an entire genetic theory, as Mendel did?

To explain the 3:1 ratio, Mendel proposed that, for each characteristic, a pea plant must contain a pair of the hereditary factors now called genes. Each gene can come in several forms or *alleles*, corresponding to variations within a trait. For example, for the stem length trait, there is a dwarf allele, d, and a tall allele, D. (Usually, we choose a small letter for a gene based on the recessive allele and use the corresponding capital letter for the dominant allele.) The true-breeding strains of pea plants contain two identical alleles and are said to be *homozygous*. The *genotypes* of these strains are dd for the dwarf strain and DD for the tall strain.

Mendel hypothesized that each parent passed along exactly one of its genes to its progeny. If a parent has genotype Dd, either a tall D allele or a dwarf d allele is passed on, rather than some sort of mix of the two. This *principle of segregation* treats the alleles associated with traits as discrete and indivisible units. A further consequence of the principle is that progeny will also have exactly two genes for a characteristic, as did their parents, and thus the number of genes does not increase in successive generations.

Chance is introduced into the model in determining which of the parental genes each descendent receives. With equal probability, either of the genes in the father will be passed to a descendent, and with equal probability, either of the genes in the mother will be passed on as well. It's as if two parental coin flips determine the outcome in the progeny.

Much more is now known about how genes segregate in the formation of *gametes* or reproductive cells. *Meiosis* is a complicated process in which gametes (egg and sperm in animals, spores in plants) carrying only one copy of each gene are formed. Modern understanding is that genes are found arranged linearly on *chromosomes*, large molecules residing in the nucleus of cells. The chromosomes come in pairs, accounting for the two copies of each gene. Indeed, in gamete formation, it is chromosomes that segregate, not genes as Mendel proposed. At fertilization, two gametes, each carrying one copy of each chromosome, join to produce new offspring.

In reality, inheritance of chromosomes is much more complicated than can be captured by our Mendelian model. The process of *crossing over*, an important source of genetic variability, makes segregation quite involved. Moreover, not all alleles fit the dominant/recessive framework that the Mendelian model supposes, and many traits are not determined by a single gene, but rather by collections of genes. Finally, whereas most familiar organisms do carry two copies of each gene in most cells, and are thus called *diploid*, there are exceptions to this.

However, we are getting ahead of ourselves by bringing up all these complications. The Mendelian model is remarkably good for predicting and understanding the inheritance of many traits and marks a first step toward understanding the biology of inheritance. We can bring modifications into the model later, after we fully understand Mendel's simpler view. So, for now, we will continue to assume segregation of parental genes and restrict our attention to the situation in which a single gene controls a single trait.

When Mendel crossed the DD true-breeding tall genotype with the dd true-breeding dwarf genotype, each descendent inherited an identical set of genes from the parents: D from the first parent and d from the second. Genotypically, these progeny are all Dd, and, because they contain two different forms of the gene, are said to be *heterozygous*.

Remember that each of the F_1 were tall pea plants. Thus, although genetically the progeny were heterozygous, the D allele was dominant over the d allele, in the sense that all the plants of F_1 resembled their tall parent. These F_1 have the same *phenotype* as their tall parents, that is, they have the same observable characteristics.

▶   What are the phenotypes of the genotypes DD, Dd, and dd?

Table 6.3. *Punnett Square for Dd × Dd*

	D	d
D	DD	Dd
d	Dd	dd

▶ If W denotes the dominant allele for round seeds and w the recessive allele for wrinkled seeds, what are the phenotypes of WW, Ww, and ww?

To understand the 3:1 phenotypic ratio in F_2, a helpful device is the *Punnett square*. Here, we place the possible gametes formed by F_1 parents as row and column headings. The entries, formed by the union of such gametes, are the F_2 genotypes. A Punnett square for the stem length gene in the self-fertilization of a pea in F_1 is shown in Table 6.3.

Because each of the gametes is equally likely, according to the model, the four entries of the square are all equally likely descriptions of the genotypes of offspring. We should thus find the three genotypes DD, Dd, and dd in a ratio of 1:2:1 in the F_2 plants.

Notice that we can also deduce the expected ratio of the phenotypes of the F_2 progeny. Since the first two of these genotypes produce the tall phenotype, we should see three tall plants (DD and Dd) for every dwarf plant (dd), giving a ratio of 3:1. Mendel's simple genetic model describes the outcome of his breeding experiments remarkably well.

We can easily extend Mendel's model to make predictions about the outcome of more complicated breeding experiments. For example, if W and w denote the alleles for round and wrinkled seeds, then we may be interested in predicting the outcome of the cross $DdWw \times ddWw$. To handle such two gene crosses, we assume, as Mendel did, that genes *assort independently*. That is, in gamete formation, the segregation of alleles of one parental gene occurs independently of the segregation of alleles for the other gene. Using the language of probability, we would say the segregations of the alleles for two different genes are independent events.

Example. To predict the outcome of the cross $DdWw \times ddWw$, we can again use a Punnett square. Because all combinations are equally likely in gametes, the parental type $DdWw$ creates four types of gametes – DW, Dw,

Table 6.4. *Punnett Square for*
DdWw × ddWw

	DW	Dw	dW	dw
dW	DdWW	DdWw	ddWW	ddWw
dw	DdWw	Ddww	ddWw	ddww
dW	DdWW	DdWw	ddWW	ddWw
dw	DdWw	Ddww	ddWw	ddww

dW, and dw – all with equal probability. Similarly, $ddWw$ creates gametes dW, dw, dW, and dw with equal probability. The resulting Punnett square is shown in Table 6.4.

Notice that there are only six different genotypes among the 16 entries in the square. However, since several of these genotypes produce the same phenotype, there are only four different phenotypes represented. Careful counting shows that, among the F_2 plants, we should expect the following fractions of the population with the given phenotypes: Tall plants with round seeds 6/16, tall plants with wrinkled seeds 2/16, dwarf plants with round seeds 6/16, and dwarf plants with wrinkled seeds 2/16.

- ▶ Which genotypes in the square produce the phenotype of tall plants with wrinkled seeds?
- ▶ Suppose you wanted to determine the phenotypes and their frequencies for a cross between $DdWwYY \times ddWWYy$, where Y represents the dominant allele for green pods and y the recessive allele for yellow pods. How big would your Punnett square be?

The size of a Punnett square grows quickly, in fact exponentially, with the number of independently assorting genes you are tracking. For n genes, the square is $2^n \times 2^n$. This makes Punnett squares impractical for all but the simplest of analyses. Ultimately, we will find that the language of probability, as introduced in Chapter 4, is a better tool for calculating the chances of different outcomes of a particular cross.

Example. To redo the example above of the cross $DdWw \times ddWw$ using probability, let's first calculate the likelihood of dwarf progeny with wrinkled seeds. Because both dwarfness and wrinkled seeds are recessive characteristics, we know that the only genotype producing these traits is $ddww$. This means that each parental strain must contribute a d and a w to such progeny.

In detail,

P(dwarf plants with wrinkled seeds)

$$= P(ddww)$$
$$= P(dw \text{ from first parent})P(dw \text{ from second parent})$$
$$= P(d \text{ from first parent})P(w \text{ from first parent})$$
$$\times \ P(d \text{ from second parent})P(w \text{ from second parent})$$
$$= \left(\frac{1}{2}\right)\left(\frac{1}{2}\right)(1)\left(\frac{1}{2}\right) = \left(\frac{1}{8}\right).$$

Naturally, this answer agrees with our result from the Punnett square.

Let's pause and examine several of the equal signs above, since they are derived from important mathematical and biological concepts. Note, for example, that the second equality, rewriting $P(ddww)$ as the product of the probabilities of inheriting alleles from each parent, is only correct if these are independent events. This assumption of independence is part of the notion of *random union of gametes*: The probability of any union of paternal and maternal gametes is the product of the proportions in which those gametes occur.

▶ Why biologically should what is inherited from one parent be independent of what is inherited from the other?

In addition, the third equality, writing the probability of inheriting dw from a parent as the product of the probabilities of inheriting d and w, is a mathematical restatement of the principle of independent assortment.

Let's try another, more involved, example.

Example. What is the probability that the progeny of $DdWw \times ddWw$ is dwarf with round seeds?

Because having round seeds is a dominant characteristic, two genotypes WW and Ww, both give rise to round seeds, and we will need to take both possibilities into consideration. Now

$$P(\text{dwarf with round seeds}) = P(ddWW \text{ or } ddWw)$$
$$= P(ddWW) + P(ddWw),$$

since $ddWW$ and $ddWw$ are disjoint events. But

$$\mathcal{P}(ddWW) = \mathcal{P}(dW \text{ from first parent})\mathcal{P}(dW \text{ from second parent})$$

$$= \left(\frac{1}{2}\right)\left(\frac{1}{2}\right)(1)\left(\frac{1}{2}\right) = \left(\frac{1}{8}\right),$$

and

$$\mathcal{P}(ddWw) = \mathcal{P}(dW \text{ from first parent})\mathcal{P}(dw \text{ from second parent})$$

$$+ \mathcal{P}(dw \text{ from first parent})\mathcal{P}(dW \text{ from second parent})$$

$$= \left(\frac{1}{2}\right)\left(\frac{1}{2}\right)(1)\left(\frac{1}{2}\right) + \left(\frac{1}{2}\right)\left(\frac{1}{2}\right)(1)\left(\frac{1}{2}\right)$$

$$= \left(\frac{1}{8}\right) + \left(\frac{1}{8}\right) = \left(\frac{1}{4}\right).$$

Finally, we add to compute

$$\mathcal{P}(\text{dwarf with round seeds}) = \left(\frac{1}{8}\right) + \left(\frac{1}{4}\right) = \left(\frac{3}{8}\right).$$

▶   Justify each step of these computations. Where have we used the fact that certain events are independent?

The last calculation was complicated enough that it is a good idea to check it a different way. Let's do this by thinking of the principal of independent assortment differently, in more probabilistic terms. When we say that the two genes assort independently, we mean that the events $E_d = \{$plant is dwarf$\}$ and $E_r = \{$plant has round seeds$\}$ are independent events. Thus, we can compute probabilities for each of these events, and then use the multiplication rule for independent probabilities to combine the answers. This focuses our attention on more manageable problems; instead of looking at the cross $DdWw \times ddWw$, we can look at the crosses $Dd \times dd$ and $Ww \times Ww$ separately.

Example. To find the probability of dwarf progeny with round seeds, we need only multiply the probabilities:

$$\mathcal{P}(\text{dwarf with round seeds}) = \mathcal{P}(E_d \cap E_r) = \mathcal{P}(E_d)\mathcal{P}(E_r).$$

For the $Dd \times dd$ cross, $\mathcal{P}(E_d) = \mathcal{P}(dd)$. This probability is $\mathcal{P}(dd) = \frac{1}{2}$, which could be found either by using a 2×2 Punnett square or by arguing that

$$P(dd) = P(d \text{ from first parent})P(d \text{ from second parent}) = \left(\frac{1}{2}\right)(1) = \frac{1}{2}.$$

For the $Ww \times Ww$ cross,

$$P(E_r) = P(WW \text{ or } Ww) = P(\text{not } ww) = 1 - P(ww) = 1 - \frac{1}{4} = \frac{3}{4}.$$

Thus,

$$P(E_d \cap E_r) = \left(\frac{1}{2}\right)\left(\frac{3}{4}\right) = \frac{3}{8},$$

just as we found before.

While we have seen several ways to calculate the frequencies of phenotypes in progeny from various crosses, we will use probability most often. It is a sophisticated tool that allows us to estimate and model genotypic and, more generally, allelic frequencies in a population. As we move into increasingly complicated genetic models, simple devices like the Punnett square cannot be usefully adapted.

Although the basic Mendelian model does not describe all genetic phenomena of interest, it is adequate to model the incidence of certain human diseases. For example, Tay-Sachs disease, a disease primarily striking children of Ashkenazi Jewish descent and usually leading to death before age 5, is developed by individuals who are homozygous with a recessive allele for a particular gene. It is estimated that roughly 1 in 31 adults in the Ashkenazi population in North America are heterozygous for the recessive allele. Recessive alleles may lie hidden for generations if most individuals marrying into a family are homozygous dominant. Tay-Sachs disease may thus occur unexpectedly and with devastating impact when two heterozygous individuals have children. For many years, estimates of the presence of the recessive allele in the population, together with extensive family medical histories when they were available, were the only means of calculating the risk that a child would develop Tay-Sachs disease. More recently, prenatal techniques such as amniocentesis are used to detect the presence of the Tay-Sachs mutation.

Problems

6.1.1. Imagine that, in a certain species, gamete formation does not occur, and instead of receiving half the genes of each parent, offspring receive the *full* set of genes from both parents. If each parent in the founding generation F_0 has two copies of a particular gene, how many copies will offspring of the nth generation F_n have?

6.1.2. Create a Punnett square for a $DdWw \times DdWw$ cross of pea plants. What proportion of the progeny has each genotype? Each phenotype?

6.1.3. In the text, probabilistic arguments are given to compute the probability of dwarf wrinkled-seed and dwarf round-seed phenotypes for the progeny of a $DdWw \times ddWw$ cross of pea plants. Complete the analysis by using probability to compute the following:

   a. The probability of a tall wrinkled-seed phenotype.

   b. The probability of a tall round-seed phenotype.

6.1.4. According to (Petersen *et al.*, 1983), the recessive allele for Tay-Sachs disease is present in 1 of 31 people in the North American Jewish subpopulation. Because of the nature of the disease, we can assume all adults with the allele are heterozygous.

   a. What is the probability that a couple drawn from this subpopulation will both have the allele?

   b. What is the probability that a child of such a couple will develop Tay-Sachs disease?

   c. What is the probability that a child, both of whose parents come from this subpopulation, will develop Tay-Sachs disease?

6.1.5. Consider three genes, each with dominant and recessive alleles, denoted A, a; B, b; and C, c.

   a. If an individual has genotype $AaBbCC$, how many different gametes might it form?

   b. If two organisms with genotype $AaBbCC$ are mated, how many different genotypes and phenotypes are possible?

6.1.6. Generalize the result of the last problem by considering N genes for a particular organism, with each gene having dominant and recessive alleles.

   a. How many different gametes can be formed by an organism that is heterozygous for n genes and homozygous for $N - n$ genes?

   b. Suppose two individuals, with identical genotypes, are heterozygous for n genes and homozygous for $N - n$ genes. How many different genotypes and phenotypes are possible if these two organisms of identical genotype are mated?

   c. Suppose one individual is heterozygous for the first k genes only and a second individual is heterozygous for first l genes only, where $k < l \leq N$. At the other loci, both organisms are homozygous recessive. How many genotypes are possible when these organisms are crossed? How many phenotypes?

6.1.7. A *testcross* is a cross between a genetically unknown organism and a homozygous recessive organism. Testcrosses can be used to determine whether an organism is heterozygous or homozygous for a particular allele.

   a. Suppose three organisms with genotypes AA, Aa, and aa are crossed with aa. What is the expected ratio of phenotypes in each of these testcrosses?

   b. Suppose that a pea plant of an unknown genotype is testcrossed with dwarf pea plants that have wrinkled seeds and yellow pods ($ddwwyy$). Of the progeny, some are tall, some are dwarf, and all have wrinkled seeds with green pods. What is the genotype of the unknown parental strain?

   c. Explain why you can determine the genotype of an unknown plant by testcrossing with another plant that is homozygous recessive for all genes of interest, but not by testcrossing with a plant that is homozygous dominant. Give both informal reasoning and quantitative justification.

6.1.8. In rabbits, two independently assorting genes affect fur. The dominant allele, B, determines black fur, and a recessive allele b determines brown fur. Normal fur length is determined by a dominant allele, R, and short fur length by a recessive allele, r. A homozygous (in both genes) black rabbit with normal-length fur is crossed with a brown, short-haired rabbit.

   a. What are the possible genotypes and phenotypes of the offspring in F_1? What is the proportion of each?

   b. If F_1 rabbits are intercrossed, what proportion of the F_2 are homozygous (both dominant and recessive) for the color gene? What proportion are homozygous for both genes? What proportion of the black rabbits are homozygous for both genes?

   c. What is the genotype ratio for black rabbits with normal length fur in F_2?

6.1.9. To test his hypothesis that two genes assorted independently, Mendel carried out another series of experiments. In one, he bred true-lines of pea plants with round, yellow seeds ($WWGG$) and wrinkled, green seeds ($wwgg$). Here, the recessive allele for green seed color is denoted by g. He crossbred these lines to get F_1, and then the F_1 plants self-fertilized to produce F_2.

   a. What are the phenotypes and genotypes of F_1?

   b. What phenotypes will be represented in F_2, and in what relative frequencies should the phenotypes occur if the genes do assort independently?

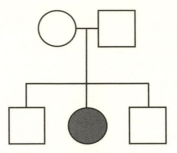

Figure 6.1. A human pedigree.

c. If Mendel's data did not exactly match that predicted in part (b), should he have doubted the assumption of independent assortment? By how much could the frequency data be off from the theoretical prediction before you would doubt the assumption? Explain informally.

6.1.10. If a certain genotype is lethal to embryos, then the expected proportions of genotypes in a new generation is, of course, affected.

In mice, an allele Y^l, known as the *yellow-lethal* mutation, is dominant for yellow fur color, but homozygotes die in the embryonic stage. Homozygotes with genotype yy have gray-brown or agouti fur.

Suppose two yellow mice are crossed. Give the genotypes, phenotypes, and expected proportions of their viable progeny, F_1.

6.1.11. Family pedigrees can be used in determining the risk of human offspring developing certain genetic diseases. One such disease is sickle-cell anemia, which occurs in individuals homozygous for a certain recessive allele.

In the pedigree of Figure 6.1, circles denote females and squares males; horizontal lines join couples, and vertical lines indicate children. Gray coloring indicates an individual has sickle-cell anemia.

a. For the relevant gene, what must the genotypes of the parents be?
b. What is the probability that a fourth child of the parents will be disease-free?
c. What are the possible genotypes of one of the sons? What is the probability of each of those genotypes?

6.1.12. Brachydactyly, or short fingers, is determined in humans by a particular gene with dominant and recessive alleles. Suppose a couple, both with brachydactyly, have two children. One child has normal length fingers and the other has short fingers.

a. Is brachydactyly a dominant or recessive trait? What are the genotypes of the parents? Of the children?

b. Suppose the couple has two more children. What is the probability that neither of them will have brachydactyly?

6.1.13. Plants heterozygous for three independently assorting genes are crossed.

a. What proportion of the progeny is expected to be homozygous for all three dominant alleles?

b. What proportion of the progeny is expected to be homozygous for all three genes?

c. What proportion of the progeny is expected to be homozygous for exactly one gene?

d. What proportion of the progeny is expected to be homozygous for at least one gene?

6.1.14. Mendel's simple model of dominant and recessive alleles does not always apply. Even when one gene with two alleles controls a trait, sometimes neither is completely dominant.

For example, in snapdragons, homozygous WW have red flowers and ww have white flowers. In heterozygotes Ww, however, both genes are expressed, and the flowers are pink. In such a case, the alleles are said to be *partially dominant*. If the heterozygote's phenotype is midway between those of the homozygotes, we say the alleles are *semidominant*.

For snapdragons, what are the phenotypic proportions in F_1 resulting from a $WW \times ww$ cross? What are the phenotypic proportions in F_2 arising from F_1 self-fertilization?

6.1.15. Some genes have multiple alleles, that is, more than two alleles exist in a population for a gene at a particular locus.

Suppose a gene has alleles a_1, a_2, and a_3, and that a_1 is dominant over a_2 and a_3, and a_2 is dominant over a_3. What are the genotype and phenotype frequencies you would expect from a cross $a_1a_3 \times a_2a_3$?

6.1.16. Mendel's model may be modified as in the last two problems to account for a gene that has more than two alleles, some of which exhibit partial or semi-dominance.

For example, the three alleles for human blood type – I^A, I^B, and I^0 – exhibit both dominance and *codominance*. Both I^A and I^B are dominant over I^0, but an individual with genotype $I^A I^B$ will have type AB blood, because both alleles are expressed equally.

    a. What are the possible genotypes for the four phenotypic blood types, A, B, AB, and O?

    b. Suppose an individual homozygous with type A blood marries an individual heterozygous with type B blood. What are the possible phenotypes of any offspring, and in what relative frequencies do these occur?

    c. Suppose parents, heterozygous with types A and B blood, have four children. How many of the children would you expect to have type O blood? Would it be possible for all of the couple's children to have type O blood? Explain, both informally and quantitatively.

6.1.17. Mendel's basic model only describes phenotypic traits that are controlled by a single gene. However, most traits are more complicated.

    A classic example is comb shape in chickens, which is determined by two independently assorting genes. There are four shapes of chicken combs: rose, pea, single, and walnut. Two genes with two alleles each are responsible for comb shape. The genotypes of the four shapes are: rose R–pp, pea rrP–, single $rrpp$, and walnut R–P–. (Here, a dash indicates either a dominant or a recessive allele is possible.)

    a. What phenotypes result from the crosses $RRpp \times rrpp$, $rrPP \times rrpp$?

    b. What phenotypes, and in what proportions, result from a $RRpp \times rrPP$ cross? If the F_1 progeny are interbred, what phenotypes, and in what proportions, are represented in F_2?

6.2. Probability Distributions in Genetics

While the Mendelian model gives a good understanding of the probability of a single child of certain parents being homozygous recessive for a particular gene, or of a single F_2 plant being tall or dwarf, often we are interested in calculating probabilities of more complicated events. For some of these, we need additional knowledge of probability, rather than genetics.

    The term "*random variable*" is sometimes used for the outcome of a measurement or count when we believe some sort of random process underlies the experiment. A few examples of random variables are

• A fair coin is flipped 10 times, and the number of heads is counted. This number is a random variable that might take on the values $0, 1, 2, \ldots, 10$.

• Parents who are heterozygous for the Tay-Sachs recessive allele have three children. The number of their children that are homozygous recessive is a random variable that might take on the values 0, 1, 2, or 3.

- Mendel's F_2 data on the progeny of the self-fertilized F_1 cross between tall and dwarf pea plants involved 1,064 plants. The proportion of the plants in F_2 that are tall was a random variable that could have taken on any of the values $0 = 0/1064$ (if all plants were dwarf), $1/1064$, $2/1064$, ..., $1 = 1064/1064$ (if all plants were tall).

The random variables listed here are built by counting or finding proportions of outcomes of simpler events. Our understanding of the simpler events should enable us to analyze these, but how?

A function that describes the probability of the various outcomes of a random variable is called a *probability distribution*. In this section, we will consider two particular distributions of use in genetics.

The binomial distribution and expected values. As a first example, suppose we flip a fair coin 3 times, and are interested in the probability of getting exactly 2 heads among the 3 flips. We can list the 8 equally likely outcomes of the 3 coin flips,

$$HHH, HHT, HTH, HTT, THH, THT, TTH, TTT,$$

each occurring with probability $1/8$. In this list, the 3 outcomes

$$HHT, HTH, THH$$

have exactly 2 heads. Thus, using the addition rule of probabilities, we find

$$\mathcal{P}(\text{exactly 2 heads in 3 flips}) = \frac{1}{8} + \frac{1}{8} + \frac{1}{8} = 3\left(\frac{1}{8}\right) = \frac{3}{8}.$$

Now suppose we wanted to find the probability of exactly 12 heads in 35 flips? We could proceed similarly, but listing cases is likely to be difficult and error-prone. However, a probability distribution called the *binomial distribution* allows us to calculate such probabilities quickly and efficiently, by associating probabilities to each of the possible outcomes $0, 1, 2, \ldots, n$ of the random variable that gives the number of heads produced by n coin flips.

To develop a formula for the binomial distribution, let's examine the example of 2 heads in 3 flips again. For each coin flip, we have probability $1/2$ of getting a head and probability $1/2$ of getting a tail. Thus, for any particular way we might get 2 heads and a tail, the probability is

$$\mathcal{P}(HHT) = \mathcal{P}(HTH) = \mathcal{P}(THH) = \left(\frac{1}{2}\right)^2\left(\frac{1}{2}\right) = \frac{1}{8}.$$

The 2 heads required two factors of 1/2, and the single tail required an additional factor of 1/2. Because the coin flips are independent, we multiply these factors.

Now why are there three scenarios in which 2 heads can be produced in the 3 flips? We have to account for which of the 3 particular flips were the ones in which the heads occur. They could occur on flips 1 and 2, flips 1 and 3, or flips 2 and 3. The number of scenarios is the number of different ways that 2 of the 3 flips can be designated as producing heads.

This simple example motivates the general formula for the binomial distribution. Suppose we perform n independent trials of a random process that has two possible outcomes. For convenience, we call one of the two outcomes a *success S* and the other a *failure F*. Suppose further that in each trial we have $\mathcal{P}(S) = p$ and $\mathcal{P}(F) = q = 1 - p$. Then, the binomial distribution calculates the probability of k successes among the n trials, as

$$\mathcal{P}(k \text{ successes in } n \text{ trials}) = \binom{n}{k} p^k q^{n-k}.$$

Here, we have introduced the notation $\binom{n}{k}$ to mean the number of different ways that the k successes might be located among the n trials.

▶   Above, we calculated the probability of 2 heads in 3 coin flips. What is a success? A failure? What is the number of trials n and the number of successes k?

Of course, for the binomial formula to be useful, we need a good way to find a value for $\binom{n}{k}$. For the 3-coin flip example, thinking of "heads" as a success, we computed $\binom{3}{2} = 3$ by listing all the cases. (Alternatively, if we think of "tails" as a success, the same list shows $\binom{3}{1} = 3$.) It really is not feasible to list all the possibilities for large n and k, however. In the exercises, you will develop the formula

$$\binom{n}{k} = \frac{n!}{(n-k)!k!}. \tag{6.1}$$

The expression $\binom{n}{k}$ is called the number of *combinations* of n objects chosen k at a time, but is usually read as "*n choose k.*" We think of it as counting how many ways we can designate (or choose) k out of the n trials to be the ones where the successes occur.

Let's consider an application of the binomial distribution to genetics.

Example. In mice, an allele A for agouti – or gray-brown, grizzled fur – is dominant over the allele a, which determines a non-agouti color. If an $Aa \times Aa$ cross produces 4 offspring, what is the probability that exactly 3 of these have agouti fur?

From the Mendelian model, we know that, for any particular offspring, the probability of the agouti phenotype is $3/4$. Although it is tempting to leap to the conclusion that this means 3 of the 4 offspring must have agouti fur, that is in fact incorrect.

We will first compute the probability that 3 of the 4 progeny have agouti fur without using the binomial distribution, working from the basic laws of probability instead. If we let A represent an agouti offspring and N non-agouti, then there are four ways in which exactly 3 of the 4 offspring could have agouti fur: in order of birth, the offspring might be $NAAA$, $ANAA$, $AANA$, or $AAAN$. Thus, we can use the multiplication and addition rules of probability to find

$\mathcal{P}$(exactly 3 of 4 offspring has agouti fur)

$$= \mathcal{P}(NAAA) + \mathcal{P}(ANAA) + \mathcal{P}(AANA) + \mathcal{P}(AAAN)$$

$$= \frac{1}{4} \cdot \frac{3}{4} \cdot \frac{3}{4} \cdot \frac{3}{4} + \frac{3}{4} \cdot \frac{1}{4} \cdot \frac{3}{4} \cdot \frac{3}{4} + \frac{3}{4} \cdot \frac{3}{4} \cdot \frac{1}{4} \cdot \frac{3}{4} + \frac{3}{4} \cdot \frac{3}{4} \cdot \frac{3}{4} \cdot \frac{1}{4}$$

$$= 4 \cdot \frac{27}{256} = .421875.$$

Now let's redo the computation using the binomial distribution. We'll call having agouti fur a success, so $p = 3/4$, $q = 1/4$. Then,

$$\mathcal{P}(\text{exactly 3 of 4 offspring has agouti fur}) = \binom{4}{3} \left(\frac{3}{4}\right)^3 \left(\frac{1}{4}\right)$$

$$= \frac{4!}{3!1!} \left(\frac{27}{256}\right) = .421875.$$

▶ If you decide to call having non-agouti fur a success, that changes the details of the work in this computation, but not the answer. How do the details change?

Notice that even though each offspring of the $Aa \times Aa$ cross has a $3/4 = .75$ chance of having the agouti phenotype, the probability that exactly 3 of 4 offspring have the phenotype is considerably lower, at around .42.

▶ Why is this statement not contradictory?

Since Mendel's studies focused on *diallelic* genes, that is, genes with exactly two alleles, the binomial distribution very naturally fits this setting. Of course, many genes have more than two alleles. Nonetheless, by grouping alleles into two categories – healthy and diseased, or dominant and recessive – the binomial distribution can often be used to make genetic predictions even when more alleles exist. For instance, in the agouti fur example, the symbol *a* actually represents a number of different alleles, each associated with different fur colorings and patterns, but all recessive to agouti. Because we are only concerned with the agouti phenotype, we can lump all others together in our analysis.

Example. What is the probability that exactly 4 of 10 mice from an $Aa \times Aa$ cross have agouti fur?

We'll use the same setup as before, only this time we are interested in determining the probability of $k = 4$ successes (agoutis) in $n = 10$ trials (births). This probability is

$$P(4 \text{ agouti in 10 births}) = \binom{10}{4}\left(\frac{3}{4}\right)^4\left(\frac{1}{4}\right)^6$$

$$= \left(\frac{10!}{4!6!}\right)\left(\frac{3}{4}\right)^4\left(\frac{1}{4}\right)^6 \approx .01622.$$

We can use the binomial distribution together with the addition rule to solve even more difficult problems.

Example. What is the probability that more than half of six progeny of a $Aa \times Aa$ cross have the agouti phenotype?

Continuing to think of an offspring with agouti fur as a success, we need to calculate $P(\text{at least 4 successes in 6 trials})$. But this is the same as

$$P(4 \text{ successes in 6 trials}) + P(5 \text{ successes in 6 trials})$$

$$+ \, P(6 \text{ successes in 6 trials}) = \binom{6}{4}\left(\frac{3}{4}\right)^4\left(\frac{1}{4}\right)^2$$

$$+ \binom{6}{5}\left(\frac{3}{4}\right)^5\left(\frac{1}{4}\right)^1 + \binom{6}{6}\left(\frac{3}{4}\right)^6\left(\frac{1}{4}\right)^0 \approx .83057.$$

Thus, it is quite likely that more than half of the six offspring have agouti fur.

Let's return to considering the number of agouti mice in a cross producing four progeny. Similar computations to those above give the probabilities that

Table 6.5. *Probabilities of Exactly i Agouti Mice Among Four Progeny of Aa × Aa Cross*

i	0	1	2	3	4
$\mathcal{P}(i)$	.00390625	.046875	.2109375	.421875	.31640625

exactly 0, 1, 2, 3, or 4 of the 4 progeny have agouti fur, as shown in Table 6.5. Of course, the entries in this table add to 1.

While Table 6.5 tells us the probability of any outcome of this four-progeny mouse cross, a useful summary of the table is the *expected value* of the number of agouti mice progeny. Informally, the expected value tells us how many agouti progeny we might expect when four offspring are produced. You should think of the expected value as an average of the outcomes, with each outcome weighted by the probability it occurs. To be more precise,

Definition. For any probability distribution describing a random variable with a finite number of possible outcome values i, the *expected value* is defined as

$$E = \sum_{\text{outcomes } i} i \cdot \mathcal{P}(i).$$

Because it is an average weighted by probabilities, the expected value of a random variable might not be an integer, even if the random variable can have only integer outcomes.

For the example described by Table 6.5, we find

$$E = 0 \cdot .00390625 + 1 \cdot .046875 + 2 \cdot .2109375 + 3 \cdot .421875$$
$$+ 4 \cdot .31640625 = 3.$$

Thus, in this example, the expected value seems to be capturing our naive belief that 3 of the 4 mice should have agouti fur, since the probability that any particular mouse does is 3/4.

As you will see in the exercises, whenever a random variable with a binomial distribution is used, something similar happens. More specifically, the expected value for the number of successes in n trials, assuming the probability of any one success is p, is

$$E = np.$$

This should seem reasonable, because it simply states that if for each trial the fraction of times you get a success is p, then of n trials, you expect np successes.

▶   What is the expected number of agouti offspring in 10 births?

Expected values for two random variables have a nice additive property. Suppose for the mouse cross above, we consider the random variables

$$X_1 = \text{the no. of agouti mice in a litter of 4,}$$
$$X_2 = \text{the no. of agouti mice in a litter of 5.}$$

Then $X_1 + X_2 =$ the no. of agouti mice in a litter of 9, since we can think of the first 4 births and the last 5 births as two separate groups. In this case, it is easy to check that

$$E(X_1 + X_2) = E(X_1) + E(X_2), \tag{6.2}$$

because the left-hand side is $9(3/4)$, and the right-hand side is $4(3/4) + 5(3/4)$. In fact, Eq. (6.2) holds for any two random variables, as you will see in the exercises.

The χ^2 distribution. Although the binomial distribution is useful in computing the probabilities of certain types of outcomes in repeated trials, many other probability distributions arise in biology. A particular useful distribution for genetics is the χ^2 distribution. Rather then predicting the likelihood of certain outcomes, the main use of the χ^2 distribution is to determine whether the outcome of an experiment fits a particular probabilistic model.

For instance, the Mendelian model predicts that all the phenotypic ratios in Table 6.2 should be 3:1. However, none of them were exactly 3:1, even though they were close. With Mendel's data the results are so close to 3:1 that few would doubt the model applies, but what if they had been further from that ratio? How far could they deviate before we might doubt the model?

In designing an experiment, a scientist ideally has a *hypothesis* to test. For Mendel's experiment, this might be: *The principal of segregation, that parents pass on each of their alleles to progeny separately and with equal likelihood, holds.* This hypothesis implies that a cross between F_1 hybrids should yield a phenotypic ratio of approximately 3:1. The larger the number of offspring, the closer we expect the experimental ratio to match the theoretical 3:1.

If data collected from the experiment is in line with the expected results, then evidence has been gathered in support of the hypothesis. If the data deviates a great deal from the expected values, then a scientist must reconsider the validity of the hypothesis; perhaps the hypothesis was wrong, or perhaps the experiment was poorly designed. An important issue for the researcher, then, is how to decide whether the data *fits* the hypothesis.

Table 6.6. *Progeny of Gg × Gg*

Phenotype	Observed No.	Expected No.
Yellow seeds	231	245.25
Green seeds	96	81.75
TOTAL	327	327

The χ^2-statistic is one way to measure *goodness of fit* of data to the hypothesis in an experiment. From the data and hypothesis, we compute a certain number according to a formula given below and denote it by χ^2. If this χ^2-statistic is large, the fit is poor. If it is small, the fit is good. An understanding of the probability distribution for this particular random variable – the χ^2-distribution – will allow us to decide how large χ^2 must be for us to consider it unlikely that the hypothesis is correct.

To illustrate how the χ^2-statistic is used, let's apply it to one of Mendel's experiments, to test the hypothesis that the principal of segregation applies to seed color. (This was one of Mendel's hypotheses, though he did not phrase it this way.) In the laboratory, we cross hybrids $Gg \times Gg$ and obtain 327 progeny. Under our hypothesis, we expect that 3/4 of these, $(.75)(327) = 245.25$, will be phenotypically dominant with yellow seeds, and the remainder, $(.25)(327) = 81.75$, will have green seeds. The experiment turns out to produce data that is a bit off from that, as shown in Table 6.6.

The χ^2-statistic is defined as

$$\chi^2 = \sum_{i=1}^{n} \frac{(O_i - E_i)^2}{E_i}.$$

Here, O_i and E_i denote the observed and expected frequencies, that is, the observed and expected numbers as in Table 6.6. Each expression $(O_i - E_i)$ measures deviation of an observation from what we expect, and because this expression is squared, any chance of positive terms canceling with negative ones is eliminated. Dividing each term by E_i gives us a sense of how large the deviation is relative to the expected number. Summing gives us a measure of total deviation.

In this experiment, we have $n = 2$ classes and find

$$\chi^2 = \sum_{i=1}^{2} \frac{(O_i - E_i)^2}{E_i} = \frac{(231 - 245.25)^2}{245.25} + \frac{(96 - 81.75)^2}{81.75} \approx 3.312.$$

If χ^2 were smaller, we would know the data fit the hypothesis better, and if it were larger, the fit would be worse. We still don't know whether 3.312 is small enough to consider the fit to be good.

Before proceeding, there is one other issue we must understand about χ^2-statistics. Because χ^2 involves adding up a number of positive terms, we would expect its value to be larger whenever there are more terms. This is captured in the idea of a parameter called the *degrees of freedom*. Counting the degrees of freedom can be quite difficult, but a rule of thumb is that there is one degree of freedom for each class whose size can vary freely. In this example, if we imagine the size of the first class (the yellow seed phenotype) varies freely (it could be any number from 0 to 327), then the size of the second class (the green seed phenotype) is obtained by subtracting the first from the total 327. This means we have one degree of freedom. More generally, if we had n classes in a test, then the first $n-1$ of them could range freely, but the last is constrained. This corresponds to $n-1$ degrees of freedom. The more degrees of freedom in a test, the larger you might find the χ^2-statistic to be, because it requires summing more positive numbers. To judge the size of a particular χ^2-statistic, we must take this into account.

With the degrees of freedom specified, statisticians have studied the χ^2 distribution. Although a formula for the distribution is too complicated to give here, information from it is incorporated in tables and in software. This makes it possible to compute, for a specified number of degrees of freedom, the probability that the χ^2-value lies in any specified range, assuming the hypothesis holds.

Keep in mind that, even when the hypothesis is true, every time we do an experiment, we will get different data and a different χ^2-statistic describing the fit. Most of these will be small, but some will be large because of chance. We would like our goodness-of-fit test to be flexible enough to accommodate this variation. So, to decide whether we consider our value of χ^2 to be too large for the data to fit the hypothesis, we pick a *significance level*, for instance $\alpha = .05$. This means we decide to view χ^2 as too large if the probability of getting a lower value is at least $1 - \alpha = 95\%$ when the hypothesis is true.

If we consult a table, such as the abbreviated Table 6.7 at the end of this section, we find that the critical value for a χ^2-statistic with one degree of freedom at the .05 level of significance is $\chi^2_{\text{critical}} = 3.841$. This means that, assuming the hypothesis is correct, only 5% of the time would we calculate a value of χ^2 that was 3.841 or larger. Thus, if our statistic is larger than 3.841, we say the data do not support our hypothesis at the .05 level of significance. However, if our statistic is less than 3.841, we find that the data do support the hypothesis at the .05 level of significance.

Since for Mendel's experiment, $\chi^2 = 3.312 < 3.841 = \chi^2_{\text{critical}}$, the value of χ^2 is not too large, and the experiment supports the hypothesis that the alleles for seed color segregate.

If, instead, our statistic had turned out to be larger than χ^2_{critical}, leading us to reject the hypothesis at the .05 level, a number of things could be responsible. It could be that the hypothesis was wrong (exactly what χ^2-statistics are trying to test), or it could be that our hypothesis is correct and we just happened to obtain extreme data through randomness.

In fact, even when the hypothesis is actually correct, this second case will happen 5% of the time. If we breed pea plants that are perfectly described by the Mendelian model again and again, as in this experiment, and calculate χ^2-statistics for each of these trials, then about 5% of the time we would expect to see χ^2-values larger than χ^2_{critical}. A χ^2 test is not capable of definitively telling us whether the hypothesis is true or not.

▶ Sometimes critical values corresponding to a level of .01 or .1 are used. Which of these makes it more likely that you will doubt the hypothesis you are testing? Which level insists on a closer fit of the data to the expected frequencies?

A significance level of .01 means that we only consider a χ^2-value to show a poor fit if it is larger than what would occur 99% of the time when the hypothesis is true. That means we are less likely to reject the hypothesis erroneously. On the other hand, the significance level .1 insists on a closer fit for us to feel the data supports the hypothesis. With $\alpha = .1$, we are more likely to reject the hypothesis erroneously.

As you can probably imagine, we are just at the tip of the iceberg in discussing χ^2-statistics. There is much more to learn about them as they are used ubiquitously in the scientific world. You will get some practice in the exercises, but a course in statistics is really necessary to delve deeper.

Table 6.7. χ^2_{critical} *Values at Significance Level* α

d.f.	$\alpha = .10$	$\alpha = .05$	$\alpha = .01$
1	2.70554	3.84146	6.63490
2	4.60517	5.99147	9.21034
3	6.25139	7.81473	11.3449
4	7.77944	9.48773	13.2767
5	9.23635	11.0705	15.2767

Note: d.f. denotes degrees of freedom.

Problems

6.2.1. List all the ways that you might have exactly 3 heads among 5 coin flips. Then compute $\binom{5}{3}$ by Eq. (6.1) to verify that it gives the correct count.

6.2.2. In the text, the binomial distribution is used to find the probability of exactly 1 of 3 coin flips producing tails. Find the probabilities of exactly none, exactly two, and exactly three tails in this situation. What is the sum of these four probabilities?

6.2.3. Verify the entries in Table 6.5.

6.2.4. Use a calculator or computer to find $\binom{10}{k}$ for each $k = 0, 1, 2, \ldots$, 10. A MATLAB command to do this type of calculation is nchoosek(10,0).

    a. For which k is $\binom{10}{k}$ smallest? For these particular values of k, explain why it has the value it does by thinking in terms of choosing objects.

    b. For which k is $\binom{10}{k}$ largest? Is this intuitively reasonable? Explain.

    c. What patterns do you notice in your calculations? Do the patterns hold if 10 is replaced by other numbers?

6.2.5. Explain the following results not by referring to formula (6.1), but in terms of choosing objects.

    a. $\binom{n}{1} = n$ and $\binom{n}{n-1} = n$ for any n.

    b. $\binom{n}{0} = 1$ and $\binom{n}{n} = 1$ for any n.

6.2.6. Suppose a family has six children.

    a. What is the probability that four are boys and two are girls?

    b. Give the probability distribution (i.e., the seven probabilities) that the family has $0, 1, \ldots$, or 6 boys. How would your answer change if you were to list the probability distribution for the number of girls in the family?

    c. What is the expected number of boys in the family?

    d. What is the probability that the family has four or more girls?

6.2.7. In the text, the binomial distribution is used to find the probability that exactly 3 of 4 offspring have agouti fur from a cross of mice heterozygous for agouti fur.

    a. Find the probabilities that exactly 30 of 40 offspring of this cross have agouti fur. Then, find the probability that exactly 300 of 400 offspring have agouti fur.

    b. Can these results be consistent with the fact that, in a large number of such offspring, we would expect 3/4 of them to have agouti fur? Explain.

6.2.8. If you roll a fair die once, what is the expected value of the outcome?

6.2.9. Suppose you roll two fair dice and add the results.

a. Calculate the expected value of the outcome by first finding the probabilities of each of the outcomes $2, 3, 4, \ldots, 12$, and then computing a weighted average of the outcomes.

b. Let X_1 and X_2 be random variables denoting the outcome of the roll of the first and second die, respectively. Find $E(X_1)$, $E(X_2)$, and $E(X_1 + X_2)$.

6.2.10. When using the binomial distribution in applications, it does not matter which of the two trial outcomes you consider a success. Use the binomial distribution to calculate the probability of 10 rolls of a die producing three sixes as follows.

a. If you call "producing a six" a success, what should p, q, n, and k be in the binomial formula for this probability? What is the resulting probability?

b. If you call "not producing a six" a success, what should p, q, n, and k be in the binomial formula for this probability? What is the resulting probability?

6.2.11. Part of the reason the formula for the binomial distribution gave the same result in both parts of the last problem was because $\binom{n}{k} = \binom{n}{n-k}$.

a. Explain in intuitive terms, in terms of choosing k or $n - k$ objects from n objects, why this formula should hold.

b. Explain why the mathematical formula (6.1) shows this formula holds.

6.2.12. One form of albinism (lack of pigment) in humans is caused by a recessive allele a. Suppose an homozygous albino marries a heterozygote, and the couple has two children.

a. What is the probability their first child will be an albino?

b. What is the probability their first child will be an albino and their second child will have normal skin pigment?

c. What is the probability exactly one of their two children will be an albino?

d. What is the probability at least one of their two children will be an albino?

e. What is the expected number of their children that will be albino?

6.2.13. Mice homozygous for a recessive allele, f, are fat. Suppose a *dihybrid* cross, $AaFf \times AaFf$, is carried out by experimenters. Here, A denotes the agouti allele.

a. How many of 25 progeny are expected to be fat with agouti fur?

b. What is the probability that exactly 4 of 25 progeny will be fat with agouti fur?

c. What is the probability that, at most, 4 of the 25 progeny will be fat with agouti fur?

d. What is the probability that at least 4 of the 25 progeny will be fat with agouti fur?

6.2.14. In a certain population of rats, the probability of an individual surviving through its first year is .5. For the rats who make it to age one, the probability of surviving a second year is .25, and for those who make it to age two, the probability of surviving a third year is also .25. All rats die before the end of their fourth year.

a. What is the probability that the age (in years, rounded down) at death of a rat is 0? 1? 2? 3? Why should these add to 1?

b. What is the expected age at death of one of these rats?

6.2.15. The yellow-lethal allele is dominant for yellow fur color, but lethal to homozygous embryos. Suppose two mice, both heterozygous for the yellow-lethal mutation, are crossed and produce 12 viable progeny.

a. What is the probability that exactly five of them will have normal coloring?

b. What is probability that 10 or more of the progeny will be yellow?

c. What is the probability that at most three of the progeny will be yellow?

6.2.16. In humans, the hereditary Huntington disease is caused by a dominant mutation. Onset of Huntington disease occurs in midlife, between 35 and 44 years of age typically, and the progressive disorder leads eventually to death. Suppose, in a married couple, one individual carries the allele for Huntington disease. They have four children.

a. What is the probability that none of their children will develop Huntington disease?

b. What is the probability that at least one of their children will develop Huntington disease?

c. What is the probability three or more of their children will develop Huntingdon disease?

6.2.17. In a trihybrid cross, $AaBbCc \times AaBbCc$, what is the probability that exactly 20 of 30 progeny will display the dominant phenotype for all three traits? What is the probability that at least two of the progeny will display the dominant phenotype for at least one of the traits?

6.2.18. The goal of this problem is to derive formula (6.1) for counting combinations. Formally, a combination of n things taken k at a time is an *unordered k* element subset of a set of n elements. However, it's better to think of it more concretely, as follows. Imagine a box of n balls with the numbers $1, 2, 3, \ldots, n$ printed on them. You pick out k of these balls and first place them in a row, in the order you picked them. Then, since you don't care about the order, you dump them in a bag. That's a combination. The number of different bags of balls you might end up with is $\binom{n}{k}$.

a. When you pick the first ball out of the box, how many different choices could you make for it? When you pick the second ball, why are there only $n - 1$ choices for it? For the lth ball, why are there $n - l + 1$ choices?

b. Why does part (a) indicate that, when the k balls are all in a row, there are $n(n - 1)(n - 2) \cdots (n - k + 1)$ possible choices you might have made? (The count of these ordered choices is sometimes called a *permutation*.)

c. Several different ordered choices might lead to the same collection of balls in the bag, so the answer in part (b) is bigger than the number of combinations. To see how much bigger, it's easiest to imagine having the balls in the bag, and (going backward in time) putting them back in some order in a row. Using reasoning similar to parts (a) and (b), explain why there are $k(k - 1) \cdots 2 \cdot 1 = k!$ choices of ways this could be done.

d. Using parts (b) and (c), conclude $\binom{n}{k} = \frac{n(n-1)(n-2)\cdots(n-k+1)}{k!}$.

e. Explain why this formula can also be written as formula (6.1).

6.2.19. The binomial distribution received its name because of a relationship to the expression $(x + y)^n$, a power of a binomial. In fact, the numbers $\binom{n}{k}$ are often called the *binomial coefficients*, because they give the coefficients in the expansion of $(x + y)^n$. That is,

$$(x + y)^n = \binom{n}{0}x^n + \binom{n}{1}x^{n-1}y + \cdots + \binom{n}{n}y^n. \tag{6.3}$$

a. Check this for $n = 2, 3$, and 4, using Eq. (6.1).

b. By thinking of $(x + y)^n$ as a product of n copies of $(x + y)$, explain why this product will produce a term $x^k y^{n-k}$ for each way we can choose k of the copies. Explain why this justifies formula (6.3).

c. What is the sum $\sum_{i=0}^{3} \binom{3}{i}$?, $\sum_{i=0}^{4} \binom{4}{i}$? Give a formula for $\sum_{i=0}^{n} \binom{n}{i}$.

6.2.20. Suppose a trial has probability of success p, so the number of successes in n trials is described by the binomial distribution. Show the expected value for the number of successes in n trials is $E = np$ as follows:

a. Express the expected value as a sum involving factorials and powers of p and q.

b. Show

$$i\frac{n!}{(n-i)!i!}p^i q^{n-i} = pn\frac{(n-1)!}{(n-i)!(i-1)!}p^{i-1}q^{(n-1)-(i-1)}.$$

c. Use part (b) to factor pn from your expression in part (a). Then use Eq. (6.3) to complete the problem.

6.2.21. The goal of this problem is to show that expected values of random variables are additive, as claimed in Eq. (6.2). Only a special case will be considered, where X_1 and X_2 are two independent random variables that, for simplicity, can take on only integer values between 1 and N.

a. Explain why the expected value of $X_1 + X_2$ is

$$E(X_1 + X_2) = \sum_{i=1}^{N}\sum_{j=1}^{N}(i+j)\mathcal{P}(X_1 = i)\mathcal{P}(X_2 = j).$$

b. Through algebra, show this can be written as

$$\sum_{i=1}^{N}i\mathcal{P}(X_1=i)\sum_{j=1}^{N}\mathcal{P}(X_2=j) + \sum_{j=1}^{N}j\mathcal{P}(X_2=j)\sum_{i=1}^{N}\mathcal{P}(X_1=i).$$

c. What are $\sum_{i=1}^{N}\mathcal{P}(X_1=i)$ and $\sum_{i=1}^{N}\mathcal{P}(X_1=i)$? Use this to conclude Eq. (6.2) holds.

6.2.22. Suppose an $Aa \times Aa$ cross produces 1,000 progeny, N with the dominant phenotype, and $1,000 - N$ with the recessive phenotype.

a. For $N = 700$, compute the χ^2-statistic to test whether this data fits the Mendelian model. Using a significance level of $\alpha = .05$, is the data in accord with the model?

b. Repeat part (a) with $N = 725$.

c. What is the smallest value of N that would be judged in accord with the model (at the $\alpha = .05$ level)? The largest value of N?

6.2.23. Explain informally why in Table 6.7, the entries get larger as you move across the rows. Explain informally why they get larger as you move down the columns.

6.2.24. The data in Table 6.8 is from Mendel's experiments with genes for seed shape and color resulting from $WwGg \times WwGg$ crosses ($W =$

Table 6.8. *Progeny of*
WwGg × *WwGg*

Phenotype	Observed No.
Round, yellow	315
Round, green	108
Wrinkled, yellow	101
Wrinkled, green	32

round; w = wrinkled; G = yellow; g = green). Use χ^2 to test if the genes for seed color and shape assort independently in pea plants. Because there are four phenotypes, there are $4 - 1 = 3$ degrees of freedom.

6.2.25. The critical value of a χ^2-statistic comes from a theoretical χ^2 distribution with appropriate number of degrees of freedom. Figure 6.2 shows a graph of a typical χ^2 distribution.

In such a graph, the values of χ^2 are along the horizontal axis, and probabilities of χ^2 falling in any interval are represented by the area above that interval and below the curve. The total area between the curve and the horizontal axis is 1 unit and corresponds to 100% or a probability of 1. The critical value χ^2_{critical} at significance level α is the value on the horizontal axis that leaves an area of α to the right. In Figure 6.2, this area is shaded for $\alpha = .05$.

a. Suppose you are performing a χ^2-test and choose a significance level of .01 (or 1%). Where, approximately, would the critical value fall on the horizontal axis in Figure 6.2?

b. Notice that the bulk of the area under the curve is just a bit to the right of the vertical axis and there is very little area under the right

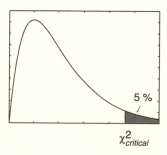

Figure 6.2. χ^2-Distribution.

tail of the curve. If your data is well explained by your experimental hypothesis, where do you expect your calculated χ^2-statistic to fall? How is the shape of this curve related to the goodness-of-fit test?

6.3. Linkage

After receiving little attention for more than 30 years, Mendel's theory of inheritance eventually became well-accepted through the efforts of the British geneticist William Bateson and others. Since Mendel only hypothesized the existence of genes, it was necessary to find the physical basis of these units of inheritance. Around the turn of the century, biologists suspected strongly that genes resided on chromosomes, large thread-like structures that could be stained and viewed under a microscope during cell division. Evidence for this was given by the American geneticist Thomas Hunt Morgan in 1910 and his coworkers. Morgan's group worked with fruit flies, *Drosophila melanogaster*, a favorite of geneticists, since they reproduce quickly and in great abundance and have some readily observable traits with simple variants.

Sex-linked genes. Let's consider one of Morgan's experiments to see how his laboratory was able to discover the important role played by chromosomes in inheritance. After 2 years of breeding *Drosophila*, a mutant male fruit fly with white eyes was born. (Normal, or *wildtype*, eye color is red.) This white-eyed male was crossed with wildtype red-eyed females, and the resulting F_1 generation had all red eyes, indicating that the new mutant allele was recessive. Then, the F_1 generation was interbred to produce F_2.

▶ Assuming Mendel's model applies, what fraction of the F_2 population should have white eyes?

The basic model predicts that, regardless of sex, $1/4$ of F_2 would be homozygous recessive, and hence have white eyes. However, when the F_1 generation were intercrossed, the F_2 were observed with phenotypes as given in the middle column of Table 6.9. In a striking departure from the expected values, there is a total absence of any white-eyed females. Also, roughly half of the males had white eyes, rather than the predicted $1/4$. While roughly $1/4$ of all progeny had white eyes, they are not distributed equally among the sexes. This is strong evidence for a connection between the determination of sex and the behavior of the eye color gene.

In a second experiment in Morgan's laboratory in which a female from F_1 was crossed with the mutant male, phenotypes of (very) roughly equal

Table 6.9. *Progeny of Two Crosses*

Phenotype	$F_1 \times F_1$	$F_1 \times$ mutant
Red-eyed, female	2459	129
White-eyed, female	0	88
Red-eyed, male	1011	132
White-eyed, male	782	86
TOTAL	4252	435

frequency occurred, as shown in the last column of Table 6.9. However, this data is *not* in contradiction with Mendelian genetics, since a $Ww \times ww$ cross would produce equal numbers of each phenotype, regardless of sex.

The first experiment points out the need for a new model consistent with its outcome. However, any new model must be capable of predicting the outcome of the second as well.

At about the same time that Morgan concluded from these experiments that the inheritance of eye color must somehow be related to the determination of sex, he also noticed a relationship between sex and chromosomes under microscopic inspection of the flies' cells. Although all chromosomes came in matching pairs in female *Drosophila*, male *Drosophila* had one nonidentical pair of chromosomes. Moreover, one of the chromosomes from the nonidentical pair in males was morphologically identical to a pair in females. Morgan suspected that this set of chromosomes, the *sex chromosomes*, must control sex determination in fruit flies and that a gene for eye color must lie on this chromosome pair.

Morgan proposed a model for this sex-linked gene behavior that used chromosomes to explain the observations from experimental data. We denote the identical sex chromosomes in females by XX, and the corresponding differing chromosomes in males by XY. In addition, we'll use w to denote the white-eye allele, and w^+ the wildtype red-eye allele. (Such notation is common for the wildtype alleles of any gene.) Hypothesizing that the eye-color gene lies on the X chromosome only, we let X^w denote a sex chromosome carrying the white-eye allele, and X^+ one carrying the wildtype allele.

In Morgan's initial experiment, the females were genotype X^+X^+ and the mutant male X^wY. Now, assuming *segregation of chromosomes* in gamete formation, in the F_1 generation we expect equal numbers of the genotypes X^+X^w and X^+Y. We continue to view the white-eyed mutation as recessive, so each female will have red eyes due to the X^+X^w genotype. Similarly, each of the F_1 males carries only a wildtype allele X^+ so they also have red eyes. The presence of the gene for eye-color on the X chromosome, with no

Table 6.10. *Punnett Square for*
$X^+ X^w \times X^w Y$

	X^+	X^w
X^w	$X^+ X^w$	$X^w X^w$
Y	$X^+ Y$	$X^w Y$

corresponding gene on the Y chromosome is consistent with Morgan's data on the phenotypic make-up of F_1.

We will leave analysis of the experiment leading to the data in the middle column of Table 6.9 to the exercises and instead consider Morgan's second experiment. When F_1 females were crossed with the mutant male, Morgan was crossing heterozygous females $X^+ X^w$ with *hemizygous* males $X^w Y$. Again, assuming segregation of chromosomes, the results of this cross are shown in the Punnett square of Table 6.10.

Now, each genotype in the table is equally likely for progeny, and each genotype gives a different phenotype. In the top row, the phenotypes are red-eyed female and white-eyed female; in the bottom row, they are red-eyed male and white-eyed male. This corresponds roughly with the approximately equal numbers in the last column of Table 6.9. (In the exercises, you are asked to perform a χ^2-test to test more rigorously if the hypothesis of X-linked inheritance of eye color meshes well with this data.)

Because males and females have a different number of X chromosomes, X-linked traits are often manifested in different proportions in the two sexes. For a female *Drosophila* to have white eyes, she must be homozygous for the mutant allele, $X^w X^w$, receiving a (possibly rare) mutant allele from each parent. However, for a male to have white eyes, he needs only one mutant allele so that his genotype is $X^w Y$. As a consequence, recessive X-linked traits are more likely to appear in males. In humans, certain types of color blindness, hemophilia, and mental retardation from fragile X syndrome are X-linked traits that are found almost exclusively in males.

Linked genes and genetic mapping. While sex-linked genes required a modification of the Mendelian model, other experiments from Morgan's laboratory pointed to additional problems with the idea of independent assortment of genes. Even when sex determination was not involved, numerous examples were found of data inconsistent with that assumption.

One such example concerns two genes in *Drosophila*. One gene affects wing shape, with the dominant allele causing straight wings and the recessive

Table 6.11. *Progeny of Cross*

Phenotype	No.
Straight wings, red eyes	520
Straight wings, purple eyes	133
Curved wings, red eyes	129
Curved wings, purple eyes	467
TOTAL	1,249

causing curved wings. The second gene affects eye color, with the dominant allele causing red eyes and the recessive causing purple eyes. Crossing a homozygote recessive for both genes (with curved-wing, purple-eye phenotype), with a heterozygote for both genes (with straight-wing, red-eye phenotype), produces data like that in Table 6.11.

▶ If the two genes assort independently, what is the expected phenotypic ratio? Is the data in line with that?

Although the basic Mendelian model, with independent assortment of the two genes, would have predicted that all four phenotypes were equally likely, the data show clear deviation from this. The inheritance of the two genes seems to be linked, in that there is a definite tendency for the progeny to have a phenotype similar to one or the other of the parents.

This linkage comes from the relationship of genes to chromosomes and the manner in which gametes are formed. The *chromosomal theory of heredity* revised and improved the Mendelian model by taking into account the physical location of genes on chromosomes and modeling such linkage.

Most cells in diploid organisms contain a set of pairs of chromosomes, with one chromosome in a pair inherited from each parent. Chromosomes are divided into two types: *autosomes* (nonsex chromosomes) and *sex chromosomes*. Chromosome number varies greatly between species and seems in no way to reflect developmental complexity; humans have 46 chromosomes, *Drosophila* 8, and cats 72.

According to the chromosomal theory of heredity, gametes are formed by the segregation of chromosomes into reproductive cells, rather than the simpler segregation of genes that Mendel imagined. Genes reside on chromosomes, arranged in a linear fashion. Somatic cells, or body cells, are diploid in that they contain the full count of $2n$ chromosomes in a species. Gamete cells have only half the number of chromosomes, n, and are called *haploid*. At fertilization, two gametes (e.g., an egg and sperm) are united to form a zygote, from which a new diploid offspring develops.

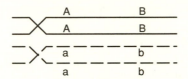

Figure 6.3. Tetrad before crossing over; four chromatids are visible.

When gametes are formed, they do not simply receive a copy of one of the chromosomes in each pair. Instead, a complicated and not completely understood process of *crossing over* provides a source of *genetic recombination*. The chromosome passed along to the gamete is not an identical copy of either one of the parental chromosomes, but instead an amalgam of the parental chromosome pair, with some genes from each.

►   If no crossing over occurred, how would the principle of independent assortment of genes need to be modified? If two genes were on different chromosomes, would they assort independently? What if they were on the same chromosome?

Let's look more closely at how crossing over works. In the process of gamete formation, chromosomes replicate forming identical *chromatids* joined at a *centromere*. Next, matching chromosomes gather together and form *homologous pairs*. This arrangement, known as a *tetrad*, can be seen in Figure 6.3.

In crossing over, two chromatids in the tetrad exchange genetic material. If the chromatids belong to different chromosomes, then this might result in an exchange of alleles. For example, suppose the solid chromosome in Figure 6.3 was inherited from the mother and contains dominant alleles for two genes AB, and that the dashed chromosome, inherited from the father, has recessive alleles for these genes, ab. (Note that the individual in this example is heterozygous for these two genes, $AaBb$.)

During crossing over, two chromatids swap DNA as shown in Figure 6.4. Since nonidentical chromatids are involved in crossing over, they exchange alleles B and b for the second gene. The *parental types* AB and ab occur in the tetrad before crossing over, but after crossing over four genotypes are represented: AB, Ab, aB, and ab. The two new genotypes, Ab and aB, the results of crossing over, are *recombinants*. In the final steps of gamete formation, the four chromatids separate, with each one going into a different gamete.

Because it is so important biologically, we point out again that only two of these gametes are identical to a parental chromosome; the two recombinant

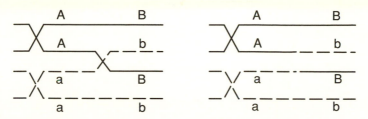

Figure 6.4. Tetrad during (L) and after (R) crossing over.

gametes, if they ultimately unite with other gametes and develop into full organisms, will introduce new genetic combinations into the population. In fact, more than one crossover can occur between homologous chromosomes, so tremendous possibilities for genetic variation are introduced. New variation is of course the raw material for evolution, since recombinants may be better adapted for survival and reproduction.

From a modeling point of view, the behavior of genes on a chromosome during gamete formation can now be captured by the probability of a crossover occurring between them. If this probability is low, then alleles for the two genes will tend to be inherited together, and parental types will dominate in the progeny. If this probability is high, then recombinants will be more common in the progeny. A probability of .5 for a crossover, so that the genes essentially behave as if on different chromosomes, would result in independent assortment. Any divergence from independent assortment is known as *linkage*.

Alfred Sturtevant, who at the time was an undergraduate student working in Morgan's laboratory, realized that the observed frequencies of crossovers could be used to create a genetic map. If we imagine that a chromosome is a long string with genes ordered along it, then it seems natural to expect that, for any little piece of the chromosome, there is some specific probability of a crossover occurring there. Sturtevant's idea was that this probabilistic behavior could be used to give an abstract notion of genetic distance, and then from that distance a map could be constructed. Specifically, he defined the *genetic distance* between two genes on a chromosome as the average number of crossovers that were observed between them during formation of many gametes. If between two genes crossovers are rare, the distance between them is small; if many crossovers typically occur, the distance is large.

Notice that genetic distance is statistical in nature. More precisely, for any stretch of a chromosome, there is a random variable giving the number of crossovers that occur on that piece in gamete formation. Its probability distribution describes the chance of 0, 1, 2, ..., crossovers occurring in that piece.

The expected value of this random variable (or more simply, the expected number of crossovers) is what Sturtevant's average was estimating. Because expected values are additive by Eq. (6.2), they will behave just like distances on a map. We formalize these ideas with a definition.

Definition. The *genetic distance* or *linkage distance* between two genes on a chromosome is the expected number of crossovers that occur between the genes in gamete formation.

Because the expected value is an average number of crossovers, theoretically a genetic distance could take on any value from 0 upward. For physically close genes, genetic distances will tend to be small, since crossovers are less likely to occur, whereas for physically distant genes, distances will tend to be larger. The type of map we will construct from crossover data is called a *linkage* or *genetic map*. This map will show the linear arrangement of the genes on a chromosome, with genetic rather than physical distances separating genes.

Let's see how a *two-point testcross* can place two genes on a linkage map. Suppose we suspect that, in *Drosophila*, the genes for curved wings c and purple eyes pr are linked. For genotypes of linked genes, we use a special notation to keep track of which alleles are on which chromosome in a given pair. For instance, we write $c\,pr/c\,pr$ for a homozygous recessive *Drosophila*, where the slash separates alleles inherited from different parents. There are now several different ways a fly could be heterozygous at both genes; $c\,pr/c^+pr^+$ and $c^+pr/c\,pr^+$ are different configurations.

As a first step in genetic mapping, we cross true-breeding, curved-wing, purple-eyed *Drosophila* with true-breeding wildtype flies: $c\,pr/c\,pr \times c^+pr^+/c^+pr^+$. Notice that all the progeny in F_1 are genotypically $c^+pr^+/c\,pr$ and phenotypically wildtype, since curved wings and purple eyes are recessive traits.

Next, we cross F_1 flies with curved-winged, purple-eyed flies to produce F_2. This testcross is $c^+pr^+/c\,pr \times c\,pr/c\,pr$, and we suppose that the data in Table 6.11 came from such an experiment. As we noticed before, there is a discrepancy between the data and the numbers predicted by Mendelian genetics. Moreover, because there are two large phenotypic classes that resemble the parents – red-eyed with straight wings and purple-eyed with curved wings, and two smaller nonparental phenotypic classes – there is evidence for linkage.

▶ What are the possible genotypes of the F_2 progeny in this second cross? Which of these are parental types and which are recombinants?

Notice how this testcross was designed to test for linkage and crossing over. In the doubly recessive homozygous parent, crossing over may occur between identical chromatids, but it has no effect on the genotype of the gamete. Such *Drosophila* only create *c pr* gametes. In contrast, the parental-type gametes from the heterozygous parent are cr^+pr^+ and $c\ pr$, and crossing over results in recombinants c^+pr and $c\ pr^+$ that will be phenotypically detectable in progeny.

▶ Why are the recombinants c^+pr and $c\ pr^+$ phenotypically observable in this cross?

Now we can estimate the average number of crossovers that occurred. Because the recombinants c^+pr and $c\ pr^+$ result from a crossover, each straight-winged, purple-eyed *Drosophila*, $c^+pr/c\ pr$, and each curvy-winged, red-eyed fruit fly, $c\ pr^+/c\ pr$, is the result of a crossover. In the testcross above, we suspect that $133 + 129 = 262$ crossovers took place.

Now, assuming all recombinants were created by a single crossover, the *recombination frequency* (no. of recombinants)/(total no. of progeny) is exactly the same as the average number of crossovers. Thus, the genetic distance is estimated by

$$\frac{\text{no. of recombinants}}{\text{total no. of progeny}} = \frac{262}{1249} \approx .21 \text{ units} \equiv 21\ cM.$$

Genetic distances are usually measured in *centiMorgans* (*cM*) in honor of Morgan.

In our calculation, we made the assumption that all recombinants were created by a single crossover. What if two crossovers occurred between the genes on a chromatid with c^+pr^+ and one with $c\ pr$? Then, the gametes produced would be of parental type, and our testcross would produce no evidence of any crossovers (see Figure 6.5). Similarly, if three crossovers occurred between the genes, that would appear to us exactly as if only 1 had occurred. Thus, our use of the recombination frequency may understate the true average. Only if we believe multiple crossovers are very rare between

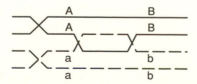

Figure 6.5. A double crossover producing no recombination of genes *A* and *B*.

Table 6.12. *Progeny of*
$gl^+d^+ws^+/gl\ d\ ws \times gl\ d\ ws/gl\ d\ ws$

Phenotype	No.
Normal leaves, tall, normal sheaf	301
Normal leaves, tall, white sheaf	146
Normal leaves, dwarf, normal sheaf	15
Normal leaves, dwarf, white sheaf	1
Glossy leaves, tall, normal sheaf	2
Glossy leaves, tall, white sheaf	17
Glossy leaves, dwarf, normal sheaf	154
Glossy leaves, dwarf, white sheaf	289
TOTAL	925

these genes can we believe our estimate of genetic distance is good. If the genes are close, and the average number of crossovers is small, then our estimation is reasonable.

Now that we have seen how testcrosses can be used as evidence for linkage and for estimating genetic distances, let's extend the method to locating three or more genes on a genetic map. Consider three genes in corn plants with recessive alleles: *d* for dwarf plants, *gl* for glossy leaves, and *ws* for white sheafs. In creating a genetic map, we now have to determine the order of the genes on the chromosome as well as find distances.

To locate the three genes, we make a three-point testcross, $gl^+d^+ws^+/gl\ d\ ws \times gl\ d\ ws/gl\ d\ ws$. (*Remember*: The order in which the genes are listed is not necessarily the correct order on the chromosome.) Sample data on phenotypes of progeny from such a cross is shown in Table 6.12.

The most numerous classes are parental types, indicating linkage of genes on a single chromosome. The remaining classes must be the result of recombination. Before counting the average number of crossovers, notice that we can now observe evidence of either one or *two* crossovers. Because we are mapping three genes, a first crossover could occur between the leftmost gene and the central gene and a second crossover between the central gene and the rightmost gene. We will use the terminology *single crossover* when only one of these is observed and *double crossover* when both are observed.

▶ From Table 6.12, what are the likely phenotypes of double crossovers? Of single crossovers?

Notice that two of the phenotypic classes are extremely rare and four of the classes are of intermediate size. Because a double crossover is much less likely than a single crossover, this identifies the phenotypic classes that correspond to double crossovers. Actually, we'll be able to figure out the gene order too now, if we examine the genotype of individual chromosomes carefully.

▶   If the genes are arranged along the chromosome in order *gl d ws*, what gametes would be produced from a double crossover in the heterozygous parent? What if they were arranged in the order *gl ws d* or *d gl ws*?

In this testcross, crossing over only effects the gametes formed by one of the parental strains. The parental-type gametes from this line are $gl^+d^+ws^+$ and *gl d ws*, with the alleles either all wildtype or all recessive. But the phenotypic classes of the double crossovers show what the chromosome inherited from this parent must have been. The class normal leaves/dwarf/white sheaf must have arisen from gametes $gl^+d\,ws$, and the class glossy leaves/tall/normal sheaf from the complementary $gl\,d^+ws^+$.

Because the outcome of a double crossover is to exchange the middle allele in the parental types, the only way a double crossover could produce the gametes here is if the genes are ordered as *d gl ws* or *ws gl d*. The *gl* gene must be in the middle. Figure 6.6 illustrates one possible configuration for a *three-strand double crossover* in which the recombinant $d^+gl\,ws^+$ is formed.

Now we are ready to estimate genetic distances. We start by finding the distance between *d* and *gl*. Four phenotypic classes result from crossovers between *d* and *gl*: tall/glossy leaves/white sheaf ($d^+gl\,ws$) and dwarf/normal leaves/normal sheaf ($d\,gl^+ws^+$) from single crossovers, tall/glossy leaves/normal sheaf ($d^+gl\,ws^+$) and dwarf/normal leaves/white sheaf ($d\,gl^+ws$) from double crossovers. Thus, the recombination frequency

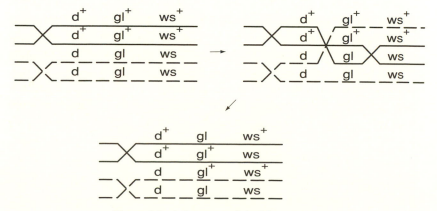

Figure 6.6. A three-strand double crossover.

between d and gl is

$$\frac{17 + 15 + 2 + 1}{925} = \frac{35}{925} \approx .04.$$

Because this is small, we estimate the genetic distance as $4\ cM$.

Similarly, single crossovers between gl and ws produce phenotypes dwarf/glossy leaves/normal sheaf ($d\ gl\ ws^+$) and tall/normal leaves/white sheaf (d^+gl^+ws). We include the double crossover phenotypic classes in our tally, too, because a crossover occurred between the genes in them also. Thus, the recombination frequency between gl and ws is

$$\frac{154 + 146 + 2 + 1}{925} = \frac{303}{925} \approx .33 \text{ or } 33\ cM.$$

Thus, we estimate the genetic distance as $33\ cM$, but since 33 is not so small, we may worry that this estimate is not so accurate.

Note that an estimation of the distance between d and ws requires that we count *all* crossovers between the genes, so double crossovers must count as two:

$$\frac{17 + 15 + 154 + 146 + 2(2) + 2(1)}{925} = \frac{338}{925} \approx .37 \text{ or } 37\ cM.$$

In particular, our estimates of genetic distance are additive, since $4\ cM + 33\ cM = 37\ cM$.

Finally, we put this together and draw the genetic map of Figure 6.7.

Return for a moment to considering only two genes, a and b. If we suspect that a and b are linked, then we might breed $a^+b^+/a\ b$ as F_1, perform a testcross with $a\ b/a\ b$, and calculate the recombination frequency. This frequency is our estimate of the genetic distance between the genes.

Notice, however, that even if a and b are located on different chromosomes, a recombination frequency can still be computed. If we did not realize they were on different chromosomes, we would count a^+b and $a\ b^+$ as single crossovers. However, because the two genes assort independently, the heterozygous parental strain produces four types of gametes, a^+b^+, a^+b, $a\ b^+$, and $a\ b$, in equal proportions. Thus, half the offspring in F_2 will show recombinant genotypes and the recombination frequency will be .5. This means we would estimate that genes on different chromosomes are $50\ cM$ apart!

Figure 6.7. A three-gene genetic map.

The error we have made is in assuming

genetic distance $\approx$ recombination frequency,

despite the fact that this approximation is only valid when the recombination frequency is small. Even for genes on the same chromosome, as recombination frequencies approach .5, the true genetic distance gets larger and larger. The approximation assumes multiple crossovers are rare, and that is only justifiable if the recombination frequency is small.

In genetic mapping, we must map genes that are close together first, and then build our map out from them. For example, if we want to find the distances in a chromosome with genes ordered $a - b - c - d$, it is better to calculate distances between a and b, b and c, c and d, than to try to use linkage information about only a and d. A reasonable rule of thumb is that recombination frequency is a good estimator of genetic distance when it is less than .25. Genes at a distance of $50\,cM$ or greater will assort approximately independently, as if they were on different chromosomes.

Performing testcrosses for genetic mapping of humans is of course neither ethical nor practical. Nonetheless, through pedigree analysis and somatic-cell hybridization techniques, genetic maps of the longest human chromosome have been built, with total length about $293\,cM$.

In addition to genetic maps, there are several other types of maps of chromosomes. A *physical map* shows markers, which might be genes or other distinguishable features, along the chromosome. Because crossover frequency does not correlate well with physical distance, such a map can look quite different, despite showing the same linear ordering of genes. *Sequencing* a chromosome to display the full structure of the DNA in terms of its constituent bases produces the highest resolution map, though genes and other features must be identified in a sequence to relate it to a genetic or physical map. Despite rapid advances in sequencing, genetic maps of the sort discussed here will remain important because of their direct applicability to problems of inheritance.

Problems

6.3.1. Three of Queen Victoria's nine children by Albert are known to have carried the X-linked allele for hemophilia. (Two of her four sons were hemophiliacs, and one of her five daughters had a hemophiliac son.) Neither she nor Albert were hemophiliacs.

  a. What must have been the genotypes of Victoria and Albert? Explain how you can rule out all other possibilities.

b. What was the probability of a son of Victoria and Albert having hemophilia? Of a daughter having hemophilia? Of a daughter being heterozygous for the allele?

c. What was the probability that exactly three of Victoria and Albert's children would carry the mutant allele?

6.3.2. Using the model Morgan developed for the X-linked, white-eyed mutant allele, compute the phenotypic ratios you would expect in the outcome of the experiment described by the data in the middle column of Table 6.9. Is the model in rough agreement with the data?

6.3.3. In another experiment, Morgan crossed white-eyed females to red-eyed males.

a. What must the genotypes of these flies be?

b. What genotypes and phenotypes would be in F_1? In what proportions?

c. If males from F_1 were crossed with females from F_1, what would be the resulting genotypes and phenotypes? In what proportion?

6.3.4. Perform a χ^2-test with $\alpha = .05$ to see if the observed data from the $X^t X^w \times X^w Y$ in the last column of Table 6.9 is consistent with expected numbers from such a cross. (Apparently, Morgan did not perform such a test.)

6.3.5. Suppose a rare disease is caused by a recessive X-linked gene, and phenotypically normal parents have a son who develops this disease.

a. If another son is born into the family, what is the probability he will develop the disease?

b. If a daughter is born into the family, what is the probability she will be a heterozygous carrier?

c. If there are two daughters in the family, what is the probability both will be carriers of the mutant allele?

6.3.6. A man with X-linked color blindness marries a woman with no history of color blindness in her family. Their daughter then marries a man with no history of color blindness and has children. What is the probability that

a. a son in the last generation will be color blind?

b. a daughter in the last generation will be color blind?

c. exactly two of three sons in the last generation will be color blind?

6.3.7. A certain allele is known to be X-linked. Determine, to the extent possible, genotypes of the parents and whether the allele is dominant or recessive if the allele is expressed in the progeny by:

a. none of the females; all of the males

b. 50% of the females; 50% of the males

c. all of the females; none of the males

d. none of the females; 50% of the males

e. 25% of the progeny

6.3.8. In a breeding experiment with flies, a particular cross produces 105 females with mutant phenotype, 98 wildtype females, and 179 wildtype males. Give a possible explanation for this outcome.

6.3.9. Vermilion eye color in *Drosophila* is caused by a recessive X-linked gene. Black body color is caused by a recessive allele on an autosome. Wildtype individuals for these genes have brick red eyes and gray body color. What phenotypic ratios are expected from the crosses:

a. gray females with brick red eyes heterozygous for both genes $\times$ black males with vermilion eyes?

b. heterozygous gray females with vermilion eyes $\times$ homozygous gray males with brick red eyes?

6.3.10. Under a hypothesis of independent assortment of genes, the cross resulting in the data shown in Table 6.11 would be expected to produce a 1:1:1:1 phenotypic ratio. Apply the χ^2-test with $\alpha = .05$ to the data to test whether the data supports rejecting a hypothesis of independent assortment.

6.3.11. Suppose a diploid organism has seven pairs of chromosomes, and each chromosome has an equal number of genes on it.

a. What is the probability that two genes chosen at random lie on distinct pairs of chromosomes?

b. Would the probability that two randomly chosen genes assort independently be greater or less than this number?

6.3.12. Suppose in the three-point test cross described by Table 6.12 you attempt to compute the genetic distance between the d and ws genes by first collapsing the table to only show information about these phenotypes.

a. Create a table like Table 6.12, but with only 4 phenotypes: tall, normal sheaf; tall, white sheaf; dwarf, normal sheaf; dwarf, white sheaf. Fill in numbers by adding appropriate entries of Table 6.12.

b. Use your table to estimate the genetic distance.

c. Why does this not agree with the estimate in the text? What is incorrect about this approach?

6.3.13. Two recessive alleles, *su* for sugary kernels and *gl* for glossy leafs, are known to exist in certain corn plants. A testcross $su^+gl^+/su\,gl \times su\,gl/su\,gl$ is performed to test for linkage. The progeny are: 198 wildtype, 228 sugary/glossy, 39 sugary, and 35 glossy. Is there evidence for linkage? If so, what is the recombination frequency between the loci for *su* and *gl*?

6.3.14. In *Drosophila*, the genes with recessive alleles *sn* for singed bristles and *m* for miniature wings are located approximately 15 *cM* apart.
   a. What sort of gametes, and in proportions, can be formed by $sn^+m^+/sn\,m$?
   b. If *Drosophila* with genotype $sn^+m^+/sn\,m$ are intercrossed, what phenotypes, and in what proportion, are the progeny?

6.3.15. Suppose two genes with alleles *a* and *b* are located 10 *cM* apart. On a different autosome, two other genes with alleles *c* and *d* are located 14 *cM* apart. Suppose individuals with genotype $a^+\,b^+/a\,b$, $c^+d^+/c\,d$ are crossed with individuals, homozygous recessive for each of these genes. What phenotypes, and in what proportions, are represented in the progeny?

6.3.16. Suppose two genes with alleles, *a* and *b*, are linked. In a heterozygote, there are two possible configurations for the chromosomes. If the genotype is $a^+b^+/a\,b$, the arrangement is called a *coupling* or *cis* configuration. If the genotype is $a^+b/a\,b^+$, the layout is known as a *repulsion* or *trans* configuration. Is it possible to use a *trans* configuration in genetic mapping? Why or why not?

6.3.17. Experimental evidence indicates that crossing over seems to be less likely near the ends or the centromere of a chromosome. Suppose two genes, *a* and *b*, are located near the centromere of a chromosome, about 5 *cM* apart. Two other genes, *c* and *d*, are located about 5 *cM* apart and about 40 *cM* away from the centromere. Which physical distance, that separating *a* and *b*, or *c* and *d*, is likely to be greater? Explain.

6.3.18. For two genes on a chromosome, give an example of a tetrad crossover configuration that results in recombinant gametes only. Why can't any tetrad configuration produce one parental-type and three recombinant gametes?

6.3.19. Suppose in a certain plant species, three genes are known to be linked. The recessive alleles for these genes are *a* for amethyst flowers, *b* for brown stalks, and *c* for curved leaves. Plants are bred with genotype

$(a^+b^+c)/(a\,b\,c^+)$, where the parentheses indicate that the order of the genes is unknown. In a testcross with $(a\,b\,c)/(a\,b\,c)$, two phenotypic classes occur in much smaller numbers: wildtype and plants with amethyst flowers, brown stalks, and curved leaves. What is the correct gene order?

6.3.20. In *Drosophila*, the genes with recessive alleles cl for clot eyes, dp for dumpy wings, and rd for reduced bristles are known to be linked.

  a. Give two different examples of appropriate testcrosses to determine the order of these genes.

  b. Suppose the phenotype wildtype eyes, dumpy wings, and reduced bristles corresponds to a recombinant from a double crossover, where the heterozygous parent had genes in the $(cl^+dp^+rd^+)/$ $(cl\,dp\,rd)$ configuration. What is the correct gene order?

6.3.21. Suppose in a certain species three genes are linked, with alleles e for enlarged eyes, h for hairy legs, and p for prickly antennae. The wildtypes for these genes are normal eyes, hairless legs, and smooth antennae. Suppose $e\,h\,p$ is the correct gene order with e and h are located 12 cM apart and h and p are located 15 cM apart. In an experiment, $e^+h\,p/e\,h^+\,p^+$ individuals are testcrossed with triply homozygous recessive individuals. What are the phenotypes of the offspring and in what frequencies should these phenotypes occur?

6.3.22. For X-linked genes, you can also analyze three-point testcrosses.

  In *Drosophila*, the alleles for cut wings ct, sable body s, and vermilion eyes v all determine recessive traits that are X-linked. The wildtype traits are long wings, gray body, and red eyes. Table 6.13 gives the results of a testcross of $(ct^+s^+v^+)/(ct\,s\,v)$ females with $(ct\,s\,v)$ males. Parentheses here denote unknown gene order.

Table 6.13. *Progeny of*
$(ct^+s^+v^+)/(ct\,s\,v) \times (ct\,s\,v)$

Phenotype	No.
Long wings, gray body, red eyes	723
Long wings, gray body, vermilion eyes	8
Long wings, sable body, red eyes	71
Long wings, sable body, vermilion eyes	125
Cut wings, gray body, red eyes	105
Cut wings, gray body, vermilion eyes	106
Cut wings, sable body, red eyes	5
Cut wings, sable body, vermilion eyes	776

   a. Notice that no data are presented on the sex of the progeny, despite the fact that X-linked genes are being investigated. Explain why it is not necessary to give that information.
   b. Does the data give evidence for linkage between the three genes? Explain.
   c. Determine the order of the three loci ct, s, and v and estimate the distances between them on a linkage map.

6.3.23. The occurrence of one crossover on a chromosome can inhibit the likelihood of a second crossover occurring nearby. This phenomenon, *interference*, typically takes places at distances less than 20 cM.

   On chromosome III in *Drosophila*, the genes cu for curled wings, Sb for stubble bristles, and e for ebony body are located at 50.0 cM, 58.2 cM, and 70.7 cM, respectively. Suppose 2,000 fruit-fly progeny result from a three-point testcross.

   a. Assuming only single crossovers can occur between consecutive pairs of these genes, and that there is no interference, what is the expected number of double crossovers between cu and e?
   b. If interference occurs, then the observed number of double crossovers is less than expected. Define the *coefficient of coincidence, c,* to be the ratio *observed number of double crossovers: expected number of double crossovers*. If only three double crossovers are observed, what is the coefficient of coincidence?
   c. The level of interference is measured by $I = 1 - c$. Explain why I is a reasonable way to quantify interference. If there is no interference, what is the value of I?
   d. What is the level of interference in the testcross above?

Projects

1. Use the outcomes of simulated experiments to map genes.

   The MATLAB program genemap will perform simulated 2- and 3-point crosses for 6 autosomal genes in *Drosophila*. (It is easily modified to simulate data for mouse genes as well.) Perform a number of such crosses and construct a genetic map from your results.

   Suggestions
   • Pick a reasonable number of progeny to produce, keeping in mind the laboratory and time resources necessary for real experiments.
   • Record all the data from each of your crosses, to present it as support for your map.
   • These genes may or may not all be on the same chromosome.

- Because in a laboratory experiment, each cross could require much time and labor, try to keep the number of crosses you do relatively small while still gathering sufficient data. Also, 3-point crosses should be viewed as more work than 2-point ones, since they would require more breeding to prepare the lines.
- Once you have produced a genetic map, use it to predict the outcome of some crossing experiments you did not do previously. Then perform the experiments. Are the results consistent with your map? Explain any discrepancies.
- If you repeat your work using crosses that produce 10 times as many progeny, how does that affect your map? Of which map would you be more confident?
- Can you back up a claim that several genes are on different chromosomes with evidence? Can you back up a claim that several genes are on the same chromosomes with evidence?

6.4. Gene Frequency in Populations

So far, we have focused on one parental cross at a time in our models of genetics. As valuable as this may be for basic biological understanding, and for medical applications, it has neglected the larger picture. In evolution, the genetic make-up of species and populations may change over time. Some traits may be lost, other new ones arise, while some persist unchanged. Though chance plays a large role in the inheritance of traits in a single parental cross, understanding how this plays out in the evolution of a population requires mathematical modeling.

Suppose several alleles of a gene are present in a population. You might imagine a gene that determines eye color, or one that can affect the fertility of its carrier. Does the proportion of each allele change over time, or does it remain fixed? The answer might depend on the particular gene, of course, since an allele decreasing fertility seems more likely to disappear from a population than one that affects a more superficial trait such as eye color. Nonetheless, alleles that seem innocuous are observed to disappear from certain breeding populations.

We'll first study a type of equilibrium of genetic composition of a population and then investigate models of two forces tending to change the composition.

Let's focus on a single gene in a large population. To describe the variability of this gene among the population members, we use *allele frequencies*. Although technically these are relative frequencies, or the proportions of all alleles that are of a certain type, we will use the simpler term "frequency" throughout this section.

The MN blood typing system in humans provides a good example of how we can estimate allele frequencies. The presence of each of the alleles M and N can be detected through antigen tests. A person with genotype MM has type M blood, and a person with genotype NN has type N blood. A heterozygote MN will test positive for both alleles and so has type MN blood. (The two alleles M and N are thus codominant, as both are equally expressed in the phenotype.)

Suppose, in a certain population, that 60 individuals have type M blood, 101 individuals type MN blood, and 53 individuals type N blood, for a total population size of 214. Because each person carries two alleles of the gene, there are a total of $2(214) = 428$ alleles in this data. To determine the frequency of M alleles, we note that each person of M blood type carries 2, those of type MN carry 1, and those of type N carry 0. Thus, the frequency of the alleles is

$$M : \frac{2(60) + 1(101)}{428} \approx .52, \quad N : \frac{1(101) + 2(53)}{428} \approx .48,$$

and of course these add to give 1.

Notice the genotype frequencies in this population are

$$MM : \frac{60}{214} \approx .28, \quad MN : \frac{101}{214} \approx .47, \quad NN : \frac{53}{214} \approx .25.$$

We can use these to calculate allele frequencies also, but because each genotype involves 2 alleles, we have to divide by 2 to account for the change in the number of objects:

$$M : \frac{2(.28) + 1(.47)}{2} = .28 + \frac{1}{2}(.47) \approx .52,$$

with a similar calculation giving the frequency of N.

Random mating and Hardy-Weinberg equilibrium. Suppose now we have a large population with the allele frequencies of the M and N blood types as calculated above. As new generations are produced, do these frequencies change?

To explore what might happen in future generations, we have to make some assumptions about the mating process. The simplest model, *random mating*, is that the genotypes of offspring are determined by the random pairing of all gametes that might be produced from current organisms. This means a given gamete is equally likely to unite with any other gamete. Because our model does not track the sex of the source of the gamete, this is of course impossible for many organisms. However, assuming allele frequencies are the same in the two sexes makes the model more reasonable.

Under the random mating model, the probability of various genotypes occurring in the next generation can be calculated simply; just multiply the appropriate allele frequencies. Since picking two gametes to unite can be viewed as two independent events, the multiplication rule of probability applies. For instance, using the previously described blood type allele frequencies, the probability that an arbitrary zygote has genotype MM is $(.52)(.52) = .2704$, because .52 is the probability of picking a gamete with the M allele. Similarly, the expected frequency of the NN genotype in the offspring is $(.48)(.48) = .2304$. Since the MN genotype can be formed in two ways, MN or NM, we find, by the addition rule for disjoint probabilities, that expected frequency is $2(.52)(.48) = .4992$.

Notice that we could have used binomial probabilities in calculating the genotype frequencies instead. For instance, if we define a success as having an M allele, then $p = \mathcal{P}(S) = .52$, $q = .48$, the number of trials is $n = 2$, and the frequency of the MN genotype is

$$\mathcal{P}(\text{one success in two trials}) = \binom{2}{1}(.52)(.48).$$

▶   Compare the genotype frequencies of the new generation, .2704, .4992, and .2304, with the original. Did they change? How?

Now, let's calculate the allele frequencies in the new generation:

$$M : .2704 + \frac{1}{2}(.4992) = .52, \quad N : \frac{1}{2}(.4992) + .2304 = .48.$$

Remarkably, these allele frequencies are exactly the same as the original ones. Although the genotype frequencies changed a bit, the allele frequencies did not change in the new generation.

▶   Repeating these calculations for a third generation would produce exactly the same allele frequencies *and* genotype frequencies as in the second generation. Explain why.

Under random mating, then, we have found that the allele frequencies are in a state of equilibrium. This equilibrium is called the *Hardy-Weinberg equilibrium*, after the British mathematician Hardy and the German physician Weinberg who independently discovered it.

Let's work more theoretically with the allele frequencies to see why such an equilibrium state exists. We continue to focus our attention on a diallelic gene, with alleles a^+ and a. Let p denote the frequency of a^+ in the population and q the frequency of a, so $p + q = 1$. The assumption of random mating is what allows us to calculate the frequency of a^+ in the next generation: each allele in a second generation individual is a^+ with probability p, or a with probability q.

In the next generation, then, the allele a^+ occurs in genotypes a^+a^+ and a^+a, which have frequencies, p^2 and $2pq$, respectively. However, only half of the alleles in a^+a genotypes are wildtype. Thus, the frequency of the allele a^+ in the progeny equals

$$p^2 + \left(\frac{1}{2}\right) 2pq = p^2 + pq = p(p + q) = p(1) = p,$$

and the allele frequencies are constant from generation to generation.

▶   Whether the assumption of random mating is reasonable for humans might depend on what gene is being considered. Give examples of some traits for which you think it is reasonable and some for which it might not be.

▶   If a population is in Hardy-Weinberg equilibrium, what sorts of things not included in our model might move it away from equilibrium?

You might have noticed in the MN blood typing examples previously described that codominance allowed us to detect heterozygotes in the population and then to compute both genotype and allele frequencies. If a gene has a completely dominant allele, however, it may be difficult to distinguish between homozygous dominant and heterozygous individuals. Nonetheless, if we assume the population is in Hardy-Weinberg equilibrium, we can still estimate allele and genotype frequencies.

For example, in the United States, approximately 1 in every 3,700 individuals suffers from cystic fibrosis, the most frequent serious genetic disease of childhood, causing severe respiratory and digestive problems. Because cystic fibrosis is caused by a recessive autosomal allele, we estimate the frequency of homozygous recessives is $1/3700$. Thus, we estimate

$$q^2 \approx \frac{1}{3700}, \quad \text{so } q \approx \frac{1}{\sqrt{3700}} \approx .0164 \quad \text{and}$$
$$p = 1 - q \approx 1 - .0164 = .9836.$$

With these values, an estimate for the proportion of heterozygotes in the population is $2pq = 2(.9836)(.0164) \approx .0323$. In other words, roughly 3% of the population carries the mutant allele without showing signs of the disease.

In nature, many alleles are not in Hardy-Weinberg equilibrium. In fact, evolution occurs through changing allelic frequency; so, if all genes were in equilibrium, there could be no evolution. Indeed, many real-life circumstances lead to non-equilibrium situations: In a certain population, mating might fail to be random with particular phenotypes preferring to mate with similar phenotypes (assortative mating), or individuals might migrate into or out of a subpopulation, disrupting an equilibrium. Differences in viability or fertility may result in certain genotypes having a higher survival rate and being more likely to reproduce. Spontaneous mutations may introduce new alleles into a population, changing allele frequencies. Even the size of a population may alter allele frequencies, because random forces may influence the genetic makeup in small populations. Although a Hardy-Weinberg equilibrium is appealing mathematically, it is not a long-term feature of the natural world.

Fitness and selection. Mutation and natural selection, two potent forces of evolutionary change, bring about changes in allele and genotype frequencies. Mutations produce new alleles, and organisms with a new genotype may have a changed ability to survive and reproduce. Only the genes of organisms that successfully produce offspring appear in future population members. Genes of organisms that are less well adapted to their environment may be passed along to the next generation in smaller numbers. Thus, the gene pool may be in constant flux as mutations introduce variability that selection may then weed out.

Geneticists use the term *fitness* for a measure of the ability of an organism to survive and reproduce. Suppose, for two alleles of a gene, A and a, an individual with genotype AA is the most fit. Then, we will define its *relative fitness*, w_{AA}, to be 1, and assign fitness values w_{Aa} and w_{aa} between 0 and 1 to the other two genotypes. For example, if relative fitness values are given by

$$w_{AA} = 1, \quad w_{Aa} = .98, \quad w_{aa} = .92,$$

then in this species the most fit genotype is AA, and heterozygotes are more fit than aa homozygotes.

▶ With these fitness values, do you think the allele frequency of A will increase or decrease over time?

Of course, there are many other possible relationships between relative fitness values. If A is completely dominant over a, and fitness depends on phenotype, then $w_{AA} = w_{Aa}$. If the homozygous recessive genotype is more fit, then we have $w_{aa} = 1$ and $0 \le w_{AA} = w_{Aa} < 1$. In the exercises, some of the many other cases will be investigated.

Although relative fitness can describe selective advantage, sometimes alternate terminology is used, focusing on the selective disadvantage of a genotype. A genotype with relative fitness w is said to have *selection coefficient* $s = 1 - w$. In our previous example, the selection coefficients are 0, .02, and .08, respectively. With a selection coefficient of .08, we see that the homozygous recessive genotype is the genotype whose members will pass on the fewest genes to progeny.

We can now model how allele frequencies change because of selection. Suppose that A occurs with frequency p in the population, so a occurs with frequency $q = 1 - p$. Our model will track how p changes with time, under the assumption that mating is random.

At fertilization, gametes randomly unite to produce genotypes AA, Aa, and aa, in proportions

$$p^2, \ 2pq, \ q^2.$$

The relative fitness values then account for the competition in survival and reproduction between the genotypes as these zygotes mature and produce new gametes. Thus, the measures of the contribution of each of these genotypes to the next collection of gametes are the products

$$w_{AA}p^2, \ w_{Aa}2pq, \ w_{aa}q^2.$$

Now, because the relative fitness coefficients are less than or equal to 1, we see

$$w_{AA}p^2 + w_{Aa}2pq + w_{aa}q^2 \le p^2 + 2pq + q^2 = (p+q)^2 = 1.$$

Therefore, we must renormalize (i.e., divide through by the quantity $w_{AA}p^2 + w_{Aa}2pq + w_{aa}q^2$) to calculate the successful contribution of gametes to the genotype proportions of the next generation, obtaining

$$\frac{w_{AA}p^2}{w_{AA}p^2 + w_{Aa}2pq + w_{aa}q^2}, \ \frac{w_{Aa}2pq}{w_{AA}p^2 + w_{Aa}2pq + w_{aa}q^2},$$

$$\frac{w_{aa}q^2}{w_{AA}p^2 + w_{Aa}2pq + w_{aa}q^2}.$$

Finally, because all the alleles contributed by the AA genotype are A, but only half the alleles contributed by the Aa genotype are, we find

$$p_{t+1} = \frac{w_{AA}p_t^2}{w_{AA}p_t^2 + w_{Aa}2p_tq_t + w_{aa}q_t^2} + \frac{1}{2}\frac{w_{Aa}2p_tq_t}{w_{AA}p_t^2 + w_{Aa}2p_tq_t + w_{aa}q_t^2}$$

$$= \frac{w_{AA}p_t^2 + w_{Aa}p_tq_t}{w_{AA}p_t^2 + w_{Aa}2p_tq_t + w_{aa}q_t^2}$$

▶   Express this in terms of p_t alone, with no q_t.

Let's consider a concrete example. Suppose, initially, 70% of the alleles are A. Thus, $p_0 = .7$ and $q_0 = .3$. If all genotypes are equally fit, then $w_{AA} = w_{Aa} = w_{aa} = 1$, and we find

$$p_1 = \frac{p_0^2 + p_0q_0}{p_0^2 + 2p_0q_0 + q_0^2} = \frac{.49 + .21}{1} = .7,$$

which illustrates the Hardy-Weinberg equilibrium. If, however, relative fitness values $w_{AA} = 1$, $w_{Aa} = .98$, and $w_{aa} = .92$ describe the genotypes, then

$$p_1 = \frac{p_0^2 + (.98)p_0q_0}{p_0^2 + (.98)2p_0q_0 + (.92)q_0^2} = \frac{.49 + (.98).21}{.9844} = .7068.$$

As you might expect, the allele frequency of A has increased slightly, from .7 to .7068, at the expense of the allele a.

Iterating the model over a few generations produces Figure 6.8. Since the genotypes are increasingly fit according to the presence of the allele A, over many generations A becomes fixed in the population and the recessive allele dies out.

This model becomes even more interesting for parameter choices where the outcome is less intuitive. What might happen if a recessive allele was the most fit? Would it be fixed eventually, or would the fact that it was only expressed in homozygotes give it too weak an influence to eventually predominate? Or, what if the heterozygotes were the most fit genotype? The outcome of such a situation is hard to predict without a mathematical model. These questions are not simply a result of mathematical curiosity, as a few biological examples show:

• In a certain species of moths, a dominant allele is associated with dark coloring. Homozygous recessives are light-colored. If a moth population lives in a forest with dark-colored trees, the light-colored moths are at a competitive disadvantage, as their predators can more easily see them. If

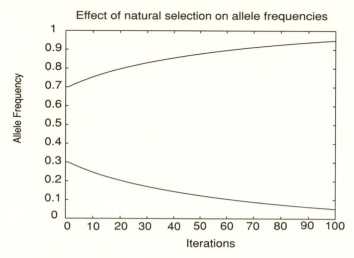

Figure 6.8. Allele frequencies of A (top) and a (bottom); relative fitness values $w_{AA} = 1$, $w_{Aa} = .98$, and $w_{aa} = .92$.

the tree bark tends to be lighter-colored, then light-colored moths are more likely to survive.

• In humans, the often-fatal disease sickle-cell anemia is associated with a homozygous recessive genotype. In certain parts of the world, the recessive allele is quite common – by some estimates about as high as 19%. Researchers have discovered that heterozygotes have an increased resistance to malaria, and thus a greater fitness in a tropical climate.

In the exercises, we will explore a number of scenarios for the effects of natural selection:

Selection for A: favors the dominant allele and associated phenotypes.
Selection against A: favors homozygous recessives.
Heterozygote Advantage or Overdominance: favors heterozygotes at the expense of homozygotes.
Homozygote Advantage: favors homozygotes, at the expense of heterozygotes.

The frequency of an allele may rise or fall, depending on the forces of selection.

Genetic drift. So far, our models addressing allele frequencies have tacitly assumed that the population under study was large. For instance, we assumed we were modeling a large population when we argued that because a certain

Table 6.14. *Probabilities That Exactly k of 4 Alleles Are A*

k	0	1	2	3	4
$\mathcal{P}(k)$	.0625	.25	.375	.25	.0625

proportion of the gametes had an allele, then the same proportion of the gametes that successfully united would have that allele. Even if half the gametes have an allele A, if we randomly pick gametes to unite, we might pick more or less than half As to form the next generation. In a small population, any deviation from half might be proportionally large, and thus proportionally greater than you are likely to have in a large population. In other words, small populations are more greatly affected by chance than are large ones.

For a concrete illustration of this, imagine a very small population of 2 individuals of genotypes Aa and Aa. Then, the alleles A and a appear in the gamete pool in proportions .5 and .5, and so random mating implies that each offspring will have genotype AA (or aa) with probability .25, and genotype Aa with probability .5.

However, if the new generation also has size 2, then to determine the alleles in this generation, we simply pick four specific gametes out of the pool. Using the binomial distribution, the probability of having exactly two of each allele in the next generation is

$$\binom{4}{2}(.5)^2(.5)^2 \approx .375.$$

This means that the probability that the allele frequencies remain stable is only 37.5%, and the more likely scenario is that allele frequencies will change. Furthermore, any change in the allele frequency must be at least .25, because there are only four alleles total in this small population. Thus, a reasonably large change is quite likely.

It might seem that this result contradicts the ideas underlying the Hardy-Weinberg equilibrium for allele frequencies. However, calculating the probabilities that exactly k of 4 alleles are A for $k = 0, 1, 2, 3$, and 4 as in Table 6.14, we see the most likely outcome is that the allele frequencies represented in the two offspring will be $p = q = .5$, the same frequencies of the parental generation and just as Hardy-Weinberg predicts. However, this most likely outcome is not very likely.

If a population is large – say 3,000 heterozygotes producing 3,000 offspring – then producing a table like Table 6.14 also shows that some change in allele numbers is likely. However, the likely size of this change is much

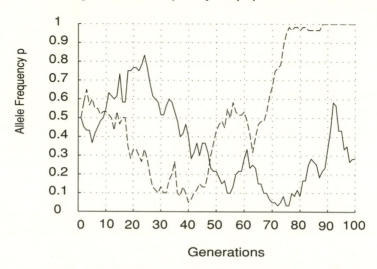

Figure 6.9. Two examples of genetic drift; population size $N = 30$.

smaller proportionally than for the two individual case. Rather than changes in allele frequencies of magnitude .25, tiny changes typically occur. Thus, the Hardy-Weinberg values are a more accurate estimate of what actually happens.

For large populations, we lose little by ignoring chance fluctuations. If a population is small, then chance fluctuations are much more important and may in fact predominate. The phenomenon of chance changes in allele frequencies dominating other factors in small populations is known as *genetic drift.*

Genetic drift may be modeled by fixing a population size N and initial allele frequencies. Then, a new generation of alleles is chosen according to the probabilities calculated by the binomial distribution. Using the new allele frequencies, this process is repeated for the next generation, and so on. Because of the random choices made at each generation, no two simulations are likely to be identical.

Figure 6.9 shows two simulations of allele frequency p over a number of generations. In both plots, the population is small, $N = 30$, and the initial value is $p = .5$. Notice the random fluctuation of the frequency p, and that whether the allele remains fixed in the population or is removed entirely is a matter of chance.

Using only concepts introduced here, it is easy to imagine a more sophisticated model that combines genetic drift with selection. But models of genes with more alleles, or of several genes that collectively determine traits affecting fitness, are also possible. Modeling the creation of new alleles through mutation, along with their possible elimination or fixation through selection, also leads to interesting insights. We have really only scratched the surface of mathematical models in population genetics.

Problems

6.4.1. An autosomal recessive allele *ct* causes curly tails in mice. Suppose, in a certain population of 450 mice, 441 mice have normal tails and 9 have curly tails, and that the allele frequencies are in Hardy-Weinberg equilibrium.

    a. Estimate the allele frequency of *ct*.

    b. What percentage of the mice population is heterozygous for this gene?

6.4.2. Color blindness is an *X*-linked trait that occurs in about 8% of human males.

    a. Give the allele frequencies for this gene. (Assume the frequencies p and q are the same in both genders, and are in equilibrium.)

    b. Approximately what percentage of the female population is color blind? What percentage of the female population with normal vision carries the mutant allele?

6.4.3. Suppose a randomly mating population segregating two alleles is in Hardy-Weinberg equilibrium.

    a. What are the allele frequencies p and q if the frequency of heterozygotes is .4? If the frequency of heterozygotes is H?

    b. Express the frequency of heterozygotes in terms of p. What values of p and q maximize this frequency? (Either graphing or calculus can be used to answer this.)

6.4.4. There is a strong connection between certain powers of polynomials and genotype frequencies in simple situations.

    a. Expand the binomial power $(p + q)^2$ and explain the meaning of each summand in terms of genotype frequencies for a diallelic gene.

    b. If a gene has multiple alleles, *multinomial expansions* are related to genotype frequencies. Suppose a gene has 3 alleles, occurring in frequencies p, q, and r. Expand $(p + q + r)^2$ and relate each term in the expansion to genotype frequencies.

    c. Does the concept of a Hardy-Weinberg equilibrium make sense for the 3 allele situation? Explain.

6.4.5. The genetics of the ABO blood typing system was explained in Problem 6.1.16.

   a. In ABO blood-typing studies in an isolated community, 32% of the population have type A blood, 15% type B blood, 4% type AB blood, and 49% type O blood. Determine the allele frequencies I^A, I^B, and I^O in this community.

   b. In the United States, approximately 40% of the population have type A blood, 11% type B blood, 5% type AB blood, and 44% type O blood. Give the system of equations that describes the blood-type frequencies in terms of the allele frequencies I^A, I^B, and I^O. Can you solve this system? If not, explain the difficulty and its biological implications.

6.4.6. Suppose a gene has 3 alleles in equilibrium in a randomly mating population. To find allele frequencies for the population, what is the minimum number of phenotype frequencies you must know? Answer the same question for n alleles.

6.4.7. Although a Hardy-Weinberg equilibrium may exist in a well-mixed population, over expansive geographic areas, natural barriers often cause variations in local equilibrium frequencies.

   Suppose two lakes separated by a short distance are populated with the same species of fish and that both lakes are in an equilibrium state. In the first lake, the frequency of a particular allele a^+ is p_1. In the second lake, the frequency of a^+ is p_2. After a flood, the two lakes are merged, and one lake is formed. Suppose both lakes contained the same number N of fish.

   a. What is the frequency p of the allele a^+ in the fish in the large lake after the flood?

   b. What are the genotype frequencies immediately after the flood? What would a Hardy-Weinberg equilibrium predict for the genotype frequencies? Explain why these two answers do not agree.

6.4.8. Show the selection model simplifies considerably if $w_{AA} = w_{Aa} = w_{aa} = 1$. Using these relative fitness values, give the simplest formula possible for p_{t+1} in terms of p_t. Explain the relationship of your formula to Hardy-Weinberg equilibrium.

6.4.9. Investigate the behavior of the selection model experimentally, using a computer program such as `onepop`, for each set of relative fitness values below. Describe your observations on the model's behavior, including likely equilibria and their stability. Are the behaviors you see biologically reasonable?

a. $w_{AA} = 1$, $w_{Aa} = .98$, and $w_{aa} = .92$ (dominant advantage)
b. $w_{AA} = .92$, $w_{Aa} = .98$, and $w_{aa} = 1$ (recessive advantage)
c. $w_{AA} = 1$, $w_{Aa} = .92$, and $w_{aa} = 1$ (homozygous advantage)
d. $w_{AA} = .92$, $w_{Aa} = 1$, and $w_{aa} = .92$ (heterozygous advantage).

6.4.10. In mice, homozygotes for the yellow-lethal allele, Y^l, die in embryonic stage, while heterozygotes have yellow fur. What are reasonable values to use in the selection model for the selection coefficients for the three genotypes? Use a computer program such as onepop to investigate the model, and describe your results. Does the allele persist in the population?

6.4.11. Relative fitness values $w_{AA} = 0$, $w_{Aa} = w_{aa} = 1$ describe a special case of the selection model.
a. Interpret these biologically.
b. Show that with these values the model is simply

$$p_{t+1} = \frac{p_t}{1 + p_t}.$$

c. Show that the explicit formula

$$p_t = \frac{p_0}{1 + t p_0}, \quad t = 1, 2, 3, \ldots$$

gives allele frequencies for this model.

6.4.12. Relative fitness values $w_{AA} = w_{Aa} = 1$, $w_{aa} = 0$ describe a special case of the selection model.
a. Interpret these biologically.
b. Give the simplest formula you can expressing p_{t+1} in terms of p_t.
c. Find an explicit formula for p_t in terms of p_0 and t.

6.4.13. Find all equilibria for the selection model as follows:
a. Express the equilibrium equation that p^* must satisfy in the form of a cubic polynomial $= 0$. This shows there are at most three equilibria.
b. Two equilibria are easy to guess. (What possible allele frequencies would not change, no matter what the relative fitness values were?) What are they?
c. Use your guesses in part (b) to help you factor the cubic polynomial in part (a) completely.
d. Use part (c) to show the third equilibrium can be written as

$$\frac{(w_{aa} - w_{Aa})}{(w_{aa} - w_{Aa}) + (w_{AA} - w_{Aa})}.$$

6.4.14. The third equilibrium for the selection model that was found in the preceeding problem is only biologically meaningful if it is a possible value for an allele frequency.

    a. Explain why the third equilibrium is only biologically meaningful if

$$(w_{aa} - w_{Aa})(w_{AA} - w_{Aa}) > 0.$$

    b. Explain why the third equilibrium is only biologically meaningful if either $w_{AA} > w_{Aa}$ and $w_{aa} > w_{Aa}$ (homozygote advantage), or if $w_{AA} < w_{Aa}$ and $w_{aa} < w_{Aa}$ (heterozygote advantage).

6.4.15. Use a program such as cobweb to investigate the stability of the selection model equilibria under the following conditions. Use a variety of parameter choices for each. Express your conclusions in biological terminology.

    a. $w_{AA} > w_{Aa}$ and $w_{aa} > w_{Aa}$ (homozygote advantage)

    b. $w_{AA} < w_{Aa}$ and $w_{aa} < w_{Aa}$ (heterozygote advantage).

6.4.16. In the selection model, the quantity

$$\overline{w}_t = w_{AA}p_t^2 + w_{Aa}2p_t q_t + w_{aa}q_t^2$$

is called the *mean fitness* of the population at time t. It is possible to show that $\overline{w}_{t+1} \geq \overline{w}_t$. Why is such a result reasonable biologically?

6.4.17. Use a computer program, such as genesim to explore the phenomenon of genetic drift. For a population of size $N = 30$, begin with equal allele frequencies and do several simulations. Repeat for $N = 300$ and $N = 3000$. Describe your observations on how population size affects drift.

6.4.18. The program genesim can model genetic drift with selection effects due to varying relative fitness levels of genotypes. For a population size that exhibits strong drift when all genotypes have the same fitness, run simulations with interesting choices of relative fitness values. Describe your observations and discuss whether they seem biologically reasonable.

6.4.19. What is the expected value of the number of A alleles in the situation described by Table 6.14? How does this fit with the idea of Hardy-Weinberg equilibrium?

6.4.20. In a population of size N, if genetic drift causes changes in allele frequencies p and q, then genotype frequencies change, too. One way to measure the effect of genetic drift is by monitoring the frequency *H* of heterozygotes, the *heterozygosity*, of a population.

a. If genetic drift tends to eliminate an allele, what will the effect be on the value of *H* over time? Explain.

b. A good model (which we will not justify here) to describe the effect of genetic drift on the heterozygosity of a population is $H_{t+1} = (1 - \frac{1}{2N})H_t$. Use the program onepop to explore the effect of population size on genetic drift and heterozygosity. Start with an initial value of $H_0 = .5$ and vary the population size *N*. What happens to *H* if $N = 100$? If $N = 1,000$? If *N* is huge? How would your answers change if the initial value was $H_0 = .2$ or $H_0 = .9$?

c. Give a formula for H_t in terms of *N*, H_0, and *t*.

Projects

1. Investigate the phenomenon of genetic drift in a simulated population.

    Study a gene with two alleles, *A* and *a*, that occur in a diploid population of size *N* in frequencies *p* and *q*. Assume that these alleles are *selectively neutral* (i.e., the resulting genotypes are all equally fit).

    Use the MATLAB program genesim to observe changes in allele frequencies in a simulated population over a number of generations. This program assumes that the population size *N* remains constant from generation to generation and that mating is random.

    Explore the effect of genetic drift on allele frequencies under a variety of assumptions.

    • The population size *N* is small, medium, or large.
    • The initial allele frequency of *A* is $p_0 = .5$, $p_0 > .5$, or $p_0 < .5$.

    The main issues to consider are:

    1. What happens to the allele frequency *p* over the long run? Is it stable? Does the allele *A* become fixed in the population? Is *A* eliminated entirely? If either of these happens, how quickly does it occur?
    2. How does the population size affect your answer to question 1 above?

    Suggestions
    • To get a feel for the effects of genetic drift, use the program genesim to explore changes in allele frequencies for lots of reasonable choices of *N* and p_0. Make a note of any unusual behavior and try to explain it.

- After a large number k of generations, how does the allele frequency p_k compare with p_0? Is there a tendency for fixation or elimination of an allele? Explore this question for different population sizes.
- If $p_0 = .5$, how likely is it that A becomes fixed in the population? For fixed N, do many simulations, record the results, and from them estimate the probability of fixation. Repeat for other N.
- Investigate the last question for specific $p_0 > .5$ and $p_0 < .5$.
- Does genetic drift tend to increase or decrease genetic variation within a population? How does the population size affect your answer?
- If one population is separated into two populations by migration (early humans leaving Africa; farm-raised fish being released into two lakes), what effect might genetic drift have on the variability between the two populations?
- If you perform many genetic drift simulations for a fixed value of p_0 and N and average the values p_{50}, what will you get? Does your answer depend on the initial value p_0? N?

2. For a gene with two alleles, A and a, both the simple selection model and genetic drift often lead to fixation of one of the alleles and elimination of the other. Why, then, do we observe so many genes with multiple alleles in real populations? Are all of them either selectively neutral, or in populations so large that drift is negligible?

    Explore and discuss one or more of the following models that offer further reasons for the stability of *polymorphic* genes.

    - *Heterozygote advantage*: In this selection model, $w_{Aa} > w_{AA}$ and $w_{Aa} > w_{aa}$. (This is the mechanism by which the persistence of the sickle cell allele is generally explained.)
    - *Frequency-dependent selection*: In this type of selection model, the fitness coefficients depend on allele frequencies. One example is

    $$w_{AA} = 1 - up^2, \ w_{Aw} = 1 - u2pq, \ w_{aa} = 1 - uq^2,$$

    for some value of u between 0 and 1. In this model, the more prevalent an allele is, the less its fitness. (In certain plants, pollen with one allele can only successfully fertilize plants with other alleles, giving rare alleles an advantage.)
    - *Mutation-selection balance*: This model modifies the classical selection model to account for recurrent mutations that continually renew the stock of an allele that might otherwise disappear. For instance, if a fraction μ of alleles that would have been A in each new generation

mutate to a, and p_t tracks the frequency of A, such a model is

$$p_{t+1} = \frac{w_{AA}p_t^2 + w_{Aa}p_tq_t}{w_{AA}p_t^2 + w_{Aa}2p_tq_t + w_{aa}q_t^2}(1 - \mu).$$

Suggestions

- Investigate these models experimentally using onepop and cobweb for a variety of parameter choices. Describe your observations and insights.

- If possible, compute equilibria for the models and discuss their stability. (If you cannot do this in general, at least do it for a few parameter choices, or by making special choices, such as $w_{AA} = w_{Aa} = 1$, $w_{aa} = 1 - s$ in the selection-mutation model.)

- *Meiotic drive*, the preferential creation of gametes of a certain type, is another mechanism that can lead to polymorphic stability. Modify the basic selection model to take meiotic drive into account and analyze your model.

7

Infectious Disease Modeling

Throughout history, devastating epidemics of infectious diseases have wiped out large percentages of the human population. In the mid-fourteenth century, the Black Death, a plague epidemic, killed roughly one-third of Europe's population. More recently, in 1918, an outbreak of the flu killed an estimated 20 million people, more people than died in all of World War I. In our own times, the acquired immune deficiency syndrome (AIDS) pandemic has brought untold personal suffering and social losses. The Centers for Disease Control (CDC) estimates that, from 1981 to 2001, approximately 21 million people died from AIDS worldwide. Millions of people all over the world are currently infected with the human immunodeficiency syndrome (HIV) virus, about 95% of them in developing countries.

Although medical advances have reduced the consequences of some infectious diseases, preventing infections in the first place is preferable to treating them. The development of vaccines gives us not only a means of protecting ourselves as individuals, but also, and perhaps more importantly from a public health view, a means of preventing sudden and widespread outbreaks.

Once a vaccine is developed, how should it be used? Should everyone in a society be required to be immunized for certain illnesses, regardless of their personal desires? Is the cost of an immunization program worthwhile if a vaccine is expensive or difficult to administer? If only those facing the highest risk of a disease are immunized, will that be sufficient to prevent epidemics? If a vaccination carries health risks to those who receive it, when are these risks worthwhile? What groups, either in terms of age or social interactions, should be targeted by vaccination programs? Questions of this sort cannot be answered simply. Individual features of diseases and societies must be taken into account. However, understanding the dynamics of disease transmission is essential to addressing them, and mathematical modeling can play a role here.

Once a model has been formulated that captures the main features of the progression and transmission of a particular disease in a population, it can be used to predict the effects of different strategies for disease eradication or control. Although political, social, and economic factors play a large role in setting public health policies, understanding the dynamics of contagion is an important step. The worldwide eradication of smallpox, through a carefully developed vaccination campaign initiated by the World Health Organization in 1967, is a remarkable example of what can be achieved with a well-designed plan. Infectious disease modeling, though often inexact, has enormous potential to help improve human lives.

Earlier in this book, we have discussed the simplest mathematical models used by biologists to understand interacting populations. Now we further develop those ideas in the particular context of infectious disease. We will focus on giving meaningful interpretations of the key parameters that appear in epidemic models. Once we have a basic framework for describing disease transmission, we'll see how various vaccination levels can prevent, or fail to prevent, an epidemic. Finally, we consider a model of a sexually transmitted disease to show how simple modeling approaches can capture the particulars of a disease with quite complicated dynamical features.

7.1. Elementary Epidemic Models

In modeling the dynamics of an infectious disease, we focus on the population in which it occurs. The models we consider assume that N, the total size of the population, is constant. We thus ignore complications that might result from new births or immigration. Although more complicated models can account for these factors as well, our assumption is often quite reasonable. For instance, if we are modeling the spread of brucellosis in a herd of cattle, during the timeframe of interest the population of cattle is unlikely to gain new members.

▶   Explain why ignoring births and immigration is a reasonable simplification for modeling the spread of chickenpox at an elementary school.
▶   What are the characteristics of a disease and population for which this would not be an appropriate assumption? Can you give an example?

We will also assume that the population under study mixes *homogeneously*; all members of the population interact with one another to the same degree. This means all uninfected individuals face the same risk of exposure to the disease by those already infected. Again, this may be quite reasonable: In a cattle herd, we would expect that all members interact roughly equally with one another.

▶ Would the homogeneous mixing assumption be reasonable in modeling chickenpox spread in a first grade class? Why or why not?

▶ How reasonable is this assumption if the population under study is all students in a particular school? All students in a particular city? All students in a country?

To begin formulating our model, at each time t, we divide the population N into three categories;

S_t: the *susceptible* class, those who may catch the disease but currently are not infected;

I_t: the *infective* class, those who are infected with the disease and are currently contagious; and

R_t: the *removed* class, those who cannot get the disease, because they either have recovered permanently, are naturally immune, or have died.

Note that we continue to count any dead individuals in R_t since it is mathematically more convenient to keep the total number of individuals constant. Moreover, because the population is constant,

$$S_t + I_t + R_t = N \qquad \text{for all } t.$$

▶ Imagine how the sizes of each of the three populations under study, S_t, I_t, R_t, must change as an outbreak of influenza at a college occurs and then subsides. If values of the three were plotted over time, how would you expect the graphs to look?

Tracking the size of the infective class I_t probably gives the clearest indication of the course of a typical disease outbreak. For an epidemic to occur, I_t must increase. A large increase in a single time step represents a rapidly spreading outbreak, while a smaller increase signals a more gradual spread. Thus, the magnitude of change in the number of infectives, ΔI, measures the virulence of the epidemic. We would expect a graph of I_t to rise, as more and more of the population becomes infected. In time, though, if it is possible for individuals to recover, the infective class I_t starts to decrease in size (i.e., $\Delta I \leq 0$). At that point, the graph of I_t turns down and the epidemic begins to subside.

Notice, however, that we have already assumed that the disease we are discussing is one from which infectives recover. Because that is not true of all infectious diseases, we'll have to pay careful attention to such assumptions when we try to model a particular disease.

The *SIR* model. The simplest epidemic model uses the three classes above. In this *SIR* model, members of the population progress through the three classes in order: susceptibles remain disease-free or become infected; infectives pass through an infectious period until they are removed permanently from the grips of the disease; and a removed individual is never at risk again. Schematically, we think of the model as:

$$\boxed{Susceptibles} \rightarrow \boxed{Infectives} \rightarrow \boxed{Removed}.$$

An outbreak of chickenpox at an elementary school can be described well by an *SIR* model. Students who have been infected will recover and will never get chickenpox again, since permanent immunity results from having been infected. Other examples of human diseases that fit the *SIR* framework, at least approximately, include the seasonal variants of influenza virus that develop each year. Once an individual has recovered from an infection, they are immune to that particular variant for life.

▶   Describe some scenarios of outbreaks of infectious diseases that can be modeled with the basic *SIR* model. Why might the model be appropriate for modeling the spread of measles, but not the spread of head lice?

Disease spreads when a susceptible individual comes in contact with an infected individual and subsequently becomes infected. Mathematically, a reasonable measure of the number of encounters between susceptible individuals and infected individuals, assuming homogeneous mixing, is given by the product $S_t I_t$. This is simply the mass action principle used for interacting populations in Chapter 3.

However, not all contacts between healthy and ill individuals result in infection. We'll use α, the *transmission coefficient*, as a measure of the likelihood that a contact between a susceptible and an infective will result in a new infection. Because the number of susceptibles S_t decreases as susceptibles become ill, the difference equation modeling the number of susceptibles is given by

$$S_{t+1} = S_t - \alpha S_t I_t.$$

▶   If $\alpha = .01$ for one disease and $\alpha = .02$ for another disease, what does this indicate about the difference between the diseases? If this is the only difference between the diseases, which will spread faster?

During one time step, the infective class grows by the addition of the newly infected. At the same time, some infectives recover or die, and so progress to the removed stage of the disease. The *removal rate* γ measures the fraction of the infective class that ceases to be infective, and thus moves into the removed class, in one time step. Clearly, the removed class increases in size by exactly the same amount that the infected class decreases. This leads to the additional equations:

$$I_{t+1} = I_t + \alpha S_t I_t - \gamma I_t,$$

$$R_{t+1} = R_t + \gamma I_t.$$

Collectively, the three above coupled difference equations form the SIR model.

Although the SIR model will serve as our basic infectious disease model, it is not appropriate for many diseases. The dynamics of tuberculosis (TB) illustrate its limitations. Many people who show a positive reaction to the TB skin test have a TB infection, but not the disease. A healthy immune system is able to fight the TB bacteria and keep it inactive. Such a person is not contagious and may never develop the disease. However, if the immune system is weakened and the infection is not medicated, a person infected may develop tuberculosis and become infectious. Proper modeling of TB will require at least one more class: those who are infected but not infective.

Almost every disease has unique features that must be incorporated into a model. The art of creating a good model is deciding which of these are important to capturing the right dynamics and which can be omitted to prevent the model from becoming too complicated to analyze. Remember that our goal in modeling disease transmission is to understand how to control it. The SIR model is a good starting model that can be refined as needed for particular diseases.

We should also remark briefly on the choice of time steps in modeling diseases. We have commented earlier that difference equations are particularly good for modeling populations with rigid life cycles. However, diseases often fail to have rigid stages of development. For that reason, and sometimes for reasons of mathematical tractability, differential equations are often used in such models. However, the modeling principles are very similar in either the difference or differential formulations. Moreover, both types of models are very schematic descriptions of the real spread of a disease, so that we have to hope our results are fairly robust to the numerous inaccuracies they involve, regardless of which approach we take.

With that said, then, how do we choose a size for our time steps? We must certainly take the particulars of the disease under study into consideration. If a person is typically sick and contagious for a week, then a time step of one day might be appropriate. It's usually better to use a shorter time step when in doubt.

Problems

7.1.1. Use the MATLAB program `sir` to investigate the behavior of the basic SIR model for a variety of parameter choices. Begin, for example, with $N = 100$, $\alpha = .001$, and $\gamma = .05$.

   a. Use a variety of choices of S_0 and I_0. How does this choice affect the behavior? Are there some choices for which the number of infectives grows in the first few time steps? Declines in the first few time steps?

   b. Holding the other parameters fixed, use larger and then smaller values for the transmission coefficient. Explain the effects on the qualitative behavior of the model.

   c. Holding the other parameters fixed, use larger and then smaller values for the removal rate. Explain the effects on the qualitative behavior of the model.

7.1.2. In using the `sir` program, initial numbers of susceptibles and infectives should always be specified by clicking with the cursor located underneath the diagonal line drawn on the phase plane. Explain what assumptions of our model require this.

7.1.3. The parasitic disease malaria is transmitted to humans through the bite of an infected mosquito. A mosquito becomes a carrier by biting an infected person. After the parasites have grown in the mosquito for about a week, the mosquito can pass the disease to another human through its bite. Is it appropriate to use an SIR model to model malaria transmission? Explain.

7.1.4. The removal rate γ can be estimated from knowing how long individuals are typically in the infective class.

   a. Suppose you know that the mean time an individual is infective is 8 days. Assuming there are a large number of infectives at various stages of the progression of the illness, what percentage of the infective class would you expect to recover each day?

   b. More generally, if m is the mean time of infectivity, measured in time steps, give a formula for γ in terms of m.

7.1.5. Obviously, the parameter α of the SIR model must be positive. However, there is also an upper bound on its size in a biologically meaningful model.

To understand this, consider a small group of 100 people in close contact, all susceptible to the disease.

a. At time 0, one individual falls ill, perhaps by exposure through contacts not included in the model. Suppose the transmission coefficient is $\alpha = 1$. What happens to the town's population after 1 day? After 10 days? What is the medical significance of $\alpha = 1$?

b. Now suppose that $\alpha = 1$ and the initial number of infectives is 5. What is the value of S_1? Why doesn't this make sense? Explain.

c. With $\alpha = .1$, what is the largest value of I_0 so that the behavior of the *SIR* model makes sense biologically, at least for the first time step?

d. Give a formula in terms of I_0 for the largest value of α so that the behavior of the *SIR* model makes sense biologically, at least for the first time step.

7.1.6. One approach to preventing disease spread is to simply quarantine infectives. Suppose a disease is modeled well by the *SIR* equations of the text, but a society decides to attempt a quarantine program, preventing a fraction q of the infectives from having contacts with susceptibles. Only $1 - q$ of the infectives will be able to spread the disease.

a. How should the mass action term, in the equation for both S_{t+1} and I_{t+1}, be changed to model this? What value of q gives the usual *SIR* model?

b. Quarantining can be viewed as a way of modifying the transmission coefficient. If an *SIR* model without quarantining had transmission coefficient α, and a fraction q of the infectives are successfully quarantined, then the model with quarantining is identical to a standard *SIR* model with some transmission coefficient α', the *effective transmission coefficient*. Give a formula for α' in terms of α and q.

c. Use the MATLAB program `sir` to investigate the behavior of your quarantine model for fixed values of N, α, and γ, and a variety of values of q. For example, let $N = 100$, $\alpha = .001$, and $\gamma = .05$ and vary q from 0 to 1. Explain the qualitative behavior you see. Can you find a value of q that prevents an epidemic from occurring, regardless of the value of I_0? Estimate the smallest such q.

7.1.7. Another approach to preventing disease spread is vaccination of susceptibles. Suppose a disease is modeled well by the *SIR* equations of the text, but a society implements a vaccination program. One simple model of this situation counts each successfully vaccinated individual in the removed class throughout the duration of the model, but otherwise still uses the basic SIR equations.

   a. Explain why this model assumes all vaccinations occur before the time $t = 0$.

   b. Suppose with $N = 100$, we have $I_0 = 1$, with the removed class composed of the fraction q of the population that was successfully vaccinated. Give formulas for S_0 and R_0. What value of q gives the usual SIR model?

   c. Use the MATLAB program `sir` to investigate the behavior of your vaccination model for a variety of values of q. Let, for example, $N = 100$, $\alpha = .001$, and $\gamma = .05$, and only vary q from 0 to 1. Explain the qualitative behavior you see. Can you find a value of q that prevents an epidemic from occurring, regardless of the value of I_0? Estimate the smallest such q.

7.2. Threshold Values and Critical Parameters

To analyze the SIR model and gain some biological insight into the parameters in the model, we'll rewrite the defining equations:

$$\Delta S = -\alpha S_t I_t,$$
$$\Delta I = \alpha S_t I_t - \gamma I_t,$$
$$\Delta R = \gamma I_t.$$

We will say an epidemic occurs if $\Delta I > 0$ for some time t (i.e., if at some time the number of infectives grows). If $\Delta I \leq 0$ for all times, then the size of the infective class does not increase and no wider outbreak of illness takes place. The first step in understanding disease dynamics, then, is to understand the sign of ΔI. Thus, we focus our attention on determining whether

$$\Delta I = \alpha S_t I_t - \gamma I_t$$
$$= (\alpha S_t - \gamma) I_t$$

is positive, zero, or negative.

    First notice from this formula that if $I_t = 0$, then $\Delta I = 0$. This is no surprise, since if the population is disease free (i.e., has no infectives), it will remain that way. Having dispensed with this easy-to-understand case, we can now assume that $I_t > 0$. This means that ΔI will be positive, zero, or negative according to whether $\alpha S_t - \gamma$ is. Because $\alpha > 0$, we can rephrase this as:

$$\text{If } S_t > \frac{\gamma}{\alpha}, \quad \text{then} \quad \Delta I > 0.$$

$$\text{If } S_t = \frac{\gamma}{\alpha}, \quad \text{then} \quad \Delta I = 0. \tag{7.1}$$

$$\text{If } S_t < \frac{\gamma}{\alpha}, \quad \text{then} \quad \Delta I < 0.$$

Notice that, from our original formulas, we have $\Delta S \leq 0$ always, so we know that S_t cannot increase. This means that, if $S_0 < \frac{\gamma}{\alpha}$, then $S_t < \frac{\gamma}{\alpha}$ for all t. Thus, if S_0 is below the value $\frac{\gamma}{\alpha}$, then $\Delta I < 0$ for all times, and the disease decreases in the population. However, when $S_0 > \frac{\gamma}{\alpha}$, the number of infectives will grow and an epidemic occurs.

For this reason, the ratio $\frac{\gamma}{\alpha}$ is an example of a *threshold* value; the relationship of S_0 to $\frac{\gamma}{\alpha}$ is an important determinant of the dynamics of the disease. Because $\frac{\gamma}{\alpha}$ represents the removal rate γ relative to the transmission coefficient α, we call it the *relative removal rate* and denote it by

$$\rho = \frac{\gamma}{\alpha}.$$

Comparing the initial number of susceptibles S_0 to the threshold value ρ, we can determine if an epidemic will occur.

▶ A larger value of γ results in a larger value of the threshold ρ. Does this make sense? Explain, in terms of the meaning of γ. What affect does a larger value of α have on ρ? Explain.

A slightly different approach to the same threshold behavior involves rewriting the equation for ΔI as:

$$\Delta I = \gamma \left(\frac{\alpha}{\gamma} S_t - 1 \right) I_t.$$

A similar sign analysis of ΔI, using the above expression, shows the important question is how the quantity $\frac{\alpha}{\gamma} S_0$ compares with 1. Mathematical epidemiologists call the expression

$$\mathcal{R}_0 = \frac{\alpha}{\gamma} S_0$$

the *basic reproduction number* of the infection. Sometime you may see this called the basic reproductive rate or basic reproductive ratio, though, so you need to be careful about terminology when reading epidemiological studies. We'll use the term "basic reproduction number" exclusively. Most importantly, if $\mathcal{R}_0 > 1$, then $\Delta I > 0$ and an epidemic occurs.

Let's consider the basic reproductive number $\mathcal{R}_0 = \frac{\alpha}{\gamma} S_0 = (\alpha S_0)(\frac{1}{\gamma})$ from a more biological viewpoint, in order to understand both its name and its conceptual importance. In the SIR model, the term $\alpha S_0 I_0$ measures the number of individuals that become infected at the outset of an epidemic. If we divide by I_0, we obtain a "per-infective" measurement: αS_0 is the number of individuals who become infected by contact with a single ill individual during the initial time step.

Actually, if we introduce one infective into an otherwise wholly susceptible population S_0, this ill individual may eventually infect many more than αS_0 others, since an infective may remain contagious for many time steps. For example, suppose a young child remains contagious with chickenpox for about 7 days. Then, using a time step of 1 day, this child would infect about $(\alpha S_0)(7)$ susceptibles over the course of a week.

Moreover, if the period of contagion lasts 7 days, then each day we expect roughly $\frac{1}{7}$ or approximately 14% of the total number of infectives to move from the infective class I_t into the removed class R_t. Because the removal rate γ measures the fraction of the infective class "cured" during a single time step, we have found a good estimate for γ; we take $\gamma = \frac{1}{7} \approx .1429$. At the same time, we have found a good interpretation for $\frac{1}{\gamma}$: it is the average duration of the infectious period. In fact, we can estimate γ for real diseases by observing infected individuals and determining the mean infectious period $\frac{1}{\gamma}$ first.

We have made progress in understanding $\mathcal{R}_0$ by thinking about this example, but we need to summarize a bit:

$$\mathcal{R}_0 = (\alpha S_0) \left(\frac{1}{\gamma} \right)$$
$$= \begin{pmatrix} \text{no. of new cases arising from one} \\ \text{infective per unit time} \end{pmatrix} \begin{pmatrix} \text{average duration} \\ \text{of infection} \end{pmatrix}.$$

Thus, $\mathcal{R}_0$ is interpreted as the average number of secondary infections that would be produced by one infective in a wholly susceptible population of size S_0.

Note that, from this point of view, the threshold value of $\mathcal{R}_0 = 1$ makes good biological sense. If $\mathcal{R}_0 > 1$, then a primary case of disease spawns more than one secondary case of the illness, the size of the infective class increases, and an epidemic results. If $\mathcal{R}_0 = 1$, then a diseased individual produces only one new case of the disease, and no epidemic can occur; there can be no growth in the number of infectives. When $\mathcal{R}_0 < 1$, the disease dies out. In short, an epidemic occurs if and only if the basic reproduction number $\mathcal{R}_0 > 1$.

Because the basic reproduction number has such a meaningful interpretation, epidemiologists try to find an expression for $\mathcal{R}_0$ for any model they propose. Although a complicated model, such as one for a sexually transmitted disease, might include many additional parameters, some combination of them should be interpretable similarly to $\mathcal{R}_0$ here. The basic reproduction number plays a role in public health decisions, because a disease prevention program will be effective in preventing outbreaks only when it ensures $\mathcal{R}_0 \leq 1$.

The severity and duration of epidemics. Once we know a model predicts that an epidemic will occur, we also want to be able to predict its severity. Suppose, for a certain disease, one infective is introduced into a population of 500 susceptible individuals. We'll assume the SIR model, and that using time steps of 1 day is adequate for describing this disease. Suppose, additionally, that data indicate that the likelihood a healthy individual becomes infected from a contact with an infective is .1% and that, once taken ill, an infective is contagious for 10 days.

▶   Justify the calculation of $\alpha = .001$ and $\gamma = .1$ in the SIR model.

For these parameter values, we find that $\rho = \frac{\gamma}{\alpha} = \frac{.1}{.001} = 100$. This means that we expect about $\frac{1}{\rho} = \frac{1}{100}$ of the susceptibles, or

$$\mathcal{R}_0 = \frac{\alpha}{\gamma} S_0 = \frac{1}{\rho} S_0 = .01 S_0 = (.01)500 = 5$$

individuals to become infected with the illness as a result of contact with the original sick person. Moreover, because $\mathcal{R}_0 = 5 > 1$, we expect an epidemic to occur. In fact, with such a large value of $\mathcal{R}_0$, we might expect a rather devastating epidemic to occur.

▶   Notice   that $I_1 = \alpha S_0 I_0 + (1 - \gamma)I_0 = .001(500)(1) + .9(1) = 1.4$. Why were there not five new cases of the disease?

▶   What would the basic reproduction number be if $S_0 = 50$? Would an epidemic occur? Explain.

Using a computer, we can trace the course of the epidemic over a series of 60 days as in Figure 7.1.

▶   Which of the curves represents S_t? I_t? R_t? How can you tell by focusing on the values at $t = 0$ and $t = 60$?

According to the graph, the number of infectives peaks at $I_t \approx 250$ at about $t \approx 21$ or 22 days. As half the population is ill at this time, this is a severe epidemic, as anticipated.

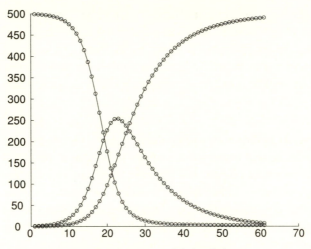

Figure 7.1. SIR model simulation.

Mathematically, we can determine information about an epidemic's peak by noting that the maximum number of infectives occurs exactly when the sign of ΔI changes from positive to negative. That is, the infective class will be largest when the number of infectives stops increasing and begins to decline. Because we have already analyzed the sign of ΔI in Eq. (7.1), we again see the importance of the relative removal rate as a critical value. When $S_t > \rho$, the infective class grows. Once $S_t < \rho$, the epidemic will subside. Returning to our example, we calculated that $\rho = 100$. Consequently, the epidemic begins to subside when $S_t = 100$, or by the time four-fifths of the population has contracted the disease. We can verify our calculation by referring to our graph – indeed, the susceptible population numbers 100 somewhere between the twenty-first and twenty-second day after the epidemic begins, and I_t peaks just at this time.

Another interesting phenomena can be detected by examining the values of S_t and I_t in Table 7.1 that were used to produce Figure 7.1.

Notice that even after the disease has ravaged the population for 100 days, there are still about two people who have remained disease free. In fact, after 150 days, there are no infectives but two disease-free citizens. Apparently, two lucky individuals escape the illness, despite the fact that they have no special immunity to the disease. We can express this long-term behavior, that as time increases S_t approaches a limiting value, by

$$\lim_{t \to \infty} S_t = 2.15.$$

Table 7.1. *SIR Model Simulation*

Day	0	1	2	3	...	20	21	22	23
S_t	500	499.50	498.80	497.82	...	135.59	102.38	76.42	56.99
I_t	1	1.40	1.96	2.74	...	244.91	253.62	254.23	248.23

Day	...	50	75	100	125	150
S_t	...	2.64	2.18	2.15	2.15	2.15
I_t	...	20.17	1.54	.12	.01	.00

Perhaps surprisingly, with the *SIR* model, it is usually the case that $\lim_{t\to\infty} S_t \neq 0$. Although the precise value of $\lim_{t\to\infty} S_t$ depends on the values of the parameters α, γ, S_0, and I_0, it is generally not zero. This means that, for a disease described well by the *SIR* model, we should expect some individuals to never fall prey to the disease, even though they lack any special immunity.

▶ Does this seem reasonable to you? Can you explain why in intuitive terms?

▶ Does an epidemic end due to a lack of susceptibles or a lack of infectives? Explain.

Since the *SIR* model is just a special case of a multiple population model, it is informative to draw a phase plane plot, just as we did earlier for other nonlinear models of interaction. Though there are three classes to track, plotting only two of the three classes is sufficient to tell how the third behaves, because $S + I + R = N$ is constant. We choose to focus on S and I, placing S on the horizontal axis and I on the vertical one. In Figure 7.2, three orbit diagrams are shown for the *SIR* model for various values of the parameters α and γ. One of the orbits $\mathcal{O}_1$ corresponds to our example above with $\alpha_1 = .001$ and $\gamma_1 = .1$. The parameter values of $\alpha_2 = .002$ and $\gamma_2 = .1$ are used for a second orbit $\mathcal{O}_2$, and $\alpha_3 = .0007$ and $\gamma_3 = .1$ for a third orbit $\mathcal{O}_3$.

▶ Which way do the trajectories go along these phase plane plots? Left to right, or right to left?

▶ Which of the plots is $\mathcal{O}_1$? $\mathcal{O}_2$? $\mathcal{O}_3$? Which epidemic is the most severe?

The plot in the phase plane gives added insight into the three epidemics and the *SIR* model. From the plot of the second orbit $\mathcal{O}_2$, the most severe

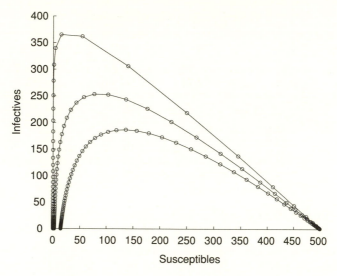

Figure 7.2. SI phase plane for the SIR model.

epidemic, you can tell that the number of infectives increases rapidly at the onset of the epidemic. In fact, just before the epidemic peaks, I_t is increasing by approximately 80 individuals per time step. In a population of only 500, this is extreme growth. Of course, a transmission coefficient of $\alpha = .002$ is quite large for a population of that size.

Note that you can approximate the relative removal rate ρ from the graphs of the three orbits, since we know ρ is the value of S_t when I_t begins to decline. Because S_0 is also easily read from the graph, once we know ρ, we can find $\mathcal{R}_0$.

▶ Determine from the graph of the orbits approximate values for ρ and $\mathcal{R}_0$. Do these values match what you would calculate from the values of the parameters γ and α?

We can make intelligent guesses about the equilibrium points of the SIR model from the phase plane, too: Each epidemic follows a wave, progressing toward a point on the horizontal axis. You will see in the problems that the SIR model has a set of equilibria, including all the points along that axis.

Problems

7.2.1. The SIR model has many equilibria.

   a. To find the equilibria, why is it not necessary to find when $\Delta R = 0$, if we find points where $\Delta S = 0$ and $\Delta I = 0$?

b. Algebraically, find the equilibrium points S^*, I^* for the SIR model. Give a common-sense explanation of why the values you find are equilibria.

c. Are these equilibria stable? Explain intuitively why they should or should not be.

7.2.2. Suppose the mean infectious period for a certain disease is 37 days.

a. What is the removal rate γ, if time steps of 1 day are used?

b. What is the removal rate γ, if time steps of 1 week are used?

7.2.3. With $\alpha = .0008$, $\gamma = .1428$, and a variety of choices of N, S_0, and I_0 in the SIR model, use a computer program to estimate the value of S_t at which this epidemic peaks. Now use the formula for the relative removal rate to determine the value of S_t at which the peak of the epidemic occurs. Do your two answers agree exactly? Explain any discrepancy.

7.2.4. In Chapter 1, the per-capita growth rate was used to understand the logistic model. In this problem, we explore the "per-infective" growth rate.

a. In the SIR model, give a formula for the per-capita growth rate of the infective class I_t.

b. Plot this relative growth rate as a function of S for the fixed values of $\alpha = .0001$ and $\gamma = .2$. Place the per-capita rate $\Delta I / I$ on the vertical axis and S on the horizontal axis. Use your graph to find the threshold value S for S_0 (i.e., find the value S such that an epidemic occurs if $S_0 > S$ and no epidemic occurs if $S_0 \leq S$).

7.2.5. An isolated island population of 100 individuals is exposed to a disease. The disease is particularly deadly; an infected individual remains contagious until overcome by death after 4 days. We want to predict the diseases's effect on the community on a daily basis. Suppose initially one individual is stricken with the disease.

a. What is the removal rate γ?

b. For what values of the relative removal rate ρ will an epidemic occur? Use this to determine for what values of the transmission coefficient α an epidemic will occur.

c. Use a computer program such as `sir` to estimate the number of days until the epidemic peaks for the values of $\alpha = .003, .005, .01$, and $.0125$, presenting your data in a table. How does the magnitude of α relate to the time until the peak?

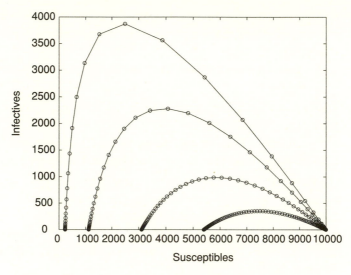

Figure 7.3. SI phase plane for the SIR model.

    d. Calculate the basic reproduction numbers and the relative removal rates for the values of α above, adding that information to your table.

7.2.6. In a population of 10,000 individuals, a new disease strikes. Once ill, an infective is contagious for $3\frac{1}{3}$ days.

    a. Give the removal rate γ, assuming time steps of 1 day.

    b. Examine the phase plane in Figure 7.3 for four possible epidemics with the value of γ determined above. Estimate the relative removal rate ρ and the transmission coefficient α for each of the four epidemics graphed. Then find the basic reproduction numbers for each of the epidemics. Give your answers in a table.

    c. Extend your table in part (b) by including the number of time steps T until the epidemic peaks, and the number of susceptibles $S_\infty = \lim_{t \to \infty} S_t$ remaining after the disease dies out. Explain the effect of increasing α on the virulence of an epidemic. Does this make biological sense?

7.2.7. A disease strikes a small town of 10,000 individuals. Suppose 10 individuals are infected initially and that the transmission coefficient has been estimated to be $\alpha = .00008$.

    a. For each of the four possible epidemics plotted in Figure 7.4, estimate the relative removal rate ρ, the removal rate γ, and the mean time of infectivity. Then calculate the basic reproduction number for each of the epidemics. Tabulate your answers.

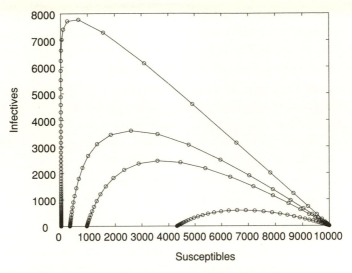

Figure 7.4. SI phase plane for the SIR model.

b. Extend your table in part (a) by including the number of tme steps T until the epidemic peaks, and the number of susceptibles $S_\infty = \lim_{t \to \infty} S_t$ remaining after the disease dies out. Explain the effect of increasing γ on the virulence of an epidemic. Does this make biological sense?

7.2.8. The text claimed that, for the SIR model, there is no epidemic of a disease if $\Delta I \le 0$. For $\Delta I = 0$, you might expect that there is no net change in the infective class, and so the number of incidences of the disease should remain constant and the disease would be *endemic* in society.

a. Investigate the $\Delta I = 0$ situation experimentally. For instance, set $\alpha = .000035$ and $\gamma = .175$, and pick values of S_0 and I_0 so that $\Delta I = 0$ initially. Then use a computer program to follow the growth or decay of the susceptible and infective classes. Record what you notice. Repeat this for some other choices of the parameters for which $\Delta I = 0$.

b. Give a common-sense explanation of why the disease ultimately dies out when $\Delta I = 0$ initially.

c. Reconsider the difference equations for ΔS and ΔI, and mathematically explain why the disease dies out if $\Delta I = 0$ initially. You will have to think about two time steps.

7.2.9. In Chapter 3, you learned to draw and interpret nullclines in a phase plane. Draw a coordinate system with S on the horizontal axis and I on the vertical axis.

    a. For the values $\alpha = .000145$ and $\gamma = .1428$, draw the nullclines for the SIR model, as well as arrows suggesting directions of orbits.

    b. Draw the nullclines and arrows suggesting directions of orbits for the general SIR model with parameters α and γ.

7.2.10. Recall the quarantine model of Problem 7.1.6 of the last section. If a fraction q of infectives is successfully quarantined, in terms of q, α, and γ what is the resulting relative removal rate? What is the effect of increasing q on the resulting relative removal rate? How does this show that a large q may prevent an epidemic?

7.2.11. As discussed in the text, to model a real disease using the SIR model, we can estimate the parameter γ from data on the length of time individuals are typically infectious. The parameter α is harder to estimate for real populations and diseases.

    Explain how you could estimate the value of the parameter α for a disease if you had sufficient data on the number of susceptibles and infectives in a population at various times throughout the course of a previous epidemic. Discuss how such data might be collected.

7.3. Variations on a Theme

Once we have understood the basic SIR model, it's not hard to make modifications to produce models that might be more appropriate for other diseases. Here, we will consider two basic variations that capture different dynamical behavior for diseases with different characteristics. We will also consider modeling large populations by tracking fractions of populations in each class, rather than absolute numbers.

The SI and SIS models. For some infectious diseases, there is no removed class. For instance, it might be that infective individuals simply cannot recover. This leads to the SI model:

$$\boxed{Susceptibles} \rightarrow \boxed{Infectives} \,.$$

The SI model assumes that $S + I = N$ and that once individuals enter the infective class, they remain there for the duration of the model. For instance, the spread of head lice in a human population with no means of effective treatment would fit the SI framework.

The equations for the *SI* model are

$$S_{t+1} = S_t - \alpha S_t I_t,$$
$$I_{t+1} = I_t + \alpha S_t I_t.$$

▶ Compare these equations with those of the *SIR* model. How do they differ?

▶ How would you expect graphs of S_t and I_t vs. time to look for the *SI* model?

Another way there could be no removed class is if once infectives recover, they are again at risk for contracting the disease. Since recovering from many diseases does not confer immunity on the former sufferer, an *SIS* model – in which individuals pass from the infective class back into the susceptible class – gives a more appropriate description of them. Syphilis and gonorrhea, for example, can be treated with antibiotics, but a patient, once cured, may become reinfected. The common cold is a disease that most of us get repeatedly. Schematically, the *SIS* model is:

$$\boxed{Susceptibles} \rightleftharpoons \boxed{Infectives},$$

and its equations are:

$$S_{t+1} = S_t - \alpha S_t I_t + \gamma I_t,$$
$$I_{t+1} = I_t + \alpha S_t I_t - \gamma I_t.$$

▶ Compare these equations to those of the *SIR* model. How do they differ?

▶ How would you expect graphs of S_t and I_t vs. time to look for the *SIS* model?

In the problems at the end of the section, you will explore these models in more detail. In particular, you will find that the *SIS* model can lead to a constant, yet nonzero number of infectives in a population. In this situation, we say the disease is endemic.

▶ Think of a few different infectious diseases you are familiar with, such as leprosy, mononucleosis, and tuberculosis. Of the three basic models, *SIR*, *SI*, or *SIS*, which if any is most appropriate for each disease?

Contact rate and contact number. Infectious disease models are often formulated a bit differently than we did above, using proportions of the population instead of absolute numbers. For large populations in particular, this may be a natural approach, because the precise size of the population may not be known, or even really matter. Such a formulation also allows us to replace the transmission coefficient α with a new parameter that has a more meaningful biological interpretation.

Working with the SIR model as an example, let's set

$$s = \frac{S}{N}, \ i = \frac{I}{N}, \ r = \frac{R}{N},$$

and rebuild the model using proportions. Notice now that

$$s + i + r = 1,$$

because we are measuring fractions of the population. We will continue to assume there are no births or immigration so that the total population size N remains constant, even though N will not appear in our equations.

In this setting, a similar thought process as used in the SIR model lead to formulas for Δs, Δi, Δr, the change in proportions of the three classes. The sir equations are:

$$\Delta s = -\beta s_t i_t,$$
$$\Delta i = \beta s_t i_t - \gamma i_t,$$
$$\Delta r = \gamma i_t,$$

where β is called the *contact rate*.

Before we explain the interpretation of β, we should note that, if we use an sir model and want to introduce a specific total population size N, we can recover SIR data by simply multiplying the sir formulas by N. For example, the net change in number of susceptibles is

$$\Delta S = (\Delta s)N = -\beta s i N = -(\beta i)(sN) = -(\beta i)S.$$

▶ Use this to show $\beta = N\alpha$ by replacing i with $\frac{I}{N}$.

▶ In the equation $\Delta S = -\alpha S_t I_t$, both S_t and I_t are measured in "individuals." What are the units of the transmission coefficient α?

▶ In the equation $\Delta s = -\beta s_t i_t$, both s and i denote "fractions of a population," and so have no units. What are the units of the contact rate β?

Let's try to understand the equation $\Delta S = -(\beta i)S$ more fully. By the term "contact," we will mean an interaction between individuals that is sufficient for disease transmission. As infectious disease spreads to a susceptible only from contact with an infective, not from contact with a healthy or immune individual, the factor βi in the equation must give the average number of contacts an individual has with infectives during a single time step. Because i denotes the fraction of the population in the infective class, we can interpret β as the average number of contacts that an individual has during a single time step. This means β measures contacts *between anyone*, whether infective or not, that *would* have caused an infection if one of the individuals was infective and the other susceptible.

Let's turn this explanation around to make sure it is clear. The contact rate β is defined as the average number of contacts an individual experiences during one time step. Then βi represents the average number of contacts with infectives that an individual experiences in a single time step. To get the number of susceptibles that fall ill from contact with infectives during a single time step, we multiply by the number of susceptibles S and obtain the expression $\beta i S$.

The value of β of course depends on the particulars of the disease under study. For example, because chickenpox is highly contagious, it seems plausible that an elementary school child might have contact, sufficient to spread chickenpox, with four other children throughout the course of a day. In this case, we would take $\beta = 4$ and $\Delta t = 1$ day. For a less contagious disease, β might be a small number such as .02.

Another important value is the *contact number* σ, the average number of contacts of a typical infective during the entire infectious period. This is a "per-infective" measurement. For the *sir* model above, we multiply the contact rate β by the mean infectious period $\frac{1}{\gamma}$ to find

$$\sigma = \frac{\beta}{\gamma}.$$

The contact number σ is closely related to the basic reproduction number $\mathcal{R}_0$. This is not surprising because both are measures of the number of secondary cases of illness produced by one infective introduced into a wholly susceptible population. In the problems, you will work out the relationship between $\mathcal{R}_0$ and σ, and see why they are essentially the same as long as S_0 is almost as big as N.

Immunization strategies. A population can be protected from disease in many ways. For example, the number of susceptible individuals can be reduced through immunizations, the contact rate can be reduced through quarantines or public health campaigns, or the removal rate can be increased through better medical treatment of the sick. In short, a society might try to change any of the parameters or initial conditions of the model describing the disease.

Vaccinations, when available, are an attractive way to control disease dynamics, but the private and public risks of immunization must be balanced. One of the main goals of any immunization program is to achieve *herd immunity*, (i.e., to ensure that no epidemic can take place even if a few cases of the disease are present). However, individuals receiving vaccinations are usually at some small risk of adverse effects. Thus, vaccination programs have goals that benefit the public welfare, possibly at the expense of a few unlucky individuals.

As a result, even immunization policies that are understood mathematically to be capable of preventing epidemics can be controversial. In the 1970's, concern over the safety of the pertussis (whooping cough) vaccine sparked public debate in Great Britain. Although the World Health Organization Smallpox Eradication Program (1967–1980) was wildly successful, the United States and Great Britain stopped routine vaccinations for smallpox in 1971, a short time after the initiation of the program. Disease surveillance in these countries indicated that more people were dying from complications arising from vaccination than would have died from smallpox itself. Moreover, discontinuing smallpox vaccinations had economic benefits; public health costs diminished.

Before public health policy is set, the likely outcome of proposed vaccination programs must be well understood so that tradeoffs in terms of public and private good can be weighed. Thus, estimating what level of vaccination confers herd immunity for a particular disease in a particular society is an essential first step.

As an example, consider a large population at risk for a disease modeled well by the sir equations. We would like to ensure that, regardless of the size i_0 of the fraction of the population that was infective, i_t never increases. Thus, we want

$$\Delta i = \beta s_t i_t - \gamma i_t < 0.$$

But, $\beta s_t i_t - \gamma i_t = (\beta s_t - \gamma)i_t$ and since $i_t \geq 0$, this means we want

$$\beta s_t - \gamma < 0, \text{ or } s_t < \frac{\gamma}{\beta} = \frac{1}{\sigma}.$$

In other words, $1/\sigma$ is the threshold value determining whether disease spreads. If the fraction of the population that is susceptible can be brought below $1/\sigma$ then an epidemic cannot occur. The fraction of the population that must be vaccinated successfully to ensure herd immunity is thus $1 - 1/\sigma$.

Of course, using this result for a disease in the real world requires estimating σ from epidemiological data. Because of uncertainties, it would also be wise to aim for immunizing a larger fraction of the population. Whether such a goal is achievable depends on many social and economic factors, but the model identifies the target.

Studying realistic immunization issues requires using more complicated models. Disease dynamics often are different among different age and social groups, and so often each of the s, i, and r groups must be broken into subgroups. (Compulsory school attendance, for instance, can have a large effect on disease transmission.) A model might break the population into several groups by age, sex, or other factors, and be used to determine which groups should be targeted in an immunization campaign.

Social and medical considerations are crucial. A vaccination campaign successful in one country may be a failure in another due either to different disease dynamics or to differing social acceptance of the program. It may, in practice, be impossible to vaccinate a high enough percentage of the population in an overcrowded city or country to avert epidemics or gain herd immunity from highly infectious diseases like measles. The best realistic policy may be to allow citizens to catch measles at a young age, when there are few complications, and gain disease-conferred immunity.

Different strategies might also be equally successful. For instance, the United States and Great Britain have adopted different vaccination programs for rubella. Rubella is not a life-threatening or dangerous disease in general, but if a pregnant woman becomes infected her infant may suffer from a serious condition know as congenital rubella syndrome (CRS). Thus, ensuring the immunity of women of childbearing age is the primary goal of any program. In the United States, all children are routinely vaccinated against rubella as part of their MMR shot at around 15 months. In Great Britain, children are allowed to contract rubella while young. Only those girls who have failed to gain disease-conferred immunity are vaccinated at around age 12.

Problems

7.3.1. Use a computer program such as `sir` or `twopop` to study the SI model. Use a variety of values of α and N. For each choice, examine the behavior of the SI model for a variety of values of S_0 and I_0. Describe your observations.

7.3.2. Investigate the SI model by doing the following:
   a. Solve for all equilibria (S^*, I^*). Are these biologically reasonable?
   b. In a phase plane, draw nullclines and arrows suggesting orbit directions. What does this tell you about the dynamics of a disease modeled by an SI model?

7.3.3. The analysis of the SI model is made easier if we note that $S_t + I_t = N$ is constant, so we can substitute the formula $S_t = N - I_t$ into the formula for I_t and only track the number of infectives.

    a. Do this and find a formula for I_{t+1} in terms of I_t.

    b. Now use a computer program such as `onepop` to investigate the model. Compare your simulations to the ones obtained with `twopop`.

7.3.4. Use a computer program such as `sir` or `twopop` to explore the SIS model. Vary the parameters α, γ, N, S_0, and I_0. Describe your observations.

7.3.5. Investigate the SIS model by doing the following:

    a. Solve for all equilibria (S^*, I^*). Are these biologically reasonable? An equilibrium with $I^* > 0$ represents the endemic occurrence of a disease. Can an SIS disease be endemic?

    b. What do the phase plane, nullclines, and orbit directions tell you about the dynamics of a disease modeled by an SIS model?

7.3.6. The analysis of the SIS model is also made easier if we note that because $S_t + I_t = N$ is constant, we can substitute the formula $I_t = N - S_t$ into the formula for S_t and only track the number of susceptibles.

    a. Do this and find a formula for S_{t+1} in terms of S_t.

    b. Now use a computer program such as `onepop` to investigate the model. Compare your simulations to the ones obtained with `twopop`.

    c. Can you find parameter values for which the approach to an endemic equilibrium is oscillatory? For which the equilibrium is not approached?

7.3.7. For the SIR model, the threshold value ρ plays an important role.

    a. Is there an analogous threshold value for the SI model? If so, find it. If not, explain why there is not.

    b. Is there an analogous threshold value for the SIS model? If so, find it. If not, explain why there is not.

7.3.8. For the SIR model, the basic reproduction number $\mathcal{R}_0$ plays an important role. How should you define $\mathcal{R}_0$ for the SIS model?

7.3.9. Suppose a disease is modeled well by the SI framework. Explain how new medical developments might make the model invalid, so that either an SIS or SIR model would be needed instead.

7.3.10. The dynamics of SIS model can be understood in terms of the logistic model of Chapter 1.

a. Show the SIS model can be expressed as

$$\Delta I = (\alpha N - \gamma)I \left(1 - I / \left(N - \frac{\gamma}{\alpha}\right)\right).$$

b. Use part (a) and your knowledge of the logistic model to determine the equilibrium, and hence a possible endemic level of the disease. What interpretation does this give to γ/α?

c. Use part (a) and your knowledge of the logistic model to determine a condition on the parameters of the model that ensures a stable equilibrium.

d. Use part (a) and your knowledge of the logistic model to determine a condition on the parameters of the model that ensures an oscillatory approach to the stable equilibrium. Do you think such behavior is likely to occur naturally?

7.3.11. Both $\mathcal{R}_0$ for the SIR model and σ for the sir model are interpreted as measures of the number of secondary cases of illness produced by one infective introduced into a wholly susceptible population. To understand their relationship better:

a. Express σ in terms of the SIR parameters α, γ, and N.

b. From your answer to part (a), compute a simple expression for $\mathcal{R}_0/\sigma$.

c. If a population is mostly susceptible initially, so $S_0 \approx N$, what will the value of $\mathcal{R}_0/\sigma$ be?

d. Which is larger: $\mathcal{R}_0$ or σ?

7.3.12. For the SIR model, we saw that $\rho = \frac{\gamma}{\alpha}$ was a threshold value for S_0, determining whether an epidemic would occur or not. What is the analogous threshold value for s_0 in the sir model? Explain your reasoning.

7.3.13. Suppose that an sir-modeled infectious disease in a certain population has an estimated contact rate of 0.1 and a removal rate of 0.05.

a. What is the contact number for this disease?

b. Assuming a vaccination is developed that is 100% effective, what percentage of the population should be immunized to achieve herd immunity?

c. Assuming the vaccination is only 90% effective (so only 9 of every 10 vaccinations confer immunity on the recipient), what percentage of the population should be immunized to achieve herd immunity?

7.3.14. For the *SIR* model with a fixed total population size N, how many individuals must be immunized to confer herd immunity? Express your answer in terms of the parameters of the model.

7.3.15. Suppose that, for an *sir* disease, we estimate the contact number is $\sigma = 0.5$. Then, according to the formula developed in the text, we should try to vaccinate a fraction $1 - \frac{1}{.5} = -1$ of the population to prevent an epidemic. Since a negative fraction of the population does not make sense, explain what this must mean.

7.3.16. Recall the approach to disease control through quarantine introduced in Problem 7.1.6.
   a. Formulate an *sir* model with a fraction q of the infectives prevented from infecting others through quarantine.
   b. In terms of the parameters of your model, find the threshold value for s_0 determining whether an epidemic will occur or not.
   c. In terms of the other parameters of the model, find the smallest value of q that will prevent epidemics from occuring for any s_0.

7.3.17. Why is there no point in asking what percentage of the population should be vaccinated to confer herd immunity for an *SIS* disease?

7.3.18. For some diseases, like tetanus and rabies, vaccination of an individual in no way contributes to achieving herd immunity.
   a. What are the features of the way these diseases are transmitted that are responsible for this?
   b. Using your understanding of such diseases, how would you design a reasonable vaccination policy for them?

7.3.19. Discuss the characteristics that an infectious disease and its vaccination might exhibit in order for a voluntary vaccination program to be worthwhile. What circumstances might make a mandatory program justified?

7.3.20. One infectious disease model that takes into account births and deaths from natural causes, as well as disease-related deaths, is given by the equations:

$$\Delta s = -\beta s_t i_t + p\mu i_t + \gamma i_t,$$
$$\Delta i = \beta s_t i_t + (1 - p)\mu i_t - (\mu + v)i_t - \gamma i_t,$$

where β is the contact rate, γ is the removal rate, μ is the birth and death rate due to natural causes (which are assumed to be equal), v

is a disease-related death rate, and p is the probability that an infant born to an infective is disease free.

a. Explain the meaning of each term in these modeling equations.

b. This model allows births and deaths. Does it assume that the total population size will remain constant?

c. Would you call this a modified SI model, a modified SIS model, or a modified SIR model? Explain.

d. The equations above do not explicitly have any terms describing deaths of susceptibles due to natural causes, or births of infants of susceptibles. Why are they not there?

7.3.21. Many of the common childhood diseases must be modeled by a more general infectious disease model, known by the acronym $MSEIR$. It allows for births and deaths, and in addition to S, I, and R uses two more classes, M and E. If a pregnant woman has immunity, either disease-conferred or from vaccination, then some antibodies are transferred across the placenta, and a newborn inherits passive immunity from its mother. This immunity is temporary, lasting possibly as long as 1 year after birth. Newborn infants with protection from maternal antibodies are placed in the M class and pass into the susceptible class after passive immunity lapses. The exposed class E consists of those who are infected but have not yet entered the infectious period where they may transmit the disease to others.

a. Make a schematic diagram for the $MSEIR$ model that shows transfer into and out of the various classes, including births and deaths. For a challenge, write down possible equations for an $MSEIR$ model.

b. The childhood diseases measles, mumps, and rubella are modeled well with an $MSEIR$ model, whereas an SIR or $SEIR$ model fails to capture key dynamics of their transmission. In the United States, the recommended age for infants to receive their first dose of measles, mumps, rubella vaccination (MMR) is between 12 and 15 months of age. Explain why the MMR vaccination age recommendation reflects the importance of including an M class when realistically modeling these childhood diseases.

Projects

1. Investigate an SIR model that also takes into account births and deaths. For many diseases, the continual introduction of newborn susceptibles is

an important contributor to the dynamics of the disease. The childhood diseases such as measles and mumps are examples. Good models of such diseases require including terms for *vital dynamics*, i.e., the addition of new susceptibles into the population by births and the removal of members of the population by deaths.

Models often assume that the birth rate μ equals the death rate μ. This has the advantage that the total population under study remains constant. One simple way to incorporate vital dynamics in an *sir* model is:

$$\Delta s = -\beta s_t i_t + \mu - \mu s_t,$$
$$\Delta i = \beta s_t i_t - (\gamma + \mu)i_t,$$
$$\Delta r = \gamma i_t - \mu r_t,$$

where s, i, r are proportions of the total population.

Investigate this model thoroughly.

Suggestions
- Explain why these three equations model the situation, by giving a rationale for each term in the modeling equations.
- What are reasonable ranges of values for β, γ, and μ?
- To get a feel for the effect of births and deaths, investigate the model numerically with `sir` for lots of reasonable choices of the parameters. You might want to start with $\beta = 1.1$, $\gamma = .2$, and $\mu = .1$, and then vary the parameters from there. Make a note of any unusual behavior, including threshold values or equilibria, and try to explain it.
- How are the dynamics of this model different from the basic *sir* model? How are they similar?
- The quantity $\frac{1}{\gamma+\mu}$ is sometimes referred to as the *mean death-adjusted infectious period*. Can you give a reasonable biological or mathematical explanation for this terminology?
- The contact number σ for this model is $\sigma = \frac{\beta}{\gamma+\mu}$. Can you give a reasonable biological explanation for this term? For what values of σ does an epidemic occur?
- Calculate the value of σ for a variety of the values for β and γ used above. Do your simulations mesh well with your theoretical predictions?
- Calculate analytically the equilibria in terms of σ, β, and μ. (Do not use γ in your answer; instead use σ.) Are these equilibria stable or unstable?
- Express the model in terms of s and i alone and consider a phase plane drawing with nullclines.

- Can you give good epidemiological interpretations to the equilibria? What do the equilibria say about the disease dynamics?
- Modify the model to account for vaccination of a percentage p of newborns at each time step. What affect does vaccination have on the behavior of the model, including any equilibria?

2. Formulate and investigate an SIRS model.

For some diseases, recovery confers temporary immunity, but over time the immunity declines. Eventually, a recovered individual is again susceptible to the disease.

Formulate an SIRS model to describe such a disease, and investigate it thoroughly.

Suggestions
- Explain each term in your modeling equations. Give names to any new parameters.
- Explain how any new parameters might be estimated for a real disease.
- Investigate the model experimentally for a variety of parameter choices. Describe the behaviors you see.
- Compute equilibria. Are they biologically reasonable?
- Draw nullclines and orbit directions in the phase plane.
- Are there any threshold values of interest?
- How might an epidemic of such a disease be prevented? What strategies might affect parameter values?

7.4. Multiple Populations and Differentiated Infectivity

Although the basic models such as the SIR, SI, and SIS provide good starts at describing disease transmission, they are not elaborate enough to capture key dynamical features of many real infectious diseases. The simple versions of these models do not take into account such things as one's age, sex, socioeconomic class, medical history, or other characteristic that may affect the likelihood of infection, time of infectivity, or transmission mechanisms. For example, the elderly and the weak are particularly susceptible to influenza. Intravenous drug users are at increased risk for a number of diseases due to their behavior. Some individuals seem to have a higher resistance to the HIV infection than others. To capture such factors in modeling disease dynamics, we need to consider several subpopulations and different parameter values for these subpopulations, using *differentiated infectivity* models.

Sexually transmitted disease in particular generally requires such models, because even the simplest model of heterosexual transmission requires considering subpopulations of males and females. We will develop such a model by studying the dynamics of gonorrhea.

Gonorrhea is caused by a bacterial infection that is spread through sexual contact. The CDC estimates that approximately 650,000 cases of gonorrhea occur each year in the United States, making it a common sexually transmitted disease. Within a few days of infection, men usually develop symptoms such as a burning sensation when urinating, a yellowish discharge from the penis, and painful or swollen testicles. Women typically have milder initial symptoms, which can be mistaken for a bladder infection. If untreated, however, women may develop pelvic inflammatory disease, which can lead to infertility. Babies born to infected woman may get the infection, which can cause blindness and be life-threatening. Untreated men may also become infertile or be left with urethral scarring. Fortunately, antibiotic treatment is effective, though penicillin is no longer used since resistant strains have developed. Once cured, an individual is unfortunately at risk of recontracting the illness; no immunity results from an infection.

While an infected female may be asymptomatic, she may still pass on the infection to a sexual partner. Moreover, it is reasonable to assume that the average time from infection to treatment might be longer for females than males, because females may be unaware of their infected state for a longer period. In addition, data indicate that, in heterosexual intercourse between infected and susceptible individuals, a new infection is roughly twice as likely to result if the male is the infective rather than the female.

Because of these sex differences, to begin modeling gonorrhea, we divide the human population into two groups: females and males. Within the two groups, we have two subclasses: susceptible females S_t^f and infective females I_t^f, and susceptible males S_t^m and infective males I_t^m. Because gonorrhea is curable, but the treatment offers no immunity from further infection, there are no removed classes.

As before, we will assume that populations remain constant, $S^f + I^f = N^f$, $S^m + I^m = N^m$, and the total population under study is $N = N^f + N^m$. If we assume, for modeling purposes, that gonorrhea is only spread through heterosexual contact, then we will need a transmission coefficient α^f to measure the rate at which gonorrhea is spread from men to women and a transmission coefficient α^m for the spread from women to men. Similarly, this model requires two removal rates γ^f and γ^m, one for each sex.

▶ From the description of gonorrhea, should α^f or α^m be larger? Should γ^f or γ^m be larger?

We will use a two-population variant of the SIS model to describe the spread of gonorrhea:

$$\Delta S^f = -\alpha^f S_t^f I_t^m + \gamma^f I_t^f,$$

$$\Delta I^f = \alpha^f S_t^f I_t^m - \gamma^f I_t^f,$$

$$\Delta S^m = -\alpha^m S_t^m I_t^f + \gamma^m I_t^m,$$

$$\Delta I^m = \alpha^m S_t^m I_t^f - \gamma^m I_t^m.$$

▶ Explain the meaning of each of the terms in the four equations above.

With four equations specifying our model, it seems too complicated for easy analysis. How we can define such things as a meaningful basic reproduction number $\mathcal{R}_0$ is not clear. Fortunately, though, we can simplify our analysis of the model by noting that since the population sizes N^f and N^m are constant, knowing the sizes of the infective classes is enough to determine the sizes of the susceptible classes. Thus, we will substitute $S_t^f = N^f - I_t^f$ and $S_t^m = N^m - I_t^m$ to get

$$\Delta I^f = \alpha^f S_t^f I_t^m - \gamma^f I_t^f$$
$$= \alpha^f (N^f - I_t^f) I_t^m - \gamma^f I_t^f,$$
$$\Delta I^m = \alpha^m S_t^m I_t^f - \gamma^m I_t^m$$
$$= \alpha^m (N^m - I_t^m) I_t^f - \gamma^m I_t^m.$$

Now our model needs only to track the sizes of the two infective classes. Still, it is hard to have an intuitive understanding of how such a model might behave.

▶ Do you think that it is possible for an epidemic to occur in the female population, but not in the male population? Could I^f be increasing while I_m is decreasing?

As you will discover in the exercises, lots of different scenarios are possible with this model. For instance, we will find that it is possible for $\Delta I^f > 0$, while $\Delta I^m < 0$; that is, the female population experiences an epidemic as the disease decreases among males. Of course, if I^f increases, we might expect that, after some time, I^m would increase too; a bit later we might have $\Delta I^f < 0$ and $\Delta I^m > 0$. Some cyclical behavior in the size of the infective classes is possible, in fact, though this SIS model usually moves toward equilibrium quickly.

In the exercises, you are asked to carry out the algebra to solve the simultaneous equations $\Delta I = 0$. You will find the equilibria, $I^f = 0$, $I^m = 0$ and

$$I^f = \frac{N^f N^m - \rho^f \rho^m}{\rho^f + N^m}, \qquad I^m = \frac{N^f N^m - \rho^f \rho^m}{\rho^m + N^f},$$

where $\rho = \frac{\gamma}{\alpha}$ denotes the respective relative removal rates.

▶ Verify that if $I^f = 0$ and $I^m = 0$, then the system above is in equilibrium. What is the medical significance of the equilibrium? Do you think it is stable or unstable?

Note that the existence of only one nonzero equilibrium shows there is at most one endemic level of the disease. This level also depends only on the sizes of the male and female populations, and the respective relative removal rate for them, but not on the initial number of infectives. Increasing the relative removal rates is the way that these endemic levels can be decreased. The precise formulas allow prediction of the endemic level expected from any relative removal rate a public health program might achieve.

Let's examine the nonzero equilibrium state more closely. For the values of I^f and I^m to make sense biologically, we must have $N^f N^m - \rho^f \rho^m > 0$, because we cannot have a negative number of infectives. In fact, this turns out to be a good test to see if a disease is endemic. If $N^f N^m - \rho^f \rho^m$ is positive, then our model predicts that gonorrhea (or a different disease modeled with this two-population SIS model) may always be present in the society. Thus, data collection for statistical estimates of the infection and removal rates can help judge whether a disease is likely to remain endemic.

Indeed, with some algebra, we can squeeze out a good interpretation of the inequality $N^f N^m - \rho^f \rho^m > 0$. Rewrite it as $\frac{N^f N^m}{\rho^f \rho^m} > 1$ and factor to get

$$\left(\frac{N^f}{\rho^f} \right) \left(\frac{N^m}{\rho^m} \right) > 1.$$

Notice the similarity between $\frac{N^f}{\rho^f} = \frac{\alpha^f}{\gamma^f} N^f$ and the basic reproduction number $\mathcal{R}_0 = \frac{S_0}{\rho} = \frac{\alpha}{\gamma} S_0$ from the one-population SIR model.

▶ Before reading ahead, use your understanding of $\mathcal{R}_0$ to interpret $\frac{N^f}{\rho^f} = \frac{\alpha^f}{\gamma^f} N^f$.

If the female population were wholly susceptible to gonorrhea, that is, if $S_t^f = N^f$, then $\frac{N^f}{\rho^f} = \frac{\alpha^f}{\gamma^f} N^f$ would give the average number of infective female – time steps caused by an infective male during a single time step. A similar interpretation holds for $\frac{N^m}{\rho^m}$. Indeed, because it is rarely the case that $N^f = S^f$, the value of $\frac{N^f}{\rho^f} = \sigma_f N^f$ gives a maximum for this number. We call the quantities $\frac{N^f}{\rho^f}$ and $\frac{N^m}{\rho^m}$ the *maximal male contact number* and the *maximal female contact number*.

In fact, our analysis above indicates that, if both $\left(\frac{N^f}{\rho^f}\right) > 1$ and $\left(\frac{N^m}{\rho^m}\right) > 1$, then the product $\frac{N^f N^m}{\rho^f \rho^m} > 1$ and the disease has an endemic equilibrium state. This property, similar to the relationship between epidemics and $\mathcal{R}_0$ for the basic SIR model, makes it tempting to refer to these quantities as reproduction numbers. Keep in mind, however, that it is at least mathematically possible for one of the maximal contact numbers to be less than 1 and the disease still to have an endemic equilibrium.

We have only taken the first step toward creating a realistic gonorrhea model. A key feature we have missed is that humans vary greatly in sexual promiscuity. More elaborate models suggest the actual disease dynamics are greatly affected by a core of highly promiscuous individuals. To build such features into a model requires introducing further subpopulations, according to both sex and number of sexual partners.

Problems

7.4.1. What diseases other than gonorrhea might be modeled with the two-population SIS model developed here?

7.4.2. A certain sexually transmitted disease is described by the model of this section. Suppose for a female population of 10,000 and a male population of 15,000, the transmission coefficients are $\alpha^f = .0000009$ and $\alpha^m = .000006$, and the removal rates are $\gamma^f = .007$ and $\gamma^m = .05$.

a. Are females more likely to catch the disease than males? Which sex is likely to recover more quickly?

b. Find the maximal female contact number and the maximal male contact number. Do you expect the disease to be endemic? Check your result with a computer simulation.

c. Calculate $N^f N^m - \rho^f \rho^m$. Then, use this expression to solve for the equilibrium values of S^f, I^f, S^m, I^m. Use a computer program such as twopop to verify your solutions.

d. Run a computer program such as twopop to determine if the nonzero equilibria are stable or unstable. Use a variety of choices for your initial infective populations.

7.4.3. Suppose when modeling a heterosexually transmitted disease, you find that $\alpha^f = .011$, $\alpha^m = .00023$, $\gamma^f = .5$, and $\gamma^m = .2$.

   a. If the time step is 1 day, what are the characteristics of this disease? Include a discussion of the meaning of the transmission coefficients and removal rates for each sex.

   b. Suppose the population is fixed with 500 people, 100 females and 400 males. Calculate the maximal female and male contact numbers. Can the disease be endemic? Explain. What if instead there are 50 females and 450 males, or if the population is evenly divided, 250 females and 250 males?

7.4.4. An infectious disease hits an isolated community. The elderly population is especially vulnerable; they catch the disease more readily and are slower to recover than are the young. Suppose you attempt to describe this using a two-population SIS model of the form:

$$\Delta S^e = -\alpha^e S_t^e (I_t^e + I_t^y) + \gamma^e I_t^e,$$
$$\Delta I^e = \alpha^e S_t^e (I_t^e + I_t^y) - \gamma^e I_t^e,$$
$$\Delta S^y = -\alpha^y S_t^y (I_t^e + I_t^y) + \gamma^y I_t^y,$$
$$\Delta I^y = \alpha^y S_t^y (I_t^e + I_t^y) - \gamma^y I_t^y.$$

   a. Assuming there are a total of N^e elderly people and N^y young people, rewrite the model to eliminate I^e and I^y.

   b. If the α values are .0003 and .0001, which is α^e and which is α^y? If the γ values are .21 and .05, which is γ^e and which is γ^y?

   c. Use a computer program such as twopop to determine what happens to the two populations using the parameter values in part (b), along with $N^e = 250$ and $N^y = 750$.

7.4.5. The text claims that the nonzero equilibrium solution for the two-population SIS model of gonorrhea occurs when $I^f = \frac{N^f N^m - \rho^f \rho^m}{\rho^f + N^m}$ and $I^m = \frac{N^f N^m - \rho^f \rho^m}{\rho^m + N^f}$. Verify this as follows:

   a. Using $\Delta I^f = 0$, show that at equilibrium

$$I^f = \frac{N^f I^m}{I^m + \rho^f}.$$

Give a similar formula for I^m at equilibrium.

b. The formulas in part (a) relate the equilibrium values of I^f and I^m. Substitute one of the formulas into the other to get an equation relating I^f at equilibrium to itself. Then solve for I^f.

c. Write down the equilibrium value for I^m by taking advantage of the symmetry of the original equations.

7.4.6. Do you expect the equilibrium state $I^f = 0$ and $I^m = 0$ for the model of this section to be stable or unstable? Does the answer to this question depend on the values of the model parameters? (You may want to use a computer to run simulations.) Linearize at the zero steady state to verify (or discredit) your guess.

8

Curve Fitting and Biological Modeling

Most of the models introduced in this text have been developed by making reasonable theoretical assumptions, which are then incorporated into a mathematical framework. However, the ultimate test of the validity of any model is that its behavior is in accord with real data. Because of the simplifications introduced in any mathematical model of a biological system, we must expect some divergence between even the most carefully collected data and well-constructed model. How can we determine if a model describes data well? How can we determine the parameter values in a model that are appropriate for describing real data? These questions are much too broad to have a single answer. There are, however, mathematical tools that can be used in addressing them.

Imagine having collected data on a population size at successive time intervals. Plotting the population values as a function of time might give a plot that appears to grow roughly exponentially. We might, therefore, think the simple Malthusian model $P_{t+1} = \lambda P_t$, introduced in Chapter 1, is adequate for describing the population growth. Then, the data points should lie approximately on a curve $P_t = \lambda^t P_0 = P_0 e^{(\ln \lambda)t}$. But what should λ be? Is there a "best" estimate of this parameter that locates the curve "closest" to the data points? How can we be confident the population is really growing exponentially, and not more slowly, with the data points actually lying on a parabola, for instance?

As this example illustrates, many questions of the correspondence between models and data can be thought of as questions of *curve fitting*. Given a number of data points, how can we choose formulas for curves that come close to all the points?

Even when no mathematical model has been formulated, curve fitting is often a good way of extracting the main features of a data set. After collecting numerical data in an experiment, a biologist might plot it and see that it appears to cluster in a roughly linear pattern, for instance. An equation for a line showing this main trend would succinctly summarize what might be the most important feature of the data, and perhaps eventually lead to a deeper understanding of the mechanism producing the pattern. Thus, fitting curves to data is useful in data-driven fields even when our understanding is too limited to produce more detailed models.

In this chapter, we will explore some of the basic ideas in curve fitting, including the most heavily used technique, called *least squares*. Though the computations necessary for basic curve fitting are readily performed by most data analysis software, understanding the mathematical ideas behind them is helpful in using such software effectively.

8.1. Fitting Curves to Data

As medical researchers develop a new drug, an important issue to be under-stood is how the concentration of the drug in the bloodstream changes as the drug is metabolized. To study this, a researcher might administer an initial dose to bring the concentration to the level of 200 mg/l, and then monitor the changing concentration over the next few days. Data such as that recorded in Table 8.1 might be obtained. Notice that no measurement was recorded for day 2; perhaps the patient missed an appointment or the laboratory work was botched.

Suppose for therapeutic value, the concentration of drug in the blood needs to be kept at a level above 100 mg/l. Then, because the table shows the level dropping below that sometime between 1 and 3 days after the initial dose, the new dose should be administered sometime in that time period. Unfortunately, the missing data for day 2 makes it hard to pin down more closely when the 100 mg/l level is crossed.

▶   Based on the available data, do you think the level that would have been measured on day 2 is greater than or equal to 100 mg/l? How would you try to persuade someone who disagreed with you?

One approach to answering this question begins with the observation that the drop in level between times 0 and 1 is much larger than that between times 3 and 4. This might indicate that each passing day produces a smaller drop

Table 8.1. *Concentration y of Drug in the Bloodstream t Days After Dosage*

t (day)	0	1	2	3	4
y (mg/l)	200	129	—	58	33

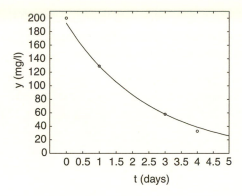

Figure 8.1. Data from Table 8.1 with an exponential decay trend.

in level, and so the measurement on day 2 would be lower than the midpoint between the day 1 and day 3 measurements. Since $(129 + 58)/2 = 93.5$, probably the day 2 measurement would have been less than 100 mg/l.

Although this sort of reasoning is fine as far as it goes, it's inadequate for answering more refined questions. For instance, what is the best estimate of the level on day 2? This is a question of *interpolating* the data to estimate values between entries in the table. If, instead, we wanted to estimate the level on day 5, then we need to *extrapolate*, because we have data entries on only one side of that day.

Plotting the above data produces the points marked in Figure 8.1. The data points appear to cluster along an exponential decay curve like the one shown. Finding a formula for that curve, or a similar one that fits the data well, would enable us to both describe the data and estimate unknown values easily. Interpolating and extrapolating could be performed by simply plugging time values into the formula for the curve. A curve that describes the data well overall, though perhaps not in all its particulars, serves as a model for the data. Because exponential decay curves are described by formulas of the form $f(t) = ae^{kt}$ with $k < 0$, our goal is to find the best choice of the parameters a and k to ensure a good fit between the data and the model.

You might imagine that just collecting more data, by taking more frequent measurements over a longer time, would be preferable to fitting a curve to the data we have. Even if we collect more data, though, we would still find it useful to fit a curve to it. Finding a formula that describes the overall trend in the data would give a succinct description of it, and might give us more insight than the raw numbers that were collected. Also, we should expect minor fluctuations in the data around its overall trend, due to measurement errors and the specifics of the patient's activities during the period of the study. Fitting a simple curve to the data is, like most models, a way of focusing attention on main features and ignoring details we consider less important.

Our first approach to finding a and k is a simple one. We can use the data points to get relationships between the parameters by plugging the points into the equation $f(t) = ae^{kt}$. For instance, the data point $(1, 129)$ gives

$$129 = ae^k.$$

With two unknowns in this equation, we cannot yet solve, so considering another data point, say $(3, 58)$, gives

$$58 = ae^{3k}.$$

Now, the first equation gives $a = 129e^{-k}$, which can be substituted into the second to obtain

$$58 = (129e^{-k})e^{3k} = 129e^{2k}.$$

Thus

$$\frac{58}{129} = e^{2k},$$

so taking a natural logarithm, we find

$$k = \frac{1}{2} \ln \frac{58}{129} \approx -.3997.$$

Now, because $a = 129e^{-k}$, using this value of k to solve for a gives $a \approx 192.4$. Thus, our first attempt at fitting an exponential curve to the data yields

$$f_1(t) = 192.4e^{-.3997t}.$$

This curve is the one that was graphed in Figure 8.1.

▶ What does this curve indicate as the amount of drug in the patient at time $t = 2$? At time $t = 5$?

Looking at the figure carefully, we notice that the graph of $y = f_1(t)$ passes through exactly two of the data points, but is only near the others. We should have expected this, because we used only two data points to solve for a and k. We have completely ignored 2 of the 4 measurements that were taken. This is, of course, a significant drawback to our approach.

▶ Suppose another researcher chooses the first two data points and solves for the constants a and k. How would the resulting curve compare with the one above? Do you think it would be a better model?

If different researchers propose different curves as good fits to the data, an objective way of measuring the fit is needed. A start at measuring goodness of fit between a curve $y = f_1(t)$ and data is to look at the difference between the y-coordinates of the data and the y-coordinates of $f_1(t)$. We can gather these differences into an *error vector* **e**. For the data and curve $y = f_1(t)$ above, we find the error is

$$\mathbf{e}_1 \approx (200, 129, 58, 33) - (192.4, 129.0, 58.00, 38.89)$$
$$\approx (7.6, 0, 0, -5.89).$$

Note that a data point below the curve produces a negative error and that one above produces a positive error. As already observed, at the two points used in fitting the curve, the individual errors are zero, or at least very close to zero due to rounding.

A major flaw in our first curve-fitting attempt is that it only used some of the data points in finding an equation. One possible way around this problem is to fit the data to a curve with more parameters. For instance, with the data above, fitting an exponential of the form $g(t) = ae^{kt} + b$ would use three data points because of the three parameters a, k, and b. The resulting curve $g(t)$ will pass exactly through those three points, making three entries in the error vector be equal to zero.

Although the idea of including more parameters in the curve seems attractive at first, it could be a real mistake. For instance, as one of the exercises will show, there is a theoretical model that justifies why a curve of the form $f(t) = ae^{kt}$ is really the appropriate one for dealing with the metabolization of a drug. Even in situations where no such theory exists, it is often better to use simple formulas to fit data rather than complicated ones. After all, some of the details in the data may be due to experimental artifacts and random variations, and are not really part of the trend we hope to capture. A simple curve that comes close to all data points may therefore be a more valuable description than a complicated curve that exactly hits all points.

Semilog and log–log graphs. As a second attempt to fitting a curve $f(t) = ae^{kt}$ to the data above, we will try to use all the data points. Of the two unknown parameters, a and k, we might think that k is more important because it indicates the rate of decay. This suggests that we should focus on a technique of finding the decay rate k using all of the data points.

A clever way to estimate k is to use a *semilog* plot. For the moment, view our four data points as approximated by ordered pairs of the form $(t, y) = (t, ae^{kt})$. If we transform the data by taking natural logarithms of the

Table 8.2. *Semilog Transformation of*
the Data in Table 8.1

t	0	1	2	3	4
ln y	5.298	4.860	—	4.060	3.497

y-coordinates, we obtain ordered pairs of the form $(t, \ln y) = (t, kt + \ln a)$. Notice that the new second coordinates of these points have a simpler pattern; they are now *linear* functions of t, because they have the form $kt + \ln a$, where k and $\ln a$ are just constants. If we could find the slope of the line relating these transformed data points, then that would be a good estimate for k, the decay rate.

In a semilog plot, we graph the transformed data $(t, \ln y)$. The name refers to the fact that we take a logarithm of only one of the coordinates. (Although it is also possible to form a different type of semilog plot using $(\ln t, y)$, that would not help us here, since our goal is to estimate k as best we can.)

The semilog transformation of the data of Table 8.1 gives Table 8.2 and Figure 8.2. The figure shows how a semilog transformation converts nearly exponential data into nearly linear data.

Although Figure 8.2 might lead us to guess that $k \approx -.5$, it's best to perform a calculation with all the data points to estimate the slope. A reasonable idea is to first find the slopes of the line segments joining adjacent transformed data points and then use the average of those three slopes as an estimate for k.

The slope between the first two transformed data points is $m \approx (4.860 - 5.298)/(1 - 0) = -.438$. Similarly, we find the slopes between the other pairs of consecutive points as -0.400 and -0.563. Finally, taking the average of these slopes, we estimate $k \approx (-.438 - .400 - .563)/3 = -.467$.

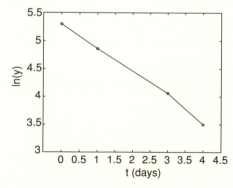

Figure 8.2. Semilog plot of the data in Table 8.1: $(t, \ln y)$.

Note that this estimate of the growth rate k is slightly different from that found in our first attempt at curve fitting in which we used only two data points. Although we still do not know if this estimate of k is better, we might suspect that it is because we used all the data in the estimation procedure.

To finish finding the equation $f_2(t) = ae^{-.467t}$ that models the data, we must pick a value for a. A quick way is to use one of the data points to solve for a. We will choose one of the middle data points, $(1, 129)$, in the hope that its central location in the data set might make the curve $f_2(t)$ fit the data the best. Substituting $t = 1$ and $y = 129$ and solving, we obtain $a \approx 205.8$, and $f_2(t) = 205.8e^{-.467t}$.

► How might you better estimate a in a way that uses all of the data?

The idea of transforming data with a logarithm was useful here because it converted exponential decay into linear behavior. A similar approach is useful when we believe a curve given by a power function $y = ax^n$ should fit our data. For this particular curve, taking a logarithm of both x and y is useful, because

$$y = ax^n \quad \text{is equivalent to} \quad \ln y = n \ln x + \ln a.$$

This means that a graph of the points $(\ln x, \ln y) = (\ln x, \ln a + n \ln x)$ will form a line, with slope n. Such a plot is called a *log–log* plot. If a log–log plot of data looks close to linear, then a good estimate of the slope of the line will be a good estimate for the degree of the correct power function to fit. Indeed, if we can find a good estimate for the equation of the line relating $\ln x$ and $\ln y$, say $\ln y \approx m \ln x + b$ for some m and b, then exponentiating this equation gives $y \approx e^b x^m$, which is a power function fitting the data.

Significantly, semilog and log–log transformations allow us to reduce the problem of fitting either exponential or power functions to data to that of fitting a line to transformed data. If we develop a means of finding good models for linear relationships between variables, then by using various transformations on our data if necessary, we will also know how to find good models of certain other types of relationships.

Measures of error. So far, we have used two *ad hoc* methods to fit an exponential curve to four data points. Both

$$f_1(t) = 192.4e^{-.3997t} \quad \text{and} \quad f_2(t) = 205.8e^{-.467t}$$

are reasonable candidates for exponential curves fitting the data, but which is better? Although we suspect that the second curve $f_2(t) = 205.8e^{-.467t}$ probably describes the data better than $f_1(t) = 192.4e^{-.3997t}$, since we at least used all the data in finding it, we need to be precise about what "better" means. Using graphical perception or vague suspicions to choose which graph is superior is too subjective; a different viewer might choose differently.

Earlier, we determined that the vector of errors for $f_1(t)$ was given by

$$\mathbf{e}_1 \approx (7.6, 0, 0, -5.89).$$

Each of these numbers measures the vertical displacement between a data point (t_i, y_i) and the point $(t_i, f_1(t_i))$ on the graph of $f_1(t)$ with the same t value. Calculating the error vector $\mathbf{e}_2$ for $f_2(t)$'s fit to the data gives

$$\mathbf{e}_2 = \begin{pmatrix} 200 \\ 129 \\ 58 \\ 33 \end{pmatrix} - \begin{pmatrix} 205.8e^{-.467(0)} \\ 205.8e^{-.467(1)} \\ 205.8e^{-.467(3)} \\ 205.8e^{-.467(4)} \end{pmatrix} \approx \begin{pmatrix} -5.8 \\ 0 \\ 7.3 \\ 1.22 \end{pmatrix}.$$

Note that only one of the entries of $\mathbf{e}_2$ is zero. Also, whereas $\mathbf{e}_1$ had two zero entries, it also has an entry larger than any of those in $\mathbf{e}_2$. Apparently, there has been a sort of trade-off, where fitting perfectly at two points produces a worse fit at others.

Instead of comparing corresponding entries in error vectors one at a time, the individual errors can be combined into a single scalar that measures the overall fit. To compute a measure of the total error for each of the fitting curves, we might try adding the components of the error vector. Unfortunately, because some of the components of the error vectors are positive and some are negative, there would be some cancellation. The number computed would give too small a measure of the total error.

A better idea is to sum the absolute value of the errors. This is called the *total deviation* for the fit of the curve to the data. For the error between f_1 and the data,

$$TD(f_1) = |7.6| + |0| + |0| + |-5.89| = 13.49,$$

whereas for f_2,

$$TD(f_2) = |-5.8| + |0| + |7.3| + |1.22| = 14.32.$$

Total deviation, therefore, gives a quantitative reason to say that f_1 fits the data better than f_2.

A second way to overcome the cancellation problem is to square each of the entries of the error vector. This is called the *sum of squares for error*.

$$SSE(f_1) = (7.6)^2 + 0^2 + 0^2 + (-5.89)^2 = 92.4521$$
$$SSE(f_2) = (-5.8)^2 + (0)^2 + (7.3)^2 + (1.22)^2 = 88.4184$$

Note that using SSE to measure total fit indicates that f_2 was a better fit than f_1.

As this example shows, SSE and TD give genuinely different criteria for determining which fit is best. Although both are reasonable measures of total error in fitting a curve to data, one must be chosen so that we have a standard way of comparing. The SSE measure of fit is the one most heavily used by scientists, and the one on which we will focus. As some of the exercises will indicate, TD has some unpleasant properties that make it a poorer choice. The use of SSE can also be grounded in statistical models of error.

But, even if we decide to use SSE to measure total fit, there might be an exponential curve that fits the data even better than f_2 does. We have found two particular curves, based on two approaches that happened to come to mind, yet there may be a still better curve that we have not thought of. How we can find the *best* curve will be a question for the next section.

Problems

8.1.1. Find a formula for the exponential $f(t) = ae^{kt}$ that passes through the first two data points in Table 8.1. Then compute the error vector, measuring its fit to the data. Is it a better or worse fit than the function $f_1(t)$ found in the text when the total error is measured by TD? Than $f_2(t)$ when total error is measured by SSE?

8.1.2. In the second approach of this section to finding an exponential curve to fit the data in Table 8.1, all data points were used to estimate k, but only one to estimate a.
   a. Invent a scheme that uses all points to estimate a (after k has been estimated) and carry it out.
   b. Use SSE to determine if the curve you found in part (a) is a better or worse fit than $y = f_2(t)$.

8.1.3. Consider the three data points: $(2, 7.6)$, $(5, 15.3)$, $(10, 32.1)$. Three candidates for best-fit line for this data are

$$y = 2.9x + 1.9, \quad y = 2.9x + 2, \quad y = 3x + 1.1.$$

   a. Plot the data points and the three lines on the same graph. (In MAT-LAB this can be done with the commands like: x=[2,5,10], y=[7.6,15.3,32.1], plot(x,y,'o'), hold on, L1= 3*x+1.1, plot(x,L1).) Which of the three appears to be the best fit?

b. For each line, compute the error vector and SSE. Which of the three lines fits the data points best by giving the smallest SSE?

c. By looking at your graphs and making informed guesses, try to find a line that produces a smaller SSE than any of the three given ones.

8.1.4. Drug levels in the bloodstream are typically observed to decay exponentially with time from the administration of a dose. A difference equation model that describes this (and gives further reason to try to fit the data of Table 8.1 to an exponential curve) is $y_{t+1} = (1 - r)y_t$, where r is the percentage of the drug that is absorbed by tissue or broken down by metabolization during one time step.

a. If the initial amount of the drug is y_0, explain why this model leads to $y_t = y_0(1 - r)^t$.

b. Letting $k = \ln(1 - r)$ and $a = y_0$, show this is equivalent to $y_t = ae^{kt}$.

c. Explain why $0 < r < 1$ for this model, and then why $k < 0$.

8.1.5. You might think that the four data points in Table 8.1 could be modeled well with a straight line.

a. Using only the two middle data points, fit a straight line $y = mt + b$ to the data. Compute the error and SSE. Is your line a better or worse fit than $y = f_2(t)$?

b. Invent a scheme to find a straight line that fits the data better than the line you found in part (a). Compute its SSE. Is it a better or worse fit than $y = f_2(t)$?

8.1.6. At times $t = 1, 2, 3, 4, 5,$ and 6 seconds, data values $y_t = 3, 7, 17, 37, 82,$ and 182 are recorded.

a. Plot the data. (In MATLAB, after storing the t and y values in vectors, use `plot(t,y,'o')`.) From this graph, do you think a linear, exponential, or power function is the best model for the data?

b. Produce a semilog plot and use it to roughly estimate the growth rate k for a model of the data given by a curve of the form $y = ae^{kt}$. (In MATLAB, `plot(t,log(y),'o')` will produce the plot.)

c. Produce a log–log plot and use it to roughly estimate the degree n of a power function, for a model of the data of the form $y = at^n$. (In MATLAB, `plot(log(t), log(y),'o')` will produce the plot.)

8.1.7. Using TD to measure total error can sometimes ignore a piece of data, as this problem will show.

Consider the three points $(0, 0)$, $(1, C)$, and $(2, 0)$, where $C > 0$, and the problem of finding the best *horizontal* line $y = b$ to fit these points.

a. Explain why any horizontal line below all three points cannot be the best fit, by drawing a plot and imagining what happens to TD as the line is moved upward.

b. Explain similarly why any horizontal line above all three points cannot be the best fit.

c. Explain why, if a horizontal line is below the middle point and above the others, then TD can be decreased by lowering the line to go through the bottom two points.

d. Conclude $y = 0$ is the best-fit horizontal line when TD is used as a measure of total error. Because this result does not depend on C, the value of C has no effect on the line.

e. For a challenge, explain why $y = 0$ is the best-fit line (horizontal or not) for the three data points.

8.1.8. Using TD to measure total error does not always produce a single best-fit curve; there can be many curves that are all equally good.

To see how this can happen, consider the four points $(0, 0)$, $(1, 1)$, $(2, 1)$, and $(3, 0)$, and the problem of finding the best *horizontal* line $y = b$ to fit these points.

a. As in the previous problem, explain why the best-fit horizontal line cannot lie above all the points or below all the points.

b. Explain why any horizontal line above the two bottom points and below the two top points will have $TD = 2$.

c. Conclude from parts (a) and (b) that there may not be a unique solution to the problem of fitting a curve to data, if total error is measured using TD. (If total error is measured by SSE, there is a unique best-fit line.)

8.2. The Method of Least Squares

While exploring the idea of fitting curves to data in the last section, we discovered that even fitting an exponential curve to data could be reformulated, through the use of semilog graphs, as a problem of fitting a straight line.

In fact, the most common curve-fitting problems experimentalists face are usually those of straight line fits. Data are collected, a plot is made (using a transformation if necessary), and the data points often appear to cluster in a roughly linear manner. Then, the best-fit line to describe the data must be chosen.

The most common means of picking the best-fit line is the *method of least squares*. The philosophy of least squares is that *the line that best fits data is the one that minimizes SSE*. Geometrically, the least-squares best-fit line is the one that minimizes the sum of squares of the vertical distances between the data points and the fitting line – of all the lines that could possibly describe the data trend, we consider as best the one with this geometric property. Note that one feature of this method is that it chooses the best line by a criterion using *all* the data points.

With this criterion, the calculation of the best-fit line ultimately turns out to be surprisingly simple. Understanding why the calculation works as it does, though, requires a bit more effort.

If there are only two data points, then finding the least-squares best-fit line through them is straightforward. We know that there is a line going exactly through any two points, and that line will have $SSE = 0$, the minimum possible value.

Although the algebra to find the line through two points can be formulated in a number of different but equivalent ways, a matrix formulation will set the stage for later work. Suppose, for instance, the data points are $(3, 2.3)$ and $(6, 1.7)$. Then, because a line has equation $y = mx + b$, we need to find m and b so that

$$2.3 = m \cdot 3 + b$$
$$1.7 = m \cdot 6 + b$$

or

$$\begin{pmatrix} 3 & 1 \\ 6 & 1 \end{pmatrix} \begin{pmatrix} m \\ b \end{pmatrix} = \begin{pmatrix} 2.3 \\ 1.7 \end{pmatrix}. \tag{8.1}$$

Solving the matrix equation gives

$$\begin{pmatrix} m \\ b \end{pmatrix} = \begin{pmatrix} 3 & 1 \\ 6 & 1 \end{pmatrix}^{-1} \begin{pmatrix} 2.3 \\ 1.7 \end{pmatrix} = \begin{pmatrix} -0.2 \\ 2.9 \end{pmatrix},$$

so the line fitting the data is $y = -0.2x + 2.9$. Because we solved the matrix equation exactly, the line goes exactly through the two data points.

Suppose now we had three data points, $(3, 2.3)$, $(6, 1.7)$, and $(9, 1.3)$. The first two are the same as above, and thus lie on the line we just found. However, the third data point is not on that line, but rather lies above it. If we are still trying to find a line $y = mx + b$ to fit this data, we would like to find a solution to

$$2.3 = m \cdot 3 + b$$
$$1.7 = m \cdot 6 + b$$
$$1.3 = m \cdot 9 + b$$

or

$$\begin{pmatrix} 3 & 1 \\ 6 & 1 \\ 9 & 1 \end{pmatrix} \begin{pmatrix} m \\ b \end{pmatrix} = \begin{pmatrix} 2.3 \\ 1.7 \\ 1.3 \end{pmatrix}. \tag{8.2}$$

▶ Why can't you attempt to solve this matrix equation by finding a matrix inverse?

Because matrix inverses can exist only for square matrices, straightforward matrix algebra is not sufficient to solve this equation. In fact, we know there is no solution to this matrix equation – if there were, then the three data points would lie exactly on a line. Since we cannot hope for an exact solution to Eq. (8.2), our aim is to instead find values for m and b that minimize SSE.

More generally, suppose we want to find the equation of a line $y = \hat{m}x + \hat{b}$ that, of all lines, best fits the data points (x_1, y_1), (x_2, y_2), ..., (x_n, y_n). We would *like* a solution, $(\hat{m}, \hat{b})$, to a system of equations:

$$y_1 = mx_1 + b$$
$$y_2 = mx_2 + b$$
$$\vdots$$
$$y_n = mx_n + b,$$

which can be written in matrix form as

$$\begin{pmatrix} x_1 & 1 \\ x_2 & 1 \\ \vdots & \vdots \\ x_n & 1 \end{pmatrix} \begin{pmatrix} m \\ b \end{pmatrix} = \begin{pmatrix} y_1 \\ y_2 \\ \vdots \\ y_n \end{pmatrix}. \tag{8.3}$$

However, this equation is unlikely to have a solution $(\hat{m}, \hat{b})$, because the original data points are unlikely to lie exactly on a line. Instead of solving this exactly, we want to find the values of $\hat{m}$ and $\hat{b}$ that "almost" satisfy it, in the precise least-squares sense.

Although we do not yet know $\hat{m}$ and $\hat{b}$, consider a line $y = mx + b$ as a candidate for the best fit one. Let

$$\tilde{y}_i = mx_i + b, \quad i = 1, 2, \ldots, n,$$

denote the y-coordinates of the points on this candidate line, with x-coordinates given by x_i. Then, the error vector for the candidate line will be

$$\begin{aligned}
\mathbf{e}(m, b) &= (y_1 - \tilde{y}_1, y_2 - \tilde{y}_2, \ldots, y_n - \tilde{y}_n) \\
&= (y_1 - mx_1 - b, y_2 - mx_2 - b, \ldots, y_n - mx_n - b).
\end{aligned}$$

The total error, using the sum of squares measure, is then

$$\begin{aligned}
SSE(m, b) &= (y_1 - mx_1 - b)^2 + (y_2 - mx_2 - b)^2 + \cdots + (y_n - mx_n - b)^2 \\
&= \sum_{i=1}^{n} (y_i - mx_i - b)^2.
\end{aligned}$$

Notice that the error vector and the total error depend on the choice of m and b for the line we consider. Our goal is to find values $\hat{m}$ and $\hat{b}$ that minimize this number among all possible choices of m and b.

We focus our attention on $\hat{m}$ first. If $SSE(\hat{m}, \hat{b})$ is minimal, then for any choice of number m, the value of $SSE(m, \hat{b})$ must be equal to or larger than it. That is,

$$SSE(m, \hat{b}) \geq SSE(\hat{m}, \hat{b}),$$

$$\sum_{i=1}^{n} (y_i - mx_i - \hat{b})^2 \geq \sum_{i=1}^{n} (y_i - \hat{m}x_i - \hat{b})^2.$$

Now consider $m = \hat{m} + \epsilon$ for some ϵ, to bring attention to the perturbation of m from its optimal value $\hat{m}$. Substituting this expression for m into the inequality and rearranging terms gives

$$\sum_{i=1}^{n} (y_i - \hat{m}x_i - \epsilon x_i - \hat{b})^2 \geq \sum_{i=1}^{n} (y_i - \hat{m}x_i - \hat{b})^2, \quad \text{or}$$

$$\sum_{i=1}^{n} \left((y_i - \hat{m}x_i - \hat{b} - \epsilon x_i)^2 - (y_i - \hat{m}x_i - \hat{b})^2 \right) \geq 0.$$

But the individual summands simplify as

$$(y_i - \hat{m}x_i - \hat{b} - \epsilon x_i)^2 - (y_i - \hat{m}x_i - \hat{b})^2$$
$$= ((y_i - \hat{m}x_i - \hat{b}) - \epsilon x_i)^2 - (y_i - \hat{m}x_i - \hat{b})^2$$
$$= (y_i - \hat{m}x_i - \hat{b})^2 - 2\epsilon x_i(y_i - \hat{m}x_i - \hat{b})$$
$$+ (\epsilon x_i)^2 - (y_i - \hat{m}x_i - \hat{b})^2$$
$$= -2\epsilon x_i(y_i - \hat{m}x_i - \hat{b}) + (\epsilon x_i)^2.$$

Therefore, the inequality above is

$$\sum_{i=1}^{n}(-2\epsilon x_i(y_i - \hat{m}x_i - \hat{b}) + (\epsilon x_i)^2) \geq 0,$$

or

$$-2\epsilon \left(\sum_{i=1}^{n} x_i(y_i - \hat{m}x_i - \hat{b}) \right) + \epsilon^2 \left(\sum_{i=1}^{n} x_i^2 \right) \geq 0.$$

In this inequality, considering a value of ϵ sufficiently close to zero, the second term will be of negligible size in comparison with the first, due to the ϵ^2. Thus, for all small values of ϵ, the first term must be nonnegative for the inequality to be satisfied. However, since ϵ might be either positive or negative, the only way the first term is always nonnegative is if

$$\sum_{i=1}^{n} x_i(y_i - \hat{m}x_i - \hat{b}) = 0. \qquad (8.4)$$

This gives us an equation $\hat{m}$ and $\hat{b}$ must satisfy to minimize SSE. Because it is an equation in only two unknowns (*Remember:* all the x_i and y_i are data values), it is more simply expressed as

$$\left(\sum_{i=1}^{n} x_i^2 \right) \hat{m} + \left(\sum_{i=1}^{n} x_i \right) \hat{b} = \sum_{i=1}^{n} x_i y_i. \qquad (8.5)$$

After much work, we have found one equation that relates $\hat{m}$ and $\hat{b}$. To find a second equation relating $\hat{m}$ and $\hat{b}$, we reason similarly focusing on $\hat{b}$. The complete argument is left for the exercises, but it yields

$$\left(\sum_{i=1}^{n} x_i \right) \hat{m} + n\hat{b} = \sum_{i=1}^{n} y_i. \qquad (8.6)$$

We now have two equations, (8.5) and (8.6), called the *normal equations*, that relate $\hat{m}$ and $\hat{b}$. With two equations in two unknowns, we can solve for $\hat{m}$ and $\hat{b}$ and so find the least-squares line $y = \hat{m}x + \hat{b}$.

Before continuing with the now routine calculation of the solutions to the normal equations, a slight detour leads to a remarkable observation about the structure of the normal equations. We need a definition first.

Definition. If M is an $m \times n$ matrix, then the $n \times m$ matrix obtained by interchanging the rows and columns of M is known as the *transpose* of M, and is denoted by M^T; for

$$
M = \begin{pmatrix} x_{11} & x_{12} & \cdots & x_{1n} \\ x_{21} & x_{22} & \cdots & x_{2n} \\ \vdots & \vdots & \ddots & \vdots \\ x_{m1} & x_{m2} & \cdots & x_{mn} \end{pmatrix}, \quad
M^T = \begin{pmatrix} x_{11} & x_{21} & \cdots & x_{m1} \\ x_{12} & x_{22} & \cdots & x_{m2} \\ \vdots & \vdots & \ddots & \vdots \\ x_{1n} & x_{2n} & \cdots & x_{mn} \end{pmatrix}.
$$

Example. If $A = \begin{pmatrix} 3 & 1 \\ 6 & 1 \\ 9 & 1 \end{pmatrix}$, then $A^T = \begin{pmatrix} 3 & 6 & 9 \\ 1 & 1 & 1 \end{pmatrix}$. Notice the first row of A becomes the first column of A^T, the second row of A becomes the second column of A^T, and so on. At the same time, the columns of A have become the rows of A^T.

Let's return to the original matrix Eq. (8.3) that we would have like to have solved to find a line through the data points. If we multiply each side of the equation on the left by the transpose of the matrix appearing in it, we obtain

$$
\begin{pmatrix} x_1 & x_2 & \cdots & x_n \\ 1 & 1 & \cdots & 1 \end{pmatrix} \begin{pmatrix} x_1 & 1 \\ x_2 & 1 \\ \vdots & \vdots \\ x_n & 1 \end{pmatrix} \begin{pmatrix} m \\ b \end{pmatrix} = \begin{pmatrix} x_1 & x_2 & \cdots & x_n \\ 1 & 1 & \cdots & 1 \end{pmatrix} \begin{pmatrix} y_1 \\ y_2 \\ \vdots \\ y_n \end{pmatrix},
$$

which, on multiplying the matrices, gives

$$
\begin{pmatrix} x_1^2 + x_2^2 + \cdots + x_n^2 & x_1 + x_2 + \cdots + x_n \\ x_1 + x_2 + \cdots + x_n & 1 + 1 + \cdots + 1 \end{pmatrix} \begin{pmatrix} m \\ b \end{pmatrix}
$$

$$
= \begin{pmatrix} x_1 y_1 + x_2 y_2 + \cdots + x_n y_n \\ y_1 + y_2 + \cdots + y_n \end{pmatrix},
$$

or, more succinctly,

$$
\begin{pmatrix} \sum x_i^2 & \sum x_i \\ \sum x_i & n \end{pmatrix} \begin{pmatrix} m \\ b \end{pmatrix} = \begin{pmatrix} \sum x_i y_i \\ \sum y_i \end{pmatrix}, \tag{8.7}
$$

where the sums range over $i = 1, \ldots, n$.

Now compare Eq. (8.7) to Eqs. (8.5) and (8.6). Amazingly, these equations are exactly the same; Eq. (8.5) is stored in the top row of Eq. (8.7), while Eq. (8.6) is in the bottom row. This observation provides a quick way to perform the method of least-squares fitting: To find the least-squares solution to a matrix equation of the form of Equation (8.3), multiply each side of the equation on the left by the transpose of the matrix and solve the resulting system for $\hat{m}$ and $\hat{b}$.

To apply this to finding the least-squares, best-fit line for the three data points $(3, 2.3), (6, 1.7)$, and $(9, 1.3)$, from Equation (8.2), we obtain the normal equations

$$\begin{pmatrix} 3 & 6 & 9 \\ 1 & 1 & 1 \end{pmatrix} \begin{pmatrix} 3 & 1 \\ 6 & 1 \\ 9 & 1 \end{pmatrix} \begin{pmatrix} m \\ b \end{pmatrix} = \begin{pmatrix} 3 & 6 & 9 \\ 1 & 1 & 1 \end{pmatrix} \begin{pmatrix} 2.3 \\ 1.7 \\ 1.3 \end{pmatrix}, \quad \text{or}$$

$$\begin{pmatrix} 126 & 18 \\ 18 & 3 \end{pmatrix} \begin{pmatrix} m \\ b \end{pmatrix} = \begin{pmatrix} 28.8 \\ 5.3 \end{pmatrix}.$$

Now multiplying both sides of the last equation by

$$\begin{pmatrix} 126 & 18 \\ 18 & 3 \end{pmatrix}^{-1} = \frac{1}{126 \cdot 3 - 18 \cdot 18} \begin{pmatrix} 3 & -18 \\ -18 & 126 \end{pmatrix}$$

gives

$$\begin{pmatrix} \hat{m} \\ \hat{b} \end{pmatrix} = \frac{1}{126 \cdot 3 - 18 \cdot 18} \begin{pmatrix} 3 & -18 \\ -18 & 126 \end{pmatrix} \begin{pmatrix} 28.8 \\ 5.3 \end{pmatrix} \approx \begin{pmatrix} -.1667 \\ 2.7667 \end{pmatrix}.$$

The least-squares best-fit line for the three data points is thus

$$y = -.1667x + 2.7667.$$

▶ Graph this line and the three data points. Does the line appear to fit the data well?

In using the least-squares approach to fit a line to data, the most important point is that you understand the criteria that you are using to choose the best line – the one with the smallest SSE. Of secondary importance is the calculation you do to actually get that line. The steps for this are:

1. Write equations you would like m and b to satisfy for all the data points to be on the line $y = mx + b$, by plugging each data point into the equation. For n data points, this gives n equations in the two unknowns, m and b, that usually cannot be solved exactly.

2. Express the equations in matrix form as

$$A \begin{pmatrix} m \\ b \end{pmatrix} = \mathbf{b}.$$

Here, A will be a matrix and $\mathbf{b}$ a vector, each with numerical entries.

3. Create the *normal equations* by multiplying on the left of both sides of the equation by A^T, giving

$$A^T A \begin{pmatrix} m \\ b \end{pmatrix} = A^T \mathbf{b}.$$

4. Solve the normal equations by computing $A^T A$, $A^T \mathbf{b}$, and $(A^T A)^{-1}$. The solution is

$$\begin{pmatrix} \hat{m} \\ \hat{b} \end{pmatrix} = (A^T A)^{-1} A^T \mathbf{b}.$$

Notice that the steps here say nothing about the real ideas behind least-squares – that appeared only in our derivation of the normal equations. However, the steps have the nice feature that they provide a simple and straightforward calculation.

Most software packages and calculators will calculate a least-squares, best-fit line (often called a *regression* line) at the touch of a button. Once you understand the idea and method of calculation, these are great labor-saving devices.

Although the matrix A was not invertible in the three data point example above, the matrix product $A^T A$ was invertible. Because of the particular form of the columns of any matrix A used in least-squares regression, the product $A^T A$ is almost always invertible, ensuring that a least-squares solution can be found using matrix algebra. In fact, provided the data has at least two points with different x-coordinates, $A^T A$ will be invertible, although a proof of this fact requires additional theory from linear algebra. Moreover, when $A^T A$ is invertible, there is one and only one solution to the normal equations. This justifies talking about *the* least-squares, best-fit line for a data set; one line is genuinely better than all others in giving a smallest value for SSE.

Problems

8.2.1. Plot the three points $(-1, 1)$, $(0, 3)$, and $(1, 4)$. Then, find the least-squares, best-fit line for them, following the four steps outlined in the text and doing all calculations without a computer. Add a graph of the line to your plot.

Table 8.3. *Population Size in Year t*

t	0	1	2	3	4	5
P	173	278	534	895	1553	2713

8.2.2. Find the least-squares, best-fit line to the data points (3, 120), (4, 116), (5, 114), (6, 109), and (7, 106) by:
  a. following the four steps given in the text, using a computer. The MATLAB command A' gives the transpose of a matrix A.
  b. following the first two steps and then using the MATLAB command A\b to find the least-squares solution $\hat{x}$ to $Ax = \mathbf{b}$.
  c. Using the MATLAB command polyfit. For instructions, type help polyfit.

8.2.3. Recall from the last section that the data of Table 8.1 showed an exponential decay that we hoped to model by an exponential formula. Table 8.2 contains transformed data that is roughly linear.
  a. Find the least-squares, best-fit line $\ln y = \hat{m}t + \hat{b}$ to the data in Table 8.2.
  b. Use your answer to part (a) to give an exponential curve $y = ae^{kt}$ fitting the data in Table 8.1.
  Note: This approach to fitting an exponential curve, using a least-squares, best-fit line to the transformed data, does not necessarily give the exponential that minimizes SSE for the untransformed data. It is, however, a standard approach to exponential curve fitting.

8.2.4. Suppose the population data in Table 8.3 is believed to be described by the model $P_{t+1} = \lambda P_t$.
  a. Produce a semilog plot and explain why it justifies the choice of the model.
  b. Find the least-squares, best-fit line to the transformed data.
  c. Use part (b) to find an exponential curve fitting the data.
  d. Use part (c) to give a good estimate of λ for this data.

8.2.5. To produce and plot simulated data points that will be nearly on the line $y = .7x + 2.1$, use the MATLAB commands

```
x=[1:10]', y=.7*x+2.1+.3*randn(10,1),
    plot(x,y,'o').
```

Then A=[x,x.^0], b=y will prepare you to perform the least-squares, line-fitting calculation.

a. Enter all these commands and recover the least-squares line. Is it exactly the line $y = .7x + 2.1$? Is it close? Perform this experiment several times and summarize your results.

b. What is the effect of the number $.3$ in these commands? If $.3$ is replaced by 3, is the line that you recover usually more or less similar to $y = .7x + 2.1$? Explain why your observation is reasonable.

c. What is the effect of using fewer or more data points on the recovery of the line? For instance, if the number 10 in these commands is replaced with 3 or 30, is the line you recover usually more or less similar to $y = .7x + 2.1$? Explain why your observation is reasonable.

8.2.6. That there is exactly one straight line through any two points is well known. However, this fact manifested itself in the fact that Equation (8.1) was solvable and had a unique solution.

a. Explain why, through any three points in the plane, you would expect to be able to find exactly one parabola of the form $y = ax^2 + bx + c$ by expressing the equations you would need to solve to find a, b, and c in matrix form. What is it about the matrix equation that suggests there is probably one and only one solution?

b. Given n points, what degree polynomial $y = p(x)$ should you consider to be likely to find one and only one such polynomial curve through the data points? Explain.

8.2.7. Consider the four data points $(-2, 8.1)$, $(0, 7)$, $(10, 5.9)$, and $(15, 5)$.

a. Use MATLAB to plot the four data points and the least-squares line fitting them.

b. Calculate the mean x-coordinate $\bar{x}$ and mean y-coordinate $\bar{y}$. Does the least-squares line pass through $(\bar{x}, \bar{y})$?

c. Perform several similar experiments by varying the data points.

8.2.8. Show that $\hat{m}$ and $\hat{b}$ for the least-squares, best-fit line to data must satisfy Eq. (8.6) as follows:

a. Explain why $SSE(\hat{m}, b) \geq SSE(\hat{m}, \hat{b})$ for any choice of b.

b. Using $b = \hat{b} + \delta$, show that for all δ

$$\sum_{i=1}^{n} \left((y_i - \hat{m}x_i - \hat{b} - \delta)^2 - (y_i - \hat{m}x_i - \hat{b})^2 \right) \geq 0.$$

c. Show that for all δ

$$-2\delta \left(\sum_{i=1}^{n} (y_i - \hat{m}x_i - \hat{b}) \right) + \delta^2 n \geq 0.$$

d. Explain why part (c) shows

$$\sum_{i=1}^{n}(y_i - \hat{m}x_i - \hat{b}) = 0.$$

e. Deduce Eq. (8.6).

8.2.9. (Calculus) The normal equations for least-squares, line fitting can also be derived using calculus. Recall that at the minimum of a differentiable function, the derivative must be zero.

a. Derive Eq. (8.4) by differentiating $SSE(m, \hat{b})$ with respect to m and setting the result equal to zero.

b. Derive Eq. (8.6) similarly.

8.2.10. Because the normal Eqs. (8.7) can be solved by inverting a 2×2 matrix, the formula for the matrix inverse leads to a formula for $\hat{m}$ and $\hat{b}$. Use this approach to deduce

$$\hat{m} = \frac{n\left(\sum x_i y_i\right) - \left(\sum x_i\right)\left(\sum y_i\right)}{n\left(\sum x_i^2\right) - \left(\sum x_i\right)^2}$$

$$\hat{b} = \frac{\left(\sum x_i^2\right)\left(\sum y_i\right) - \left(\sum x_i\right)\left(\sum x_i y_i\right)}{n\left(\sum x_i^2\right) - \left(\sum x_i\right)^2}.$$

8.2.11. Use the result of the last problem to show that, if $\bar{x}$ an $\bar{y}$ are the mean x- and y-coordinates of the data points, then $(\bar{x}, \bar{y})$ lies on the least-squares, best-fit line.

8.2.12. The least-squares solution to the equation $Ac = b$, for a vector of unknowns c, is given by $\hat{c} = (A^T A)^{-1} A^T b$. You might think this formula could be simplified as

$$\hat{c} = (A^T A)^{-1} A^T b$$
$$= A^{-1}(A^T)^{-1} A^T b$$
$$= A^{-1} I\, b$$
$$= A^{-1} b.$$

Explain the error in this reasoning.

8.3. Polynomial Curve Fitting

While least-squares fitting of straight lines to data is very commonly done, fitting higher degree polynomials is often useful as well. If we have decided to model some data by an nth degree polynomial $y = f(t)$, then finding the polynomial of that degree minimizing SSE can be done by a procedure very much like that outlined in the last section. Although we will not give a proof of why the calculation works to produce the least-squares, best-fit polynomial, an argument similar to the one for the best-fit line can be made. Instead, we will look at an interesting example.

Modeling the growth of AIDS. In the early stages of epidemics of most infectious diseases, the total number of cases often grows exponentially. More precisely, suppose at times $t = 0, 1, 2, \ldots$, the total number of infected individuals in a population is counted, yielding $I_0, I_1, I_2, \ldots$. Then, a plot of the points (t, I_t) typically shows the data points clustered around a curve that looks much like an exponential growth curve, at least for small values of t. As the exercises will show, this type of behavior is predicted by standard infectious disease models, such as those of Chapter 7. Figure 7.1 of that chapter shows an example of this, where the plot of the number of infectives grows roughly exponentially in the early time steps.

The CDC began collecting data on the AIDS epidemic in the United States on a monthly basis in 1982, with reporting of new cases required by law. Although there are many flaws in this data, such as time lags between infection, diagnosis, and reporting, it still provides our best picture of the spreading epidemic. A change in the surveillance definition in 1987, and again in 1993, further complicate analysis. Nonetheless, studying this data might give useful insights into the mechanisms of the disease spread.

Recorded in Table 8.4 are some of the data on the cumulative number of AIDS cases in the United States, as reported by the end of the calendar year. Note that because data are cumulative, all cases reported by the end of a year appear in the count for subsequent years as well.

When a plot of this data is made, it does show a marked increase over time. However, unlike most epidemic data, it does *not* seem to grow approximately exponentially.

► If the data were growing approximately exponentially, what would a semilog plot of the data look like?

Table 8.4. *Cumulative No. of AIDS Cases Reported to CDC*

Year	1	2	3	4	5	6	7
Cases	158	767	2787	7198	15454	28629	50280

Note: "Year" is year since 1980.

As you will see in the exercises, a semilog plot of this data does not produce the approximately linear behavior that an exponential model would lead to. More surprisingly, a log–log plot of the data shows the transformed points clustering along a line with slope approximately 3.

▶ If a log–log plot is linear, with slope 3, then what curve is likely to be a good model for the data?

The log–log plot suggests that fitting the data to a polynomial of the form $y = at^3$ might be appropriate. However, we will use a slightly more general curve produced by the general cubic polynomial

$$y = c_3 t^3 + c_2 t^2 + c_1 t + c_0.$$

Here, t represents year since 1980, and y the cumulative number of AIDS cases reported.

For each of the seven data points, we have an equation relating the unknown coefficients c_3, c_2, c_1, and c_0. For example, from the points $(1, 158)$ and $(7, 50280)$, we obtain the equations

$$158 = c_3(1)^3 + c_2(1)^2 + c_1(1)^1 + c_0,$$
$$50{,}280 = c_3(7)^3 + c_2(7)^2 + c_1(7)^1 + c_0.$$

Instead of writing each of the seven equations down individually, we can express the system in matrix form as

$$\begin{pmatrix} 1^3 & 1^2 & 1 & 1 \\ 2^3 & 2^2 & 2 & 1 \\ 3^3 & 3^2 & 3 & 1 \\ 4^3 & 4^2 & 4 & 1 \\ 5^3 & 5^2 & 5 & 1 \\ 6^3 & 6^2 & 6 & 1 \\ 7^3 & 7^2 & 7 & 1 \end{pmatrix} \begin{pmatrix} c_3 \\ c_2 \\ c_1 \\ c_0 \end{pmatrix} = \begin{pmatrix} 158 \\ 767 \\ 2787 \\ 7198 \\ 15454 \\ 28629 \\ 50280 \end{pmatrix}.$$

More compactly, $A\mathbf{c} = \mathbf{b}$, where $\mathbf{b}$ is the 7×1 column vector containing the cumulative numbers of cases, A is the 7×4 matrix constructed from powers of the t-values, and $\mathbf{c}$ is the vector of unknown coefficients we hope to find.

▶ Explain why this matrix equation could only have an exact solution if all seven points lie on the graph of a cubic.

Of course, we do not expect this equation to have an exact solution, because we do not expect any cubic to pass exactly through seven data points. Instead, we are interested in finding an approximate solution $\hat{\mathbf{c}}$, which minimizes SSE.

Fortunately, the algorithm for finding this least-squares solution is just as for fitting a line: Multiply each side of the equation on the left by A^T to get the normal equations and solve them exactly.

In full detail,

$$
\begin{pmatrix} 1^3 & 2^3 & 3^3 & 4^3 & 5^3 & 6^3 & 7^3 \\ 1^2 & 2^2 & 3^2 & 4^2 & 5^2 & 6^2 & 7^2 \\ 1 & 2 & 3 & 4 & 5 & 6 & 7 \\ 1 & 1 & 1 & 1 & 1 & 1 & 1 \end{pmatrix}
\begin{pmatrix} 1^3 & 1^2 & 1 & 1 \\ 2^3 & 2^2 & 2 & 1 \\ 3^3 & 3^2 & 3 & 1 \\ 4^3 & 4^2 & 4 & 1 \\ 5^3 & 5^2 & 5 & 1 \\ 6^3 & 6^2 & 6 & 1 \\ 7^3 & 7^2 & 7 & 1 \end{pmatrix}
\begin{pmatrix} \hat{c}_3 \\ \hat{c}_2 \\ \hat{c}_1 \\ \hat{c}_0 \end{pmatrix} =
$$

$$
\begin{pmatrix} 1^3 & 2^3 & 3^3 & 4^3 & 5^3 & 6^3 & 7^3 \\ 1^2 & 2^2 & 3^2 & 4^2 & 5^2 & 6^2 & 7^2 \\ 1 & 2 & 3 & 4 & 5 & 6 & 7 \\ 1 & 1 & 1 & 1 & 1 & 1 & 1 \end{pmatrix}
\begin{pmatrix} 158 \\ 767 \\ 2787 \\ 7198 \\ 15454 \\ 28629 \\ 50280 \end{pmatrix} .
$$

Using a computer to perform the matrix multiplication yields

$$
\begin{pmatrix} 184820 & 29008 & 4676 & 784 \\ 29008 & 4676 & 784 & 140 \\ 4676 & 784 & 140 & 28 \\ 784 & 140 & 28 & 7 \end{pmatrix}
\begin{pmatrix} \hat{c}_3 \\ \hat{c}_2 \\ \hat{c}_1 \\ \hat{c}_0 \end{pmatrix} =
\begin{pmatrix} 25903869 \\ 4024191 \\ 639849 \\ 105273 \end{pmatrix} .
$$

Finally, solving this, we find

$$
\mathbf{c} = (A^T A)^{-1} A^T b \quad \text{or}
$$

$$
\mathbf{c} \approx \begin{pmatrix} 266.5 \\ -1189.6 \\ 2673.8 \\ -1708.6 \end{pmatrix} .
$$

Thus, the equation

$$
y = 266.5t^3 - 1189.6t^2 + 2673.8t - 1708.6
$$

is the least-squares, best-fit cubic to the cumulative AIDS data.

The plot in Figure 8.3, showing the AIDS data together with the fitted curve illustrates how good a fit this cubic is. Indeed, the fit is extraordinarily good. In the exercises, we will quantify this and see in a different way why a cubic is a particularly good choice of curve for this data.

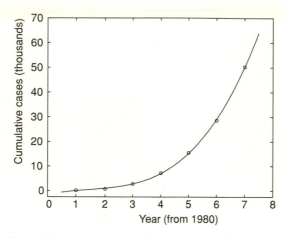

Figure 8.3. Least-squares cubic fitting data of Table 8.4.

Why the AIDS epidemic in the United States appears to have grown cubically, rather than exponentially, in its early years is an interesting question. Of course AIDS, with its complicated pattern of transmission through different behaviors and in various subpopulations, requires a much more complicated model than those of Chapter 7 to even begin to capture its dynamics. For instance, transmission through sexual contact, intravenous drug use, or blood transfusions is not likely to satisfy the homogeneous mixing assumption of basic infection models. Still, even without homogenous mixing, one might expect exponential growth within "well-mixed" subpopulations, and more detailed data analysis fails to show that. In (Colgate *et al.*, 1989), a possible explanation for the cubic growth is proposed through a mathematical model that includes variation in behavior placing individuals at risk.

We could do an even better job of fitting a polynomial to this data if we allowed ourselves to use one of a higher degree. Because there are only 7 data points in our example, attempting to fit a sixth-degree polynomial to the data would lead to a matrix equation $A\mathbf{c} = \mathbf{b}$, where A is a square 7×7 matrix. That system can be solved exactly, and so we can find a polynomial of degree 6 whose graph goes through all the data points exactly.

However, fitting the data exactly in this way is not desirable if a simpler cubic curve already does such a good job of capturing the main data trend. We expect data to not conform exactly to a model and do not want to *overfit* the data with a complicated curve. We instead choose our model so that it balances the competing demands of simplicity and providing a good fit. Some of the exercises will investigate this issue more.

Problems

8.3.1. Produce regular, semilog, and log–log plots of the data in Table 8.4. Why do your plots indicate that an exponential curve is probably not an appropriate model, whereas a cubic polynomial might be?

8.3.2. In using MATLAB to perform polynomial fitting, we often need to enter a matrix like $A = \begin{pmatrix} 1^2 & 1^1 & 1 \\ 2^2 & 2^1 & 1 \\ 3^2 & 3^1 & 1 \end{pmatrix}$. One way to produce it is with the commands x = [1,2,3]', A = [x.^2, x, x.^0].

Practice by using MATLAB to check all steps of the calculation of the least-squares cubic fitting the data in Table 8.4. The cumulative numbers of cases in the table can be loaded by running aidsdata.

8.3.3. Consider the following table of data:

x	1	2	3	4	5
y	1.1	8.7	19.8	39.5	64.7

a. What matrix equation would be solvable if the data points all lay on a quadratic, $y = ax^2 + bx + c$?

b. What are the normal equations that should be solved to find the least-squares, best-fit quadratic?

c. What is the least-squares, best-fit quadratic for this data?

8.3.4. Consider the four points $(1, 2)$, $(2, 9)$, $(3, 1)$, and $(4, 4)$.

a. If all these points were on a line, what matrix equation $A\mathbf{c} = \mathbf{b}$ would have a solution? What are the associated normal equations for the least-squares, best-fit line? What is the best-fit line? What is SSE for this line?

b. If all these points were on a quadratic curve, what matrix equation $A\mathbf{c} = \mathbf{b}$ would have a solution? What are the associated normal equations for the least-squares, best-fit quadratic? What is the best-fit quadratic? What is SSE for this quadratic?

c. If all these points were on a cubic curve, what matrix equation $A\mathbf{c} = \mathbf{b}$ would have a solution? What are the associated normal equations for the least-squares best-fit cubic? What is the best-fit cubic? What is SSE for this cubic?

d. Produce a plot of the four data points and the best linear, quadratic, and cubic curves fitting the points.

8.3.5. Suppose six data points were collected in an experiment: (1, 40.2), (3, 29.4), (5, 27), (7, 18.2), (8, 18), and (9, 14).

    a. Use MATLAB to graph the data points. What degree polynomial do you think might be a good fit for these data?

    b. Use built-in MATLAB commands to fit polynomials of varying degrees to the data. To get started, try using the following commands:

```
xdata = [1 3 5 7 8 9],
ydata = [40.2 29.4 27 18.2 18 14]
plot(xdata, ydata, 'ro'), axis([0 10 0 50])
hold on
L1 = polyfit(xdata,ydata,1);
x = [0:.1:10]; y = polyval(L1, x);
plot(x,y)
```

Modify the commands above to graph the best least-squares polynomial for degrees $n = 2, 3, 4, 5$. For example, you will want to use the command L3 = polyfit(xdata, ydata, 3) to get the coefficients for the least-squares cubic.

    c. What degree polynomial best captures the tendency of the data? Although the degree 5 polynomial passes through all six data points, why might it be a poor choice to describe the data?

8.3.6. We can find least-squares, best-fit polynomials of various degrees fitting a set of data. The larger the degree of the polynomial, the better fit we can get. Generally, it is desirable to fit data with as simple a function as possible that does a good job. Investigate the best-fit polynomials of degree 1 through degree 6 for the data in Table 8.4 by running aidsdata and then using the MATLAB commands

```
x=[1:7]', y=cml1981_1987
plot(x,y,'ro'), hold on
xx=[1:.1:7];
for i=1:6
    c=polyfit(x,y,i); yf=polyval(c,x); e=y-yf;
    sse=e'*e; yy=polyval(c,xx); plot(xx,yy)
    disp(['Degree = ', num2str(i), ', SSE = ',...
      num2str(sse)])
    pause
end
```

    a. From the graphs these command produce, why does it seem most reasonable to fit the data with a cubic?

    b. From SSEs computed by these commands, why does it seem reasonable to fit the data with a cubic?

8.3.7. Running `aidsdata` creates a variable `cmlJan1982_Dec1987` with monthly data on cumulative AIDS cases that is more detailed than that in Table 8.4. Modify the commands in the last problem to fit polynomials of various degrees to this data. Based on both the graphs and SSE, what degree polynomial do you think is a good one to fit this data? Is a cubic polynomial still a reasonable choice to model it?

8.3.8. Consider the three data points $(1, 3)$, $(2, 5)$, and $(3, 10)$, and the problem of fitting a *horizontal* line of the form $y = c$ to them.

    a. What matrix equation would have a solution if all these points were on a horizontal line? What is the associated normal equation? What is the least-squares, best-fit horizontal line?

    b. Show the result in part (a) could be found by averaging the y-coordinates of the data.

    c. Show that, for any set of data points, the least-squares, best-fit horizontal line is always given by $y = \bar{y}$, where $\bar{y}$ is the average y-coordinate of the data.

8.3.9. The CDC's AIDS data provides a good example of why caution is necessary in extrapolating. In MATLAB, type `aidsdata` to define the variables `cmlJan1982_Dec1987`, the monthly cumulative number of AIDS cases from January 1982 to December 1987, and `cmlJan1982_Dec2000`, the monthly cumulative number of cases from January 1982 to December 2000.

    a. Use MATLAB to plot `cmlJan1982_Dec1987`, find the best-fit cubic modeling that data, and plot the cubic with the data. Does the cubic seem to be an adequate fit?

    b. Plot the data `cmlJan1982_Dec2000`, along with the prediction of the data given by the cubic you found in part (a). Are they close?

    c. What biological, medical, or social factors might be responsible for what you observed in part (b)?

    d. What degree polynomial is needed to provide a reasonable model of the data `cmlJan1982_Dec2000`? Find a good polynomial model and graph it along with the data.

8.3.10. Simple infectious disease models result in approximately exponential growth of the number of infectives in the early stages of an epidemic.

To see this, first consider an SI model, where $\Delta I = \alpha S I$ for some parameter α.

a. If the total population is N, and I_t is small relative to N, explain why

$$I_{t+1} = I_t + \alpha I_t (N - I_t) \approx (1 + N\alpha)I_t.$$

b. Explain why the approximation $I_{t+1} \approx (1 + N\alpha)I_t$ leads to approximately exponential growth.

c. Show similarly that an SIR or SIS model will show approximately exponential growth in the number of infectives in the early days of an epidemic.

Appendix A
Basic Analysis of Numerical Data

Often, the goal of an experiment is the taking of some measurement or a series of measurements. Although it may seem that, with such data in hand, the important work has been done, and all that remains is the mopping up of data analysis, the interpretation of the raw numbers may be as involved and difficult as any experimental setup. Numbers by themselves tell you nothing and extracting meaning from them is an art.

In this appendix, we look at some of the basic ideas involved in interpreting numerical data. We will not focus on any particular type of experiment, but rather imagine the likely outcomes of many measurements and learn the simplest ways of extracting information from large batches of numbers. Although not all data are numerical in nature (you might record qualitative information such as color, for example), it is only numerical data that will be discussed here.

We also focus primarily on questions of the interpretation of data and do not attempt to discuss points of experimental design. This is actually a rather artificial distinction, since when designing an experiment, a scientist must be sure that once data are obtained they will be amenable to analysis. Thus, what may appear as an after-the-fact analysis of data in this discussion should really be an analysis that the experimenter intended to do from the start.

A.1. The Meaning of a Measurement

To be concrete, suppose we are interested in investigating the effects of a certain nutrient on the growth of plants. We prepare two pots of soil, adding a certain amount of the nutrient to one (the experimental pot), but not to another (the control pot). In each pot, we plant a bean and then measure the height of the bean after 20 days. Suppose at that time the control bean is 10 cm tall, and the experimental bean is 15 cm tall.

An important goal of an experiment is to be able to draw conclusions that you can then apply in other situations.

345

▸ What can we conclude from this experiment? Does the nutrient cause beans to grow more? Does it cause them to be 150% taller? Would you feel comfortable making predictions about other beans based only on this experiment's result? Would you be surprised if another experimenter got different results? Why or why not?

A cautious scientist would be hesitant to conclude anything from this experiment. In part, this is because of background knowledge that plant growth seems to be highly variable, even under seemingly similar conditions. Also, humans often make mistakes, and the experimenter may have unwittingly botched the experiment.

We would probably feel better drawing a conclusion if the experiment were repeated many times. (Of course, it might be best to design the experiment so all the repetitions are done at the same time, because that would cut down on variation in other factors such as temperature, length of day, etc.) Perhaps the control and experimental groups should have 10 beans each, or 100, or 1,000.

▸ How does an experimenter decide how many repetitions of an experiment should be done? What trade-offs must be made? If the bean experiment were repeated with pine trees that were to be grown for 20 years, would it be reasonable to use the same number of repetitions as for beans?

Pretend we redid the experiment, this time using five beans in each group (five is used only to keep the amount of data small for illustrative purposes). The heights in centimeters are measured and found to be

Control: 9.3, 14.2, 11.7, 10.2, 9.8

Experimental: 12.1, 16.3, 13.2, 13.5, 14.9

Notice there is variability in the data, in that not all control plants reached exactly the same height, nor did the experimental ones. In fact, one of the control plants is actually taller than three of the experimental ones.

▸ Does the original data for one bean seem to fit with this data? If it does, would you draw the same general conclusions you might have before? Would there be any subtle differences in your conclusions? Should you feel more confident about your conclusion and if so how much more confident?

▸ How would you briefly summarize, in words and not numbers, your conclusions based on this data?

▶ If another researcher repeated this experiment with only one plant in each group and found the control plant was 13.6 cm tall and the experimental one was 13.4 cm tall, would that data be surprising? What conclusion would that experimenter be likely to draw based solely on that data? Is your data compelling enough to say the other researcher's conclusion is wrong?

Clearly, an important issue in analyzing this data is understanding the variability within each group. In fact, you have probably already made some hypotheses as to why the numbers might be so varied within each group.

▶ Give as many reasons as you can for the variability in the data.

It's worth distinguishing two main reasons why data might vary. The first, called experimental error, is due to mistakes made (perhaps unavoidably) on the part of the experimenter. For instance, the ruler used for measuring height might be inaccurate, or the location of the top of the plant might have been misjudged, or the nutrient may not have been applied in exactly the amount claimed.

The second reason is that, in dealing with a very complicated system such as a living organism, there are simply more variables than we can possibly control at once. For instance, the beans may differ genetically, and the conditions of soil, light, and air that each plant is exposed to are not identical no matter how hard we try to make them so. One could argue that this is all a form of experimental error, in that we have not been able to carry out our experiment carefully enough. That misses the point, though, because if the experiment could be carried out so that none of this variability were present, then our results might actually be *less* meaningful. Knowing how all clones of a specific bean would respond to certain very specific conditions may well be less valuable than knowing how a random sampling of beans will behave in a less tightly controlled setting.

In studying anything complicated (and biological system are all complicated), we should expect variability in measurements. Experimental error should, of course, be minimized, but variability in the data often will remain. We should take lots of measurements to be sure we have a good idea of the nature of this variability, so that the variability within the data does not obscure the effects we are trying to measure. The more data we have, the better conclusions we should be able to draw.

We have now arrived at the central problem the discipline of statistics is designed to address. Too little data can be misleading, so that we draw incorrect conclusions, but too much data becomes incomprehensible to us.

How can we boil down the large quantities of information we need to prevent mistakes into a simpler, yet meaningful nugget of information? Given that variability is not just due to error but actually an important part of the systems we are studying, how can we quantify the natural variability of what we are experimentally investigating? In what follows, we begin to address these questions.

A.2. Understanding Variable Data – Histograms and Distributions

People often find pictorial tools useful in understanding data – perhaps because vision is a more basic function of our brain than numerical reasoning. Thus, our first step in understanding variable data will be based on a graphical device, the histogram.

Let's consider just the control group of beans. Suppose we had 20 beans in this group, and the heights we measured in centimeters were:

9.3, 9.7, 10.1, 10.2, 10.4, 10.6, 10.7, 10.7, 10.9, 11.0,

11.1, 11.1, 11.3, 11.3, 11.6, 11.7, 11.9, 12.3, 12.4, 13.4.

We begin by grouping the data together in intervals of some convenient size. For instance, here we see there are two data points between 9 and 10, seven data points between 10 and 11, eight between 11 and 12, two between 12 and 13, and one between 13 and 14. Thus, we draw the histogram on the left of Figure A.1.

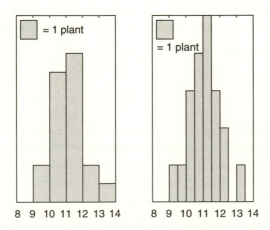

Figure A.1. Two histograms describing the same data.

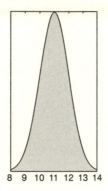

8 9 10 11 12 13 14

Figure A.2. A normal distribution approximating the histograms in Figure A.1.

Notice that the *area* of each bar tells us the number of data points in that interval. Here, the width of each bar is 1, so the height also tells us the number of data points, but it will be the area that matters.

Grouping the data using a finer scale, with intervals having width 0.5 instead of 1, produces the histogram on the right of Figure A.1. Because area represents the number of data points, and the bars are now 0.5 wide, a single plant is denoted by a bar twice as high as what would have appeared in the left figure.

By insisting on using area to denote the number of data points in an interval, we have kept the total shaded area in the each of these histograms the same. If we had used vertical heights to denote the number of data points, the second graph would look much flatter than the first, since there are fewer points in each of its intervals. Although the two histograms are very similar, the smaller interval size in the second one makes it appear a little less step-like.

▶   How would the histogram change if an interval of .25 was used to group the data? An interval of .05?

▶   Suppose we had 100 data points to work with to construct histograms in this manner? How would the histograms be different? How would the histogram change if the grouping interval, or *bin size*, was made smaller?

With enough data points, and a sufficiently fine bin size, the histograms appear to look more and more like the graph in Figure A.2, which we call a *distribution*. The jaggedness of the original histogram is smoothed out.

This particular distribution has a bell-shaped curve and is called a *normal* distribution. "Normal" here is a technical term that you should not think of as

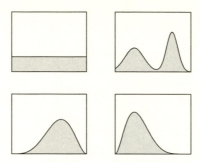

Figure A.3. Four distributions.

having anything to do with meanings like "natural," "common," "good," or "expected." Although it is common for data to follow a normal distribution, it is also common that it does not.

A good way of thinking of the distribution is that if you repeated your experiment endlessly, taking measurements that were fully accurate, and plotted histograms with finer and finer groupings, then the histograms would look more and more like the graph of the distribution. Of course, this means the distribution is not something you can really ever determine exactly. It is a mathematical idealization of how we believe the data would appear if we had more data than we do.

Thus, when someone makes a claim that certain data fits a certain distribution, what they mean is that the data appears to fit such a distribution. In other words, they are saying that a certain distribution is a good statistical *model* of their data.

A few other examples of distributions that might appear to describe data are shown in Figure A.3.

The distribution in the upper left is said to be *uniform*, because it describes data that is equally spread over the interval. The upper right distribution, which describes data that tends to fall in one lower interval and one higher one, is said to be *bimodal*. Notice the bottom two distributions look vaguely like the normal distribution, but are *skewed* one way or another. There are, of course, many other distributions. Those that have been found to be particularly useful for describing data have been named and studied extensively by statisticians.

▶ Consider the following hypothetical data sets and draw reasonable distribution curves for them. How would you describe the shape of each distribution in words?

a: The age at death of a large number of a certain bird are recorded; mortality is particularly low for young adults.

b: The number of puppies in a litter is recorded for a large number of dogs.

c: Downed trees in a certain forest are located, and the angles the trunks make with due north are measured.

d: The ages of all individuals on a university campus during the workday are recorded.

Probably you drew (a) as bimodal, with peaks showing the deaths of the very young and very old. For (c), if there is no prevailing wind, a uniform distribution is reasonable; otherwise you might have a distribution that looks more normal, with a peak at the angle the wind typically blows. Because university campuses tend to be populated mainly by individuals in their late teens and 20's, for (d) you should imagine a skewed distribution, with a peak to the left and a "tail" stretching off to the right describing faculty and staff ages.

Notice that (b) is a little different from the others in that any measurement of the number of puppies in a litter will give an integer value such as 3 or 7, but never a number like 4.7. When only certain values, separated by gaps, are possible for data, we say the data is *discrete*. When data values do not have to be separated in this way, as in (a) and (c), we say the data is *continuous*.

For discrete data, it is not too useful to group data in intervals that are very small, so it is really best to think of something shaped like a histogram, with step-like features, as giving the distribution. Thus, the distribution for (b) might be a stepped version of a normal distribution, with peak located at the average litter size.

▶ Would the data in (d) be discrete or continuous?

Now that the concept of a distribution is clear, let's turn things around. Suppose before doing an experiment, we hypothesize that the data we will obtain will fit the normal distribution of Figure A.4.

▶ Before continuing, decide if each of the following data sets seems consistent with that hypothesis and state how confident you feel about each answer.

a: 2.0, 3.6, 3.8, 4.1, 4.3, 6.9

b: 6.9

c: 3.8, 3.9, 3.9, 4, 4, 4.1

d: 1.1, 1.4, 2.1, 6.3, 6.5, 7.1

You should feel the given distribution fits (a) well. For (b), with only a single number, we do not have much to go on. The distribution shows it is

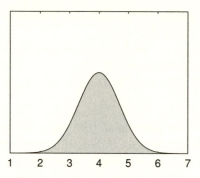

Figure A.4. A normal distribution.

possible to get numbers around 6.9, but they are not as likely as those around 4. The data in (c) seems like it would be better described by a distribution with a much narrower and higher peak around 4. For (d), a bimodal distribution seems more reasonable. Still, with so few data points, any of these might in fact come from an experiment described by the distribution in the figure.

▶   Can you make up a set of a few data points that you can say with complete certainty do *not* arise from the normal distribution of Figure A.4?

Because the graph in Figure A.4 lies above the horizontal axis for all values, there is some likelihood of the experiment producing any number you might mention. While a large number of data points whose histogram is not similar to the figure makes it very likely the distribution does not describe the experiment, you cannot completely rule out the possibility.

A.3. Mean, Median, and Mode

When approaching a new set of data, drawing a histogram to understand the nature of the distribution is the best place to begin. Then, there are several numerical ways of describing the key features of the distribution.

The first question to be addressed is how do we locate the *central tendency* of the data. Does the data tend to cluster around some single value? If it does appear to cluster, how do we determine that value?

A quick glance at the bimodal distribution shows that there is not always a single central tendency to locate. If we are faced with a bimodal distribution, then often we should expect that we are really dealing with a data set that should be broken down into two smaller sets that should be analyzed separately. Perhaps we have missed some important experimental variable and need to rethink our investigation.

► Can you think of some data that might be bimodally distributed? Can it be naturally broken into two smaller data sets?

For simplicity, assume we have a data set producing a histogram with a single major hump, perhaps not too pronounced and maybe not symmetrical, but that is at least visually the dominant feature.

There are three distinct ways of choosing what might be called the central tendency of the data, each with its own strength and weaknesses.

The Mode: The mode is simply the data value that occurs most frequently. That definition must be modified a bit in practice, though, since if you look back at the list of 20 bean heights at the beginning of Section 2, you will see that three of them occurred twice and all the others once. It is perhaps better to do some grouping and say that with an interval of 1 the mode for that data was between 11 and 12, and with a grouping interval of 0.5, it was between 11 and 11.5.

► How can you tell by glancing at a histogram what the mode of the data is?
► Can the mode change if you use a different grouping interval?
► Can a change in a few data values change the mode by much? Is the effect different if the largest or smallest data points are changed, or those midsized?

The Median: This is the data value that occurs in the precise middle of all the data values when they are arranged in order. For instance, in the data of 20 bean heights in Section 2, since we have an even number of data points, we find 11.0 and 11.1 are the middle values. The best we can do is average the two and report 11.05 as the median.

► How can you tell by a glance at a histogram where the median is located?

Because in a histogram area is used to denote the number of data points, the median will be the value where a vertical line splits the total area in half. Since distributions are idealizations of histograms, this also allows the median to be located on the graph of a distribution.

► Could the median change much if a few of the data points were changed? How sensitive is the median to changes in extreme data points vs. those near the middle?

The median is insensitive to changes of the more extreme data points, unless of course these values, are changed so much that they jump from one side of the median to the other. The median may move if data points closest to the median are changed.

The Mean: The mean of a set of data is just the usual average. To calculate it, you add up all the data values and then divide this sum by the number of values you have added. If we have n data values denoted by x_1, x_2, x_3, ..., x_n, then the mean μ is simply

$$\mu = \frac{1}{n}(x_1 + x_2 + x_3 + \cdots + x_n) = \frac{1}{n}\sum_{i=1}^{n} x_i$$

▶ What is the mean of the 20 bean heights in Section 2?

To guess the mean of a data set by looking at a histogram is harder than estimating the mode or median. Although we will not explain why here, because that would involve a detour into physics, the mean is located at the *center of mass* of the histogram along the horizontal axis. This is the point along the horizontal axis at which the histogram would balance if you imagine it as cut out of a piece of cardboard. This interpretation will help you in getting a rough idea of the location of a mean from a histogram, but do not expect to be able to pin down a mean very accurately without doing a calculation.

▶ Does the mean you calculated for the bean data appear to be at this balance point on the histogram of the data?

▶ How sensitive is the mean to changes in only a few of the data points? Does it matter whether these points are near the extremes or in the middle?

Changing an extreme value can have a large effect on a mean. If you think in terms of a balance point for a histogram, moving an extreme value outward is likely to cause the histogram to tip in that direction, just as a see-saw does if a weight far out on an arm is moved farther out. That means the new balance point must be found by moving in the same direction. On the other hand, changing a data value near the mean has little effect on the mean – just as on a see-saw weights near the pivot have little effect.

▶ Can you change just one value in the bean height data set so that while the mode and median do not change, the mean does?

Notice for a normal distribution the mode, median, and mean would all be the same. In general, though, the three are not the same. Figure A.5 provides a

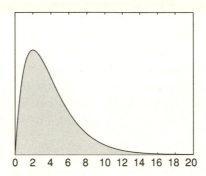

Figure A.5. Mean $= 4$, median ≈ 3.3, and mode $= 2$.

good illustration of a skewed distribution. The median, which splits the area, is around 3.3 on the horizontal axis. The mean, which is the balance point, is further to the right, at 4, due to the rightward stretching tail. The mode, which is at the peak, is around 2.

▶ Look back over all the histograms and distributions in this appendix and estimate where the mode, median, and mean are on each.

In practice, all three of these concepts are used to describe data. The mean is probably used most frequently, the median next most, and the mode the least, but this varies depending on what is being studied.

For instance, in reporting incomes and housing prices, governments tends to emphasize the median as being the most important of the three. This is simply because the median is less sensitive to extreme values. Relatively few very large values can cause the mean to be much larger than the median.

▶ Would you be more interested in knowing the mean, median, or mode of life spans for your society? Which do you think is most optimistic?

A.4. The Spread of Data

The concepts of mean, median, and mode are useful in that they allow us to represent the most important feature of a data set with a single number. The drawback of reporting only them is that we draw attention away from the fact that the data has variation in it. It is useful to have an easily reportable measure of the spread of the data as well.

When the median is chosen to represent the central tendency of the data, the most natural way of specifying the spread of the data is through reporting *quartiles*. Just as the median is the value that divides all the data points in half (half being above the median and half being below the median), the *first quartile point* is the value that has one quarter of the data points below it and three quarters above. The second quartile point is just the median, and the third quartile point has three quarters of the data values below it, with one quarter above. The *interquartile range* is the interval from the first quartile point to the third quartile point. It always contains the median and 50% of the data. Finding all of these is easily done from an ordered list of data.

- ▸   For the 20 bean heights of Section 2, find the interquartile range.
- ▸   From looking at a histogram or distribution, how can you judge where the first and third quartiles lie? Estimate them for all the distributions in this appendix.

Of course there is nothing special about quartiles. Sometimes quintiles (fifths), or deciles (tenths), or even percentiles (hundredths), are used. Specifying the full *range* of the data, by giving the smallest and largest values, also gives the reader a better understanding of the variability of the data.

If the mean is chosen as the way of specifying the data's central tendency, then it is usual to also report the *standard deviation* as the measure of data spread. To develop the idea of the standard deviation, consider an example. Suppose the height (in centimeters) of five bean plants is our data:

$$12.1, \ 16.3, \ 13.2, \ 13.5, \ 14.9.$$

The mean of this data is $\mu = 14.0$. Notice that two of the data values are larger than the mean, and three are smaller, as is reasonable.

A first approach to understanding the spread of the data would be to see how far each data point is from the mean. Hence, we calculate

$$12.1 - 14.0 = -1.9$$
$$16.3 - 14.0 = 2.3$$
$$13.2 - 14.0 = -0.8$$
$$13.5 - 14.0 = -0.5$$
$$14.9 - 14.0 = 0.9.$$

We get negative values when the data point is smaller than the mean and positive values when it is larger than the mean.

A seemingly good idea would be to average these *differences from the mean* that we have just calculated. So, we should add them up and divide by 5:

$$\frac{(-1.9 + 2.3 - 0.8 - 0.5 + 0.9)}{5} = \frac{0}{5} = 0.$$

Unfortunately, zero is not a good measure of spread. In fact, this calculation of average differences will always give zero, as a little algebra can show. The crux of the matter is that some of the differences will be positive and some will be negative, and adding them up always results in cancellation.

The next natural idea would be to just make all these differences from the mean positive before averaging them (i.e., average their absolute values). If we do this we get:

$$\frac{(1.9 + 2.3 + 0.8 + 0.5 + 0.9)}{5} = \frac{6.4}{5} = 1.28.$$

This looks a bit better, and means that, on average, our data points differ from the mean of 14.0 cm by 1.28 cm. This quantity 1.28 is referred to as the *mean deviation* and is a reasonable measure of data spread. To summarize for a set of n data points,

$$\text{mean deviation} = \frac{1}{n} \sum_{i=1}^{n} |x_i - \mu|.$$

Another way of handling the problem of cancellations is to square the differences and then average the squares:

$$\frac{((-1.9)^2 + (2.3)^2 + (-0.8)^2 + (-0.5)^2 + (0.9)^2)}{5} = \frac{10.6}{5} = 2.12.$$

This quantity is called the *variance* or *mean square deviation* of the data. Note that, since our data was in centimeters (cm), the calculation produced a quantity whose units should be squared centimeters (cm^2). To have the same units as our original data, we take the square root and get

$$\sqrt{2.12} \approx 1.46.$$

This last quantity is called the *standard deviation* of the data.

To summarize for a set of n data points, the standard deviation, denoted usually by σ, is

$$\sigma = \sqrt{\frac{1}{n} \sum_{i=1}^{n} (x_i - \mu)^2}.$$

The variance, σ^2, is calculated by just leaving the square root out of this formula.

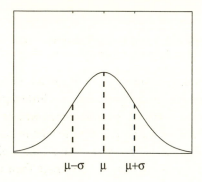

Figure A.6. Normal distribution, with mean μ and standard deviation σ.

▶ Calculate the standard deviation for the 20 bean heights of Section 2.

The standard deviation is the most commonly used measure of data spread, though the reasons for this require more statistical theory than can be explained here. When the mean of data is reported, the standard deviation usually should be as well.

For the theoretical model of data given by the normal distribution, there is a good graphical interpretation of the standard deviation, as shown in Figure A.6. It is the horizontal distance from the peak of the normal curve (the mean) to the *inflection points* (the points where the curve changes from being concave up to concave down or vice versa). The larger the standard deviation, the wider the bell curve.

It can be shown that, for a normal distribution, approximately 34% of the area under the graph is between the mean and the inflection point to the left of the mean. Because the graph is symmetrical, this holds for the area between the mean and the inflection point to the right as well. Because area corresponds to the number of data points, that means that if your data is normally distributed, about 68% of your data should be within one standard deviation of the mean. Similarly, about 95% will be within two standard deviations from the mean, and more than 99% within three standard deviations.

A more precise definition of the normal distribution can now be given. The normal distribution with mean μ and standard deviation σ, where μ is any number and $\sigma > 0$, is

$$f(x) = \frac{1}{\sigma\sqrt{2\pi}} e^{-\frac{(x-\mu)^2}{2\sigma^2}}.$$

Although other curves appear to be bell-shaped, only the particular curves given by this formula are called normal.

A.5. Populations and Samples

There is one last point to be made concerning these basic statistical concepts. In the discussion so far, the focus has been on having some data (a list of numbers) and finding ways of describing that data. But we might want to make conclusions that go beyond the particular data we have collected.

Again, thinking of bean heights as our data set, we can adopt a slightly more sophisticated viewpoint. While we only have the heights of 20 beans recorded, we can certainly imagine performing our experiment on all beans in the world. We will consider all beans as the *population* we are trying to study, and the 20 beans we actually experimented with as being a *sample* from that population. Although a histogram is a good way of graphically treating the data from our sample, the distribution curve is what describes the population as a whole. Of course, we cannot know exactly what the distribution curve for the population really is without experimenting on every bean, but we can make well-informed guesses based on histograms for data sets involving some reasonably large number of beans.

With this viewpoint, there is a change in what we would like to get from our bean data. Although we understand the mean and standard deviation of the data, our real interest is the mean and standard deviation of the entire population. But, without data on the entire population, we can't find these exactly. We will, however, be able to estimate them. Let

$$\mu = \text{mean of population}$$

$$\sigma = \text{standard deviation of population}$$

be the two quantities we would like to estimate.

Not surprisingly, the best estimate you can give for the mean μ of the entire population is simply the mean of the sample. More formally, the mean of the sample is

$$\overline{x} = \frac{1}{n}(x_1 + x_2 + x_3 + \cdots + x_n) = \frac{1}{n}\sum_{i=1}^{n} x_i$$

and

$$\mu \approx \overline{x}.$$

Although the standard deviation of the sample is not a bad estimate of the standard deviation of the population, there is a better one. Statistics texts prove that if

$$s = \sqrt{\frac{1}{n-1}\sum_{i=1}^{n}(x_i - \overline{x})^2}$$

then

$$\sigma \approx s$$

and that s is the best estimator σ.

▶   Calculate the estimate for the population standard deviation for all beans based on the 20 bean heights of Section 2 and compare it to the standard deviation of that data set that you calculated before.

▶   Compare this formula to the formula for standard deviation of a data set and find all the differences. What effect will these differences have on the value you would obtain?

There are, of course, two differences between the formulas for s and σ. First, rather than use μ, which we do not know exactly, we use $\bar{x}$, which estimates it. Then, after adding up the squared deviations from the data mean, we divide by one less than the number of data points, rather than by the number of data points. Since we divide by a smaller number, we end up with a bigger number. Taking the square root afterward still leaves us with a bigger number. Thus, the estimate for the population standard deviation will be inflated a bit from the data's standard deviation.

The informal reason why this inflation is desirable is subtle: In the formula for s, we use not only our data points, but also the mean $\bar{x}$ of the data as an estimate of the unknown mean μ of the population. The data points are likely to be clustered more closely around their own mean $\bar{x}$ than they are around the population's mean μ. Thus if we do not modify the formula, we would get a standard deviation that was smaller than σ. Inflating the standard deviation slightly gives a better estimate. The full argument why replacing n by $n - 1$ is precisely the right thing to do can be found in statistics books.

If we have a large sample, so that n is big, then n and $n - 1$ are really about the same size, so whichever one we use should not matter too much. This is reasonable, since if the sample size is large, then the sample mean $\bar{x}$ is likely to be quite close to the population mean μ, so little adjustment is necessary.

A.6. Practice

Returning to beans for one last time, suppose we grow 10 beans under a set of conditions we will refer to as condition A, and 10 beans under a different set of conditions which we will refer to as condition B. We count the number of leaves on each plant and get the following data:

> Condition A: 4, 6, 5, 6, 8, 4, 6, 5, 10, 5
>
> Condition B: 7, 5, 9, 6, 10, 8, 9, 7, 8, 7

Analyze this data using all the concepts in this appendix (histograms and distributions, mean, median, mode, interquartile range, standard deviation, sample vs. population).

Appendix B

For Further Reading

For further study, there are many textbooks focusing on mathematical models in biology. They generally assume a solid knowledge of calculus and some differential equations and linear algebra, though sections may be read by those with less mathematical background. Among the books covering a variety of biological topics are:

- Leah Edelstein-Keshet. *Mathematical Models in Biology*. McGraw-Hill, New York, 1988.
- Frank C. Hoppenstaedt and Charles S. Peskin. *Modeling and Simulation in Medicine and the Life Sciences*. Springer, New York, second edition, 2002.
- J. Mazumdar. *An Introduction to Mathematical Physiology and Biology*. Cambridge University Press, Cambridge, second edition, 1999.
- James D. Murray. *Mathematical Biology I: An Introduction* and *Mathematical Biology II: Spatial Models and Biomedical Applications*. Springer, New York, third edition, 2002.
- Clifford Taubes. *Modeling Differential Equations in Biology*. Prentice Hall, Upper Saddle River, NJ, 2001.
- S. I. Rubinow. *Introduction to Mathematical Biology*. John Wiley, New York, 1975.
- E. Yeargers, R. Shonkwiler, and J. Herod. *An Introduction to the Mathematics of Biology: With Computer Algebra Models*. Birkhauser, Boston, 1996.

For linear models, including ones using differential equations, recommended books are:

- Hal Caswell. *Matrix Population Models: Construction, Analysis, and Interpretation*. Sinauer Associates, Sunderland, MA, 1989.
- Michael R. Cullen. *Linear Models in Biology*. Ellis Horwood, Chichester, England, 1985.

In addition to sections of the books above, infectious disease models have been the focus of a number of texts and survey papers:

- L. J. S. Allen. Some discrete-time SI, SIR, and SIS epidemic models. *Math. Biosci.*, 124:83–105, 1994.
- Roy M. Anderson and Robert M. May. *Infectious Diseases of Humans: Dynamics and Control*. Oxford University Press, Oxford, England, 1992.
- Fred Brauer and Carlos Castillo-Chavez. *Mathematical Models in Population Biology and Epidemiology*. Springer, New York, 2001.
- Herbert W. Hethcote. The mathematics of infectious diseases. *SIAM Rev.*, 42(4):599–653, 2000 (electronic).

Material on molecular evolution and phylogenetic tree construction has not yet appeared in other texts at this level. Several good surveys exist, directed at researchers and advanced students, as does low-cost or free software:

- W.-H. Li. *Molecular Evolution*. Sinauer Associates, Sunderland, MA, 1997.
- J. Felsenstein. *PHYLIP (Phylogeny Inference Package), Version 3.5c*. Department of Genetics, University of Washington, 1993.
- D. L. Swofford. *PAUP* (Phylogenetic Analysis Using Parsimony *and Other Methods), Version 4*. Sinauer Associates, Sunderland, MA, 2002.
- David L. Swofford, Gary J. Olsen, Peter J. Waddell, and David M. Hillis. *Phylogenetic Inference*, in *Molecular Systematics*. Sinauer Associates, Sunderland, MA, second edition, 1996.

More on the classical genetics topics can be found in:

- J. F. Crow and M. Kimura. *An Introduction to Population Genetics Theory*. Harper and Row, New York, 1970.
- Electronic Scholarly Publishing. *Foundations of Classical Genetics*, a collection of important papers in the development of classical genetics [http://www.esp.org].
- Daniel L. Hartl and Andrew G. Clark. *Principles of Population Genetics*. Sinauer Associates, Sunderland, MA, second edition, 1989.

Books providing a solid background on some of the mathematical and statistical topics introduced here include:

- David C. Lay. *Linear Algebra and Its Applications*. Addison-Welsey, Boston, third edition, 2002.

- Marcello Pagano and Kimberlee Gauvreau. *Principles of Biostatistics*. Duxbury, Pacific Grove, CA, second edition, 2000.
- Sheldon Ross. *A First Course in Probability*. Prentice Hall, Upper Saddle River, NJ, fifth edition, 1997.
- Gilbert Strang. *Introduction to Linear Algebra*. Wellesley-Cambridge Press, Wellesley, MA, second edition, 1993.
- Dennis D. Wackerly, William Mendenhall III, and Richard L. Scheaffer. *Mathematical Statistics with Applications*. Duxbury, Pacific Grove, CA, sixth edition, 2002.

References

Altman, L. (1994). *AIDS mystery that won't go away: Did a dentist infect 6 patients?* N.Y. Times. July 5, C3.

Anderson, S., Bankier, A. T., Barrell, B. G., de Bruijn, M. H. L., Coulson, A. R., Drouin, J., Eperon, I. C., Nierlich, D. P., Roe, B. A., Sanger, F., Schrier, P. H., Smith, A. J. H., Staden, R., and Young, I. G. (1981). *Sequence and organization of the human mitochondrial genome.* Nature, **290**, 457–465.

Andersson, J., Doolittle, W., and Nesbø, C. (2001). *Are there bugs in our genome?* Science, **292**, 1848–1850.

Baker, C. and Palumbi, S. (1994). *Which whales are hunted? A molecular genetic approach to monitoring whaling.* Science, **265**, 1538–1539.

Brown, W. M., Prager, E. M., Wang, A., and Wilson, A. C. (1982). *Mitochondrial DNA sequences of primates: Tempo and mode of evolution.* J. Mol. Evol., **18**, 225–239.

Cann, R., Stoneking, M., and Wilson, A. (1987). *Mitochondrial DNA and human evolution.* Nature, **325**, 31–36.

Chapela, I., Rehner, S., Schultz, T., and Mueller, U. (1994). *Evolutionary history of the symbiosis between fungus-growing ants and their fungi.* Science, **266**, 1691–1694.

Colgate, S. A., Stanley, E. A., Hyman, J. M., Qualls, C. R., and Layne, S. P. (1989). *Aids and a risk-based model.* Los Alamos Science, **18**(5), 2–40.

Crouse, D. T., Crowder, L. B., and Caswell, H. (1987). *A stage-based population model for loggerhead sea turtles and implications for conservation.* Ecology, **68**(5), 1412–1423.

Cullen, M. R. (1985). *Linear models in biology.* Ellis Horwood, Chichester, England.

Cushing, J. M., Henson, S. M., Desharnais, R. A., Dennis, B., Costantino, R. F., and King, A. (2001). *A chaotic attractor in ecology: Theory and experimental data.* Chaos Solitons Fractals, **12**(2), 219–234 [Chaos in ecology].

Farris, J. S. (1972). *Estimating phylogenetic trees from distance matrices.* Am. Nat., **106**, 645–668.

Felsenstein, J. (1993). *Phylip (phylogeny inference package), Version 3.5c.* Department of Genetics, University of Washington.

Fitch, W. and Margoliash, E. (1967). *The construction of phylogenetic trees.* Science, **155**, 279–284.

Gibbons, A. (1992). *Mitochondrial Eve: Wounded but not dead yet.* Science, **257**, 873–875.

365

Hafner, M. S., Sudman, P. D., Villablanca, F. X., Sprading, T. A., Demastes, J. W., and Nadler, S. A. (1994). *Disparate rates of molecular evolution in cospeciating hosts and parasites.* Science, **265**, 1087–1090.

Hayasaka, K., Gojobori, T., and Horai, S. (1988). *Molecular phylogeny and the evolution of primate mitochondrial DNA*, Mol. Biol. Evol., **5**, 626–644.

Hillis, D., Moritz, C., and Mable, B. (eds.) (1996). *Molecular Systematics.* Sinauer Associates, Sunderland, MA.

Hinkle, G., Wetterer, J., Schultz, T., and Sogin, M. (1994). *Phylogeny of the attine ant fungi based on analysis of small subunit ribosomal RNA gene sequences.* Science, **266**, 1695–1697.

Keyfitz, N. and Flieger, W. (1968). *World Population; an analysis of vital data.* University of Chicago Press, Chicago.

Keyfitz, N. and Murphy, E. M. (1967). *Matrix and multiple decrement in population analysis.* Biometrics, **23**, 485–503.

Li, W.-H. (1997). *Molecular Evolution.* Sinauer Associates, Sunderland, MA.

Ludwig, D., Jones, D. D., and Holling, C. S. (1978). *Qualitative analysis of insect outbreak systems: The spruce budworm and forest.* J. Anim. Ecol., **47**(1), 315–332.

May, R. M. (1978). *Simple mathematical models with very complicated dynamics.* Nature, **261**, 459–567.

Mendel, G. (1866). *Versuche über pflanzenhybriden.* Verh. naturforsch Ver. Brünn, **4**, 3–47 [Translation: Experiments in Plant Hybridization (1865), 1–39 (electronic: http://www.esp.org)].

Nellis, C. H. and Keith, L. (1976). *Population dynamics of coyotes in central Alberta, 1964–68.* J. Wildlife Management, **40**(3), 389–399.

Ou, C., Cieselski, C. A., Meyers, G., Bandea, C. I., Luo, C., Korber, B. T. M., Mullins, J. I., Schochetman, G., Berkelman, R., Economou, A. N., Witte, J. J., Furman, L. J., Satten, G. A., MacInnes, K. A., Curran, J. W., and Jaffe, H. W. (1992). *Molecular epidemiology of HIV transmission in a dental practice.* Science, **256**, 1165–1171.

Petersen, G. M., Rotter, J. I., Cantor, R. M., Field, L. L., Greenwald, S., Lim, J. S., Roy, C., Schoenfeld, V., Lowden, J. A., and Kaback, M. M. (1983). *The Tay-Sachs disease gene in North American Jewish populations: Geographic variations and origin.* Am. J. Hum. Genet., **35**(6), 1258–1269.

Ricker, W. E. (1954). *Stock and recruitment.* J. Fish. Res. Bd. Canada, **11**(5), 559–623.

Salzberg, S., White, O., Peterson, J., and Eisen, J. (2001). *Microbial genes in the human genome: Lateral transfer or gene loss?* Science, **292**, 1903–1906.

Studier, J. and Keppler, K. (1988). *A note on the neighbor-joining algorithm of Saitou and Nei.* Mol. Biol. Evol., **5**, 729–731.

Swofford, D. L. (2002). *Paup* (phylogenetic analysis using parsimony *and other methods), Version 4.* Sinauer Associates, Sunderland, MA.

Tateno, Y., Nei, M., and Tajima, F. (1982). *Accuracy of estimated phylogenetic trees from molecular data. I. Distantly related trees.* J. Mol. Evol., **18**, 387–404.

Vogel, G. (1997). *Phylogenetic analysis: Getting its day in court.* Science, **275**, 1559–1560.

Vogel, G. (1998). *HIV strain analysis debuts in murder trial.* Science, **282**, 851–853.

Yerushalmy, J., Harkness, J. T., Cope, J. H., and Kennedy, B. R. (1950). *The role of dual reading in mass radiography.* Am. Rev. Tuber., **61**, 443–464.

Index

absolute value, 73
Africa, Out of, 171, 179, 209
agouti fur, 231
AIDS, 209, 279, 336, 338
albinism, 239
Allee effect, 37
allele, 217
  codominant, 227, 262, 264
  dominant, 216, 217
  fixation of, 267
  frequencies, 261
  multiple, 227, 276
  mutation, 265
  partially dominant, 227
  recessive, 216, 217
  semidominant, 227
  wildtype, 244
autocatalytic model, 19
autosome, *see* chromosome

base substitution, 115, 138
bases, 114
basic reproduction number, 287, 299
Bateson, William, 244
bifurcation diagram, 25
bin size, 349
binomial coefficients, 241
blood type
  ABO system, 227, 272
  MN system, 262
bootstrapping, 207
brachydactyly, 226

cannibalism, 108
carrying capacity, 12
Centers for Disease Control (CDC), 212, 279, 336
centromere, 248
chaos, 26–28, 39
characteristic equation, 79
chickenpox, 27, 282, 288, 299
χ^2-statistic, 235
chromatid, 248
chromosome, 218, 244
  autosome, 247

homologous, 248
sex, 245, 247
cobweb diagram, 15–16, 88
codominance, *see* allele
codon, 114, 142
color blindness, 246, 256, 271
combinations, 230, 240
competition
  contest, 36
  model, 85, 105–106, 109–110
  scramble, 36
competitive exclusion, 110
complex numbers, 73–74
  absolute value of, 73
contact number, 299
  maximal male and female, 311
contact rate, 298, 299
crossing over, 218, 248, 251, 252
  interference, 260
curve fitting, 315
  least squares, 316, 325
    line, 325, 331
    polynomial, 335
cystic fibrosis, 264

deletion, 115
Demography, Fundamental Theorem of, 71
density dependence, 11
determinant, 61, 79, 167
  and inverse of matrix, 61
deviation
  mean, 357
  mean square, 357
  standard, 356–359
  total (TD), 322
difference equation(s), 3, 5, 39
  coupled, 42, 86
  vs. differential equations, 9, 39, 283
differential equation(s), 9, 38, 39, 283
  logistic, 40
diffusion, 30
diploid, 218, 247
disjoint, 120

distance
  additive and symmetric, 160–162, 168, 176
  genetic, 249, 250, 255
  Jukes-Cantor, 157–159, 176, 180
  Kimura, 159–160, 166, 180
  linkage, *see* distance, genetic
  log-det, 160–162, 167–180
  methods of tree construction, 180
  phylogenetic, 114, 155–170
  physical, 250
distribution, *see also* random variable, 349
  bimodal, 350, 352
  binomial, 229–233
    expected value, 233, 241
  central tendency, 352
  χ^2, 234–237, 243
  continuous, 351
  discrete, 351
  normal, 349, 351, 358
  probability, 229
  skewed, 350
  uniform, 350
DNA, 113–116
  aligned sequences, 116
  coding, 115, 138
  junk, 115
  mutation, 115–116, 138
dominance, *see* allele
Drosophila melanogaster, 244

edge, 172
eigenvalue and eigenvector, 65–83, 142, 145, 167
  complex, 73, 102
  computation of, 78–83
  dominant, 70
  power method, 81
  strictly dominant, 70
emigration, 8
equilibrium, 7, 20–21, 44, 65, 88, 94
  saddle, 102
  stable, 21, 90, 102
  unstable, 21, 102
Euler's method, 39
event(s), 118
  complementary, 122, 126
  independent, 123
    definition of, 132
  mutually exclusive, 120, 129
expected value, *see* random variable
exponential model, 5
extrapolation, 317

fecundity, 2, 55
F_i, 216
Fick's law, 30
Fitch-Margoliash
  algorithm, 183–186, 191, 192
  method, 189–190
fitness, 265
  mean, 274
  relative, 265
fixed point, *see* equilibrium

Florida dentist AIDS cluster, 209, 212
4-point condition, 193
fragile X syndrome, 246

gametes, 218
  random union of, 221
GenBank, 209
gene, 114, 215, 217
  linkage, 246–255
    cis and *trans* configurations, 258
  polymorphic, 276
  sex-linked, 244–246
gene transfer, lateral, 173, 209
genetic code, 114
genetic drift, 268–271
genotype, 217
  parental type, 248
  recombinant, 248, 249
geometric model, 5
gonorrhea, 297, 308
growth rate
  finite, 3, 11
  finite intrinsic, 13
  intrinsic, 11, 70
  per capita, 11
  relative, 37

haploid, 247
Hardy-Weinberg equilibrium, 263
hemizygote, 246
hemophilia, 246, 255
herd immunity, 300
heterozygosity, 275
heterozygote, 218
  advantage, 268, 272, 276
histogram, 348
HIV, 209, 307
hominoid, 171, 208, 210
homozygote, 217
  advantage, 268, 272
Huntington disease, 240
hypothesis test, 234

immigration, 8
immune system model, 106–107
immunization, 279, 285, 299
independent assortment
  of chromosomes, 248
  of genes, 219, 221, 222, 225, 246, 249
infectious disease
  endemic, 295, 297, 310
  epidemic, 279, 281, 286
  model
    differentiated infectivity, 307
    MSEIR, 305
    SI, 296, 343
    SIR, 281, 343
    sir, 298
    SIRS, 307
    SIS, 297, 343
    sexually transmitted (STD), 307
infective class, 281

influenza, 282
informative site, 202
inheritance
  chromosomal theory, 247
    Mendelian model, 217–218
initial condition, 3
insertion, 115
interpolation, 317
interquartile range, 356
intersection, 123
inversion, 115

Jacobian matrix, 104
Jukes-Cantor model, 176, 180

leaf, 173
least squares, *see* curve fitting
leprosy, 297
lice, head, 282, 296
likelihood, 206
linear algebra, 44, 61
linearization, 21–24, 99–101
logistic model, 12, 24–27, 33, 86

malaria, 284
Malthus, Thomas, 5
map
  genetic, 249, 250, 254
  linkage, *see* map, genetic
Markov
  matrix, 142
  model, 57, 141
mass action, 87, 106, 282
mating
  assortative, 265
  random, 262
matrix
  addition, 50
  characteristic equation of, 79
  definition, 45
  identity, 59
  inverse, 59
    and determinant, 61
    formula, 61
  multiplication, 46, 48–49
  projection, 46
  scalar multiple, 50
  singular, 62
  transition, 46, 140, 141
  transpose, 167, 330
Maximum Likelihood method, 206, 207
Maximum Parsimony method, 198–202, 207
  assumptions of, 202
mean, 354, 358, 359
mean infectious period, 288
  death adjusted, 306
measles, 27, 282, 301, 305, 306
median, 353
meiosis, 218
meiotic drive, 277
Mendel, Gregor, 215
mitochondria, 171, 179, 208

mixing, homogeneous, 87, 280, 282, 339
mode, 353
model
  linear, 5, 43
  nonlinear, 11, 87
molecular clock, 144, 158, 176, 183
molecular evolution, 113
  model, 138–155, 176, 206
    equilibrium base distribution, 145
    general Markov, 148, 160–162
    Jukes-Cantor, 143–147, 155–159
    Kimura, 147–148, 159–160
    protein, 151
mononucleosis, 297
Morgan, Thomas Hunt, 244
multinomial coefficients, 271
mumps, 27, 305, 306
mutation, 113, 115
  back, 116, 143
  hidden, 116
mutation-selection balance, 276
mutualism model, 85, 107–108, 110–111

Neighbor Joining algorithm, 191–195, 207
normal equations, 329, 332, 338
nucleotides, 114
nullclines, 95, 97

operational taxonomic unit (OTU), 172
orbit, 89
orthologous sequences, 171
outgroup, 187, 198, 200
overdominance, 268

parallel evolution, 208
parasites, 208
parsimony score, 198
partial derivative, 101
pattern, 204
pedigree, 226
permutations, 241
perturbation, 21, 100
pertussis, 300
phase plane, 89
phenotype, 218
physiology models, 39
plot
  log–log, 321
  semilog, 319
population genetics, 261–277
population model
  density dependent, 11, 33–38
  discrete *vs.* continuous, 39–40
  discrete logistic, 12
  harvesting, 30–31
  interacting, 85–111
  Leslie, 53–55, 72, 149
  linear, 5, 41–78, 315
    intrinsic growth rate, 70
    stable age/stage distribution, 71
  Malthusian, 2–12
  Markov, 57

population model (*cont.*)
　nonlinear, 11–38, 85–111
　　Ricker, 34, 37
　　structured, 41
　　Usher, 55–56, 109
power method, 81
predator–prey model, 85–105
primate, 171, 210
probability, 116–138
　addition rule, 119, 121, 125–127, 129
　conditional, 130–133
　　definition of, 132
　frequency interpretation, 117
　multiplication rule, 124–127
Punnett square, 219–220, 223
purine, 114, 115, 126
pyrimidine, 114, 115, 126

quarantine, 285, 296, 299, 304
quartiles, 355

rabies, 304
random variable, 228
　expected value, 233
　　additive property of, 234, 242, 250
recessive, *see* allele
recombination frequency, 255
regression, 332
removal rate, 283, 288, 299, 308
　relative, 287, 290, 310
removed class, 281
RNA, 114
root, 173
Rosco, 144
rubella, 301, 305

sample, 359
scalar, 50
segregation
　of chromosomes, 218, 245, 247
　of genes, 217, 219
selection, 265, 268, 272
　coefficient, 266
　frequency-dependent, 276
sensitivity, 134–135
sensitivity analysis, 77
sickle-cell anemia, 226
significance level, 236
smallpox, 280, 300
specificity, 134–135
spruce budworm, 38
stability, 21, 88
　analysis, 21–24, 99–103
　　by calculus, 22, 104
　local *vs.* global, 24, 103
stable age/stage distribution, 71
statistics, 345
steady state, *see* equilibrium
Stirling's formula, 179
Strong Ergodic Theorem, 71, 81–83,
　　148

structurally unstable model, 91
Sturtevant, Alfred, 249
sum of squares for error (SSE), 322
susceptible class, 281
symbiosis, 107, 209
syphilis, 297

T cells, 106
taxon, 172
Tay-Sachs disease, 223, 224, 228
testcross, 225
　3-point, 252
　2-point, 250
tetanus, 304
tetrad, 248
3-point formulas, 183
threshold value, 287, 300
transient, 20, 94
transition, 115, 126, 130, 147, 164
transmission coefficient, 282, 298, 308
transversion, 115, 126, 130, 147, 164
tree, 172
　bifurcating, 173
　construction
　　algorithms *vs.* optimality criteria, 207
　　methods, 180–208
　metric, 175
　neighbors, 192
　parsimony score, 198
　phylogenetic, 171, 172
　　applications of, 208
　rooted, 173–175
　rooting, 186
　topological, 173
　　number of, 175, 177–179
　unrooted, 173, 174, 200
tribolium, 28, 108
tuberculosis, 283, 297
turbidity, 8

union, 120
UPGMA, 181–183, 186, 192

vaccination, *see* immunization
variability in data, 208, 347, 355
variance, 357
vector
　addition, 50
　definition, 45
　multiplication by matrix, 46
　scalar multiple, 50
vertex
　interior, 173
　terminal, 173

whale hunting, 209

yellow-lethal allele, 226, 240,
　　273

zygote, 247